Proceedings of the 8th International Workshop on Complex Structures and Vector Fields

Topics in Contemporary Differential Geometry, Complex Analysis and Mathematical Physics

Proceedings of the 8th International Workshop on Complex Structures and Vector Fields

Topics in Contemporary Differential Geometry, Complex Analysis and Mathematical Physics

Institute of Mathematics and Informatics, Bulgaria 21 – 26 August 2006

Editors

Stancho Dimiev
Bulgarian Academy of Sciences, Bulgaria

Kouei Sekigawa
Niigata University, Japan

World Scientific

NEW JERSEY • LONDON • SINGAPORE • BEIJING • SHANGHAI • HONG KONG • TAIPEI • CHENNAI

Published by

World Scientific Publishing Co. Pte. Ltd.

5 Toh Tuck Link, Singapore 596224

USA office: 27 Warren Street, Suite 401-402, Hackensack, NJ 07601

UK office: 57 Shelton Street, Covent Garden, London WC2H 9HE

British Library Cataloguing-in-Publication Data
A catalogue record for this book is available from the British Library.

TOPICS IN CONTEMPORARY DIFFERENTIAL GEOMETRY, COMPLEX ANALYSIS AND MATHEMATICAL PHYSICS
Proceedings of the 8th International Workshop on Complex Structures and Vector Fields

ISBN-13 978-981-270-790-1
ISBN-10 981-270-790-5

Printed in Singapore by World Scientific Printers (S) Pte Ltd

PREFACE

This book is the Proceedings of the 8th International Workshop on Complex Structures and Vector Fields held at Bulgarian Academy of Sciences, Institute of Mathematics and Informatics (Sofia) from August 21 to August 2, 2006. The first Workshop was held at the same place on 1992. We are aiming at the higher achievement of the studies of current topics in Complex Analysis, Differential Geometry, Mathematical Physics and also of the intermediate ones among them including their applications. It is notable that many new specialists in Mathematical Physics attended the present Workshop besides regular participants in the previous Workshops and also that a new tendency to expand our subject matters is adopted in the present Workshop, and places especially emphases on the further development of the studies in Differential Geometry, Complex Analysis, Partial Differential Equations and Integrable System, and also on the expansion of the research areas including new ones in Mathematical Physics in the forthcoming Workshops.

This book is dedicated to the memory of three distinguished scientists, Professor Shigeru Ishihara who is regarded as a teacher of many Japanese participants, Professor Shozo Koshi who was an active participant of the Workshop, and Professor Sawa Manoff who was an active participant of the Workshop and also made much effort for the development of the same Workshop.

The editors express their deepest gratitude to Professor T. Oguro for his outstanding co-operation and efforts in the arrangements of this volume.

Editors

THE 8TH INTERNATIONAL WORKSHOP ON COMPLEX STRUCTURES AND VECTOR FIELDS

21-26 August 2006, Sofia – Bulgaria

ORGANIZING COMMITTEE

S. Dimiev (Chairman)	– IMI, Sofia, Bulgaria
St. Donev	– INRNE, Sofia, Bulgaria
V. Gerdjikov	– INRNE, Sofia, Bulgaria
B. Iliev	– INRNE, Sofia, Bulgaria
R. Lazov	– IMI, Sofia, Bulgaria

SCIENTIFIC COMMITTEE

V. Georgiev (Chairman)	— Pisa, Itary
S. Dimiev (Chairman)	— Bulgaria, Bulgaria
D. Alekseevski	— Hull, UK
K. Sekigawa	— Niigata, Japan
H. Hashimoto	— Nagoya, Japan
T. Adachi	— Nagoya, Japan
S. Maeda	— Matsue, Japan
Y. Mathsushita	— Shiga, Japan
S. Marchiafawa	— Rome, Italy
M. Pontecorvo	— Rome, Italy
V. Rostov	— Nice, France
A. Sergeev	— Moscow, Russia
O. Mushkarov	— Sofia, Bulgaria
J. Davidov	— Sofia, Bulgaria
G. Ganchev	— Sofia, Bulgaria
I. Pestov	— Dubna, Russia
St. Donev	— Sofia, Bulgaria
V. Gerdjikov	— Sofia, Bulgaria
B. Iliev	— Sofia, Bulgaria
R. Lazov	— Sofia, Bulgaria

CONTRIBUTED COMMUNICATIONS

1. Yasuo Matsushita
   Almost Kähler-Einstein structures on 8-dimensional Walker manifolds
2. Peter Popivanov
   On the hypoellipticity of complex valued vector fields
3. Toshiaki Adachi
   Killing helices on a complex projective space and canonical magnetic fields on geodesic spheres
4. Vestislav Apostolov
   Generalized Kähler manifolds, commuting complex structures and split tangent bundles
5. Rossen Dandoloff
   Anholonomy of a moving space curve and the Schrödinger equation
6. Hideya Hashimoto
   On almost Hermitian structures of 6-dimensional submanifolds in the octonions
7. Norio Ejiri
   Complex submanifolds and Lagrangian submanifolds associated with minimal surfaces in tori
8. Julian Lawrinowicz
   Clusters and fractals in mathematics and physics
9. Vladimir Georgiev
   Nonexistence of nonzero resonances for Schrödinger operators with singular perturbation
10. Vladimir Gerdjikov
    Hamiltonian aspects of soliton equations with deep reductions
11. Tihomir Valchev
    Bäcklund transformations and Riemann-Hilbert problem for N-wave equations with additional symmetries
12. Georgi Grahovski
    The Caudrey-Beals-Coifman systems and the gauge group action
13. Nikolay Kostov
    The Manakov model as two moving interacting curves
14. Stoil Donev
    Integrability, curvature and description of photon-like objects

15. Tadashi Sugiyama
Characterization of totally umgilic immersions by curves of order 2
16. Kouei Sekigawa
Some critical almost Kähler structures
17. Sadahiro Maeda
Characterization of parallel isometric immersions of space forms into space forms in the class of isotropic immersions
18. Milen Hristov
On the real hypersurfaces of Hermitian manifolds
19. Galya Nakova
On some non-integrable almost contact manifolds with Norden metric of dimension 5
20. Bozhidar Iliev
Normal frames and linear transports along paths in line bundles. Applications to classical electrodynamics
21. Velichka Milousheva
On the geometric structure of hypersurfaces of conullity two in Euclidean space
22. Mancho Manev
On quasi-Kähler manifolds with Norden metric
23. Dobrinka Gribacheva
Lie groups as four-dimensional Riemannian product manifolds
24. Marta Teofilova
Lie groups as several complex manifolds with Norden metric
25. Vladimir Balan
KCC and linear stability for the Parkinson tremor model
26. Stancho Dimiev
Bi-complex analytic pseudo-Euclidean geometry and applications
27. Nikolai Nikolov
On the definition of Kobayashi-Buseman pseudometric
28. Petar Stoev
Differential forms of bi-complex variables
29. Branimir Kiradjiev
Quadratic forms on the bi-complex algebras
30. Lilya Apostolova
Real analyticity of the hyper-Kähler almost Kähler manifolds

31. Assen Kyuldjiev
Symmetries of the Manev problem and its real Hamiltonian form

32. Rossen Ivanov
Camassa-Holm equation as a geodesic flow of the right invariant metric

33. Hiroshi Matsuzoe
Information geometry and affine differential geometry

34. Mihail Konstantinov
Effects of finite arithmetic — myths and realities

35. Ilona Zasada
Phase transitions in binary alloy thin films

36. Angel Ivanov, George Venkov
Global existence of the solution to the Hartree equation

37. Vanya Markova
Application of unscented and extended Kalman filtering for estimating quaternion motion

38. Marin Marinov
Homogenization of nonlinear parabolic operators of high order

CONTENTS

MODULI SPACE OF KILLING HELICES OF LOW ORDERS ON A COMPLEX SPACE FORM

T. ADACHI*

Department of Mathematics, Nagoya Institute of Technology
Gokiso, Nagoya, 466-8555, Japan
E-mail: adachi@nitech.ac.jp

We give a report on the moduli space of helices of proper order less than 5 which are generated by some Killing vector fields on a complex space form from the viewpoint of the length spectrum.

1. Introduction

In this note we give a summary of my work concerning essential Killing helices of low orders on a non-flat complex space form, which is either a complex projective space or a complex hyperbolic space. A smooth curve γ parameterized by its arclength on a Riemannian manifold M is said to be a *helix of proper order d* if it satisfies the following system of ordinary differential equations

$$\nabla_{\dot{\gamma}} Y_j = -\kappa_{j-1} Y_{j-1} + \kappa_j Y_{j+1}, \quad 1 \leqq j \leqq d, \tag{1.1}$$

with positive constants $\kappa_1, \ldots, \kappa_{d-1}$ and an orthonormal system $\{Y_1 = \dot{\gamma}, Y_2, \ldots, Y_d\}$ of vector fields along γ. Here $\kappa_0 = \kappa_d = 0$, and Y_0, Y_{d+1} are null vector fields along γ. These constants $\kappa_1, \ldots, \kappa_{d-1}$ are called the *geodesic curvatures* of γ and the system $\{Y_i\}$ the *Frenet frame* of γ. We call a helix *Killing* if it is generated by some Killing vector field on M. On real space forms, which are standard spheres, Euclidean spaces and real hyperbolic spaces, all helices are Killing and lengths of closed helices are given by their geodesic curvatures. But on a complex space form the situation is different. We study the difference on laminations on the moduli spaces of Killing helices which are induced by the length spectrum.

*The author is partially supported by Grant-in-Aid for Scientific Research (C) (No. 17540072) JSPS, and Scientist Exchange Program between JSPS and MES.

2. Moduli space of Killing helices and length spectrum

We say two helices γ_1, γ_2 on a Riemannian manifold M are *congruent* to each other if there exist an isometry φ of M and a constant t_0 satisfying $\gamma_2(t) = \varphi \circ \gamma_1(t+t_0)$ for all t. We denote by $\mathcal{K}_d(M)$ the set of all congruence classes of Killing helices of proper order d on M. We put $\mathcal{K}(M) = \bigcup_{d=1}^{\infty} \mathcal{K}_d(M)$ and call it the *moduli space* of Killing helices. On a real space form $\mathbb{R}M^n$, as helices are classified by their geodesic curvatures, we see $\mathcal{K}_d(\mathbb{R}M^n)$ is bijective to $(0,\infty)^{d-1}$, the $(d-1)$ product of half lines, when $d \le n$. But on a non-flat complex space form, as isometries are either holomorphic or anti-holomorphic, the moduli space of Killing helices is not so simple. For a helix on a Kähler manifold (M, J) with Frenet frame $\{Y_i\}$, we define its *complex torsions* τ_{ij} $(1 \le i < j \le d)$ by $\tau_{ij} = \langle Y_i, JY_j \rangle$. As was pointed out in [8], on a non-flat complex space form $\mathbb{C}M^n$ a helix γ is Killing if and only if all its complex torsions are constant along γ.

We call a helix γ *closed* if there is positive t_c with $\gamma(t+t_c) = \gamma(t)$ for all t. The minimum positive t_c with this property is called the length of γ and is denoted by length(γ). When γ is not closed we say it is *open* and put length$(\gamma) = \infty$. The *length spectrum* $\mathcal{L} : \mathcal{K}(M) \to (0,\infty]$ is defined by $\mathcal{L}([\gamma]) = \text{length}(\gamma)$, where $[\gamma]$ denotes the congruence class containing a helix γ. For the sake of simplicity we denote a restriction of $\mathcal{L}$ onto a subset of $\mathcal{K}(M)$ also by $\mathcal{L}$.

3. Moduli space of helices on a real space form

For the sake of comparison, we here show some properties on length spectrum of helices on a real space form $\mathbb{R}M^n(c)$ of constant sectional curvature c. The length spectrum $\mathcal{L} : \mathcal{K}_2(\mathbb{R}M^n(c)) \cong (0,\infty) \to (0,\infty]$ of circles of positive geodesic curvature, which are helices of proper order 2, is given as $\mathcal{L}(\kappa) = 2\pi/\sqrt{\kappa^2+c}$, where we read it infinity when $\kappa^2 + c \le 0$. Thus if we induce the canonical Euclidean differential structure on $\mathcal{K}_2(\mathbb{R}M^n(c))$, we see the length spectrum is smooth on this moduli space.

For about the moduli space of helices of proper order 3 on $\mathbb{R}M^n(c)$ $(n \ge 3)$, the feature depends on sectional curvature c. All helices of proper order 3 on a Euclidean space $\mathbb{R}^n$ are unbounded. For a standard sphere $S^n(c)$ of constant sectional curvature c, we have a canonical foliation $\{\mathcal{G}_\alpha\}_{\alpha \in (1,\infty)}$ on $\mathcal{K}_3(S^n(c))$ which is related with the length spectrum and is given as

$$\mathcal{G}_\alpha = \{ [\gamma_{\kappa_1,\kappa_2}] \mid \kappa_1^2 + (\kappa_2 - \alpha\sqrt{c}/2)^2 = c(\alpha^2-1)/4 \},$$

where $[\gamma_{\kappa_1,\kappa_2}]$ denotes the congruence class of helices of proper order 3 on $S^n(c)$ with geodesic curvatures κ_1, κ_2 (see Figure 1).

Theorem 3.1. *The length spectrum* $\mathcal{L} : \mathcal{K}_3\big(S^n(c)\big) \to (0,\infty]$ *is constant on each leaf. Each leaf is set theoretically maximal with respect to this property. A leaf* $\mathcal{G}_\alpha$ *consists of congruence classes of closed helices if and only if* α *and* $\sqrt{\alpha^2-1}$ *are rational.*

For a real hyperbolic space $H^n(c)$ of constant sectional curvature c, we have a canonical foliation $\{\mathcal{G}_\alpha\}_{\alpha\in(-\infty,\infty)}$ on $\mathcal{K}_3\big(H^n(c)\big)$ which is given as

$$\mathcal{G}_\alpha = \big\{\, [\gamma_{\kappa_1,\kappa_2}] \;\big|\; \kappa_1^2 + (\kappa_2 - \alpha\sqrt{|c|}/2)^2 = -c(\alpha^2+1)/4 \,\big\}.$$

We should note that the moduli space $\mathcal{BK}_3\big(H^n(c)\big)$ of bounded helices of proper order 3 on $H^n(c)$ is given as $\big\{\, [\gamma_{\kappa_1,\kappa_2}] \;\big|\; \kappa_1^2+(\kappa_2-\sqrt{|c|}/2)^2 > -c/2 \,\big\}$. On this space the foliation $\{\mathcal{G}_\alpha\}_{\alpha\in(1,\infty)}$ satisfies the same property as of the foliation on $\mathcal{K}_3\big(S^n\big)$. In both cases of a standard sphere and of a real hyperbolic space, these foliations can be naturally extend to a foliation or a lamination on $\mathcal{K}_2\big(\mathbb{R}M^n(c)\big) \cup \mathcal{K}_3\big(\mathbb{R}M^n(c)\big)$.

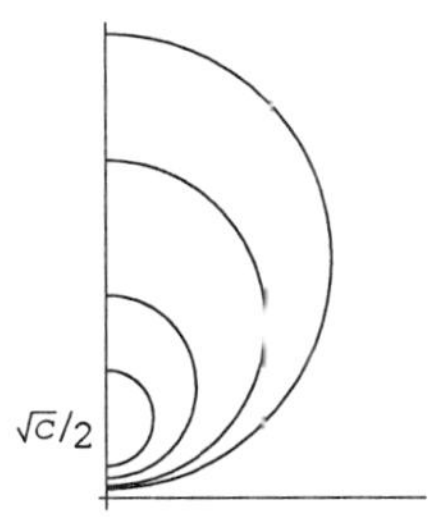

Fig. 1. Foliation on $\mathcal{K}_3\big(S^n(c)\big)$

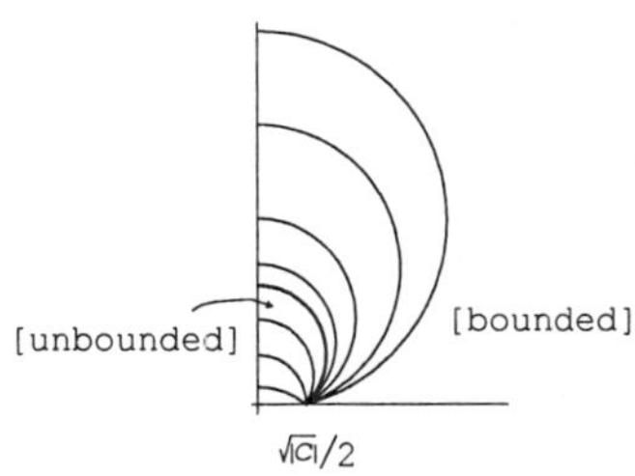

Fig. 2. Foliation on $\mathcal{K}_3\big(H^n(c)\big)$

4. Moduli space of circles on a complex space form

We now study the moduli space of helices on a non-flat complex space form. On a Kähler manifold, the complex torsion τ_{12} of each circle γ is always constant along γ, because

$$\tau_{12}' = \langle \nabla_{\dot\gamma}\dot\gamma, JY_2\rangle + \langle \dot\gamma, J\nabla_{\dot\gamma}Y_2\rangle = \kappa_1\big(\langle Y_2, JY_2\rangle - \langle\dot\gamma, J\dot\gamma\rangle\big) = 0.$$

Therefore we see the moduli space $\mathcal{K}_2\big(\mathbb{C}M^n\big)$ of circles of positive geodesic curvature on a non-flat complex space form is set theoretically bijective to the product $(0,\infty)\times[0,1]$ when $n \ge 2$. In this section we suppose $n \ge 2$ and we shall denote by $[\gamma_{\kappa,\tau}]$ the congruence class of circles with geodesic curvature κ and complex torsion $\tau_{12} = \tau$ on a complex space form $\mathbb{C}M^n(c)$ of constant holomorphic sectional curvature c.

For a complex projective space $\mathbb{C}P^n(c)$, we have a lamination structure $\{\mathcal{F}_\mu\}_{\mu\in[0,1)\cup\{\star\}}$ on $\mathcal{K}_2(\mathbb{C}P^n(c))$ defined by

$$\mathcal{F}_\mu = \begin{cases} \{[\gamma_{\kappa,0}] \mid \kappa > 0\}, & \text{if } \mu = 0, \\ \{[\gamma_{\kappa,\tau}] \mid 3\sqrt{3}c\kappa\tau(4\kappa^2+c)^{-3/2} = \mu\}, & \text{if } 0 < \mu < 1, \\ \{[\gamma_{\kappa,1}] \mid \kappa > 0\}, & \text{if } \mu = \star. \end{cases}$$

Theorem 4.1. *The length spectrum $\mathcal{L} : \mathcal{K}_2(\mathbb{C}P^n(c)) \to (0,\infty]$ is smooth on each leaf with respect to the canonical induced Euclidean differential structure. Each leaf is maximal with respect to this property.*

1) *The leaf $\mathcal{F}_\star$ consists of congruence classes of closed circles satisfying $\mathcal{L}([\gamma_{\kappa,1}]) = 2\pi/\sqrt{\kappa^2+c}$.*
2) *The leaf $\mathcal{F}_0$ also consists of congruence classes of closed circles satisfying $\mathcal{L}([\gamma_{\kappa,0}]) = 4\pi/\sqrt{4\kappa^2+c}$.*
3) *The leaf $\mathcal{F}_\mu$ $(0 < \mu < 1)$ consists of congruence classes of closed circles if and only if $\mu = q(9p^2-q^2)(3p^2+q^2)^{-3/2}$ with some relatively prime positive integers p, q satisfying $p > q$. On this leaf $\mathcal{L}([\gamma_{\kappa,\tau}]) = 2\delta(p,q)\pi\sqrt{(3p^2+q^2)/\{3(4\kappa^2+c)\}}$, where $\delta(p,q) = 1$ when the product pq is odd and $\delta(p,q) = 2$ when pq is even.*

For a complex hyperbolic space $\mathbb{C}H^n(c)$, we also have a lamination structure $\{\mathcal{F}_\mu\}_{\mu\in[0,\infty]\cup\{\star\}}$ on $\mathcal{K}_2(\mathbb{C}H^n(c))$ defined by

$$\mathcal{F}_\mu = \begin{cases} \{[\gamma_{\kappa,0}] \mid \kappa > 0\}, & \text{if } \mu = 0, \\ \{[\gamma_{\kappa,\tau}] \mid 3\sqrt{3}|c|\kappa\tau|4\kappa^2+c|^{-3/2} = \mu\}, & \text{if } 0 < \mu < \infty, \\ \{[\gamma_{\sqrt{|c|}/2,\tau}] \mid 0 < \tau < 1\}, & \text{if } \mu = \infty, \\ \{[\gamma_{\kappa,1}] \mid \kappa > 0\}, & \text{if } \mu = \star. \end{cases}$$

This lamination has the same properties as of the lamination on $\mathcal{K}_2(\mathbb{C}P^n)$ if we restrict ourselves on the moduli space

$$\mathcal{BK}_2(\mathbb{C}H^n(c)) = \{\, [\gamma_{\kappa,\tau}] \mid 0 \le \tau < \nu(\kappa) \,\} \cup \{\, [\gamma_{\kappa,1}] \mid \kappa > \sqrt{|c|} \,\}$$

of bounded circles on $\mathbb{C}H^n(c)$ (see [2]). Here $\nu : (0,\infty) \to [0,1]$ is given by

$$\nu(\kappa) = \begin{cases} 0, & \text{if } 0 < \kappa \le \sqrt{|c|}/2, \\ (4\kappa^2+c)^{3/2}/(3\sqrt{3}\,|c|\kappa), & \text{if } \sqrt{|c|}/2 < \kappa < \sqrt{|c|}, \\ 1, & \text{if } \kappa \ge \sqrt{|c|}. \end{cases}$$

In view of the features of these laminations on the moduli spaces of circles, we find the set $\{\, [\gamma_{\kappa,1}] \mid \kappa > 0 \,\}$ of congruence classes of *trajectories for Kähler magnetic fields* is quite different from other part of $\mathcal{K}_2(\mathbb{C}M^n(c))$. Since each trajectory lies on some totally geodesic $\mathbb{C}M^1$ and other circles do not lie on $\mathbb{C}M^1$, we shall classify helices by this property. We call a helix on $\mathbb{C}M^n$ of proper order $2k-1$ or $2k$ *essential* if it lies on some totally geodesic $\mathbb{C}M^k$. We denote by $\mathcal{EK}_d(\mathbb{C}M^n(c))$ the set of all congruence classes of essential Killing helices of proper order d on $\mathbb{C}M^n(c)$.

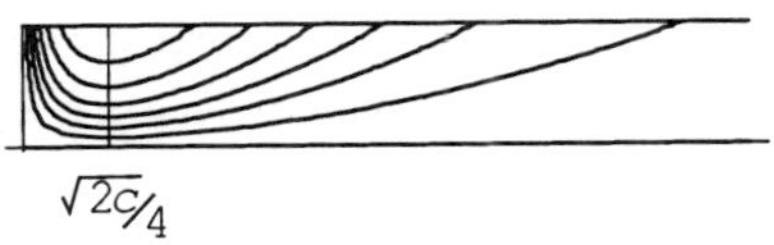

Fig. 3. Lamination on $\mathcal{K}_2(\mathbb{C}P^n(c))$

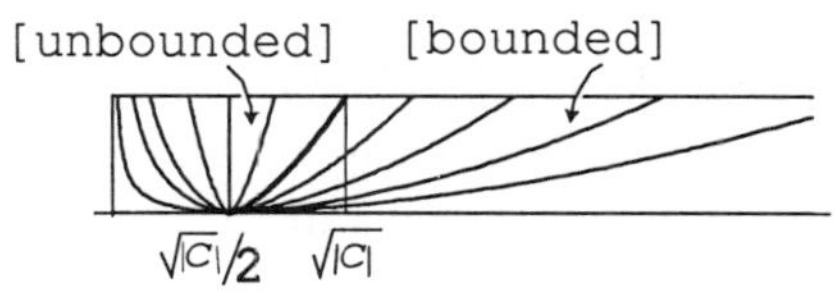

Fig. 4. Lamination on $\mathcal{K}_2(\mathbb{C}H^n(c))$

5. Moduli spaces of Killing helices of orders less than 5 on a complex space form

Though all circles on a non-flat complex space form are Killing, helices of proper order greater than 2 are not necessarily Killing. Computing τ'_{ij} by using (1.1) we see a helix of proper order d on $\mathbb{C}M^n$ is Killing if and only if

$$-\kappa_{i-1}\tau_{i-1j} + \kappa_i\tau_{i+1j} - \kappa_{j-1}\tau_{ij-1} + \kappa_j\tau_{ij+1} = 0, \quad 1 \le i < j \le d, \qquad (5.1)$$

where we set $\tau_{0k} = \tau_{kk} = \tau_{kd+1} = 0$ ([7]). Applying these relations to a helix of proper order 3 on $\mathbb{C}M^n$ $(n \ge 2)$, we find it is Killing if and only if its geodesic curvatures and complex torsions satisfy $\tau_{13} = 0$ and $\kappa_1\tau_{23} = \kappa_2\tau_{12}$. If we consider the initial frame we find the following.

1) A helix is essential Killing if and only if $\tau_{12} = \pm\kappa_1/\sqrt{\kappa_1^2+\kappa_2^2}$, $\tau_{13} = 0$, $\tau_{23} = \pm\kappa_2/\sqrt{\kappa_1^2+\kappa_2^2}$, where the double signs take the same signature.
2) When $n \ge 3$, a helix is Killing if and only if its complex torsions satisfy $\tau_{12} = \kappa_1\tau$, $\tau_{13} = 0$, $\tau_{23} = \kappa_2\tau$ with some τ satisfying $|\tau| \le 1/\sqrt{\kappa_1^2+\kappa_2^2}$.

Thus we see the moduli space $\mathcal{EK}_3(\mathbb{C}M^n)$ of essential Killing helices of proper order 3 is bijective to a quater of a plane $(0,\infty)^2$ and the moduli space $\mathcal{K}_3(\mathbb{C}M^n)$ is bijective to the set $(0,\infty)^2 \times [0,1]$.

When we consider Killing helices of proper order 4, the relations (5.1) turn to $\kappa_1\tau_{23} + \kappa_3\tau_{14} = \kappa_2\tau_{12}$, $\kappa_3\tau_{23} + \kappa_1\tau_{14} = \kappa_2\tau_{34}$ and $\tau_{13} = \tau_{24} = 0$. Considering the initial frame we find a helix of proper order 4 on $\mathbb{C}M^n$ $(n \ge 2)$ is essential Killing if and only if its complex torsions satisfy one of the following conditions:

i) $\tau_{12}=\tau_{34}=\pm(\kappa_1+\kappa_3)\Big/\sqrt{\kappa_2^2+(\kappa_1+\kappa_3)^2},\quad \tau_{13}=\tau_{24}=0,$

$\tau_{23}=\tau_{14}=\pm\kappa_2\Big/\sqrt{\kappa_2^2+(\kappa_1+\kappa_3)^2},$

ii) $\tau_{12}=-\tau_{34}=\pm(\kappa_1-\kappa_3)\Big/\sqrt{\kappa_2^2+(\kappa_1-\kappa_3)^2},\quad \tau_{13}=\tau_{24}=0,$

$\tau_{23}=-\tau_{14}=\pm\kappa_2\Big/\sqrt{\kappa_2^2+(\kappa_1-\kappa_3)^2}.$

Here the double signs take the same signatures in each case. Thus the moduli space $\mathcal{EK}_4(\mathbb{C}M^n)$ of essential Killing helices of proper order 4 is bijective to the set $(0,\infty)^2\times(\mathbb{R}\setminus\{0\})$. When $\kappa_3>0$, a point $(\kappa_1,\kappa_2,\kappa_3)$ corresponds to the congruence class of Killing helices with geodesic curvatures κ_1, κ_2, κ_3 and complex torsions in the condition i), and when $\kappa_3<0$, it corresponds to the congruence class of Killing helices with geodesic curvatures κ_1, κ_2, $-\kappa_3$ and complex torsions in the condition ii). Thus we find moduli spaces of essential Killing helices of proper order less than 5 on $\mathbb{C}M^n$, which are

$$\mathcal{EK}_1(\mathbb{C}M^n)=\{0\},\qquad \mathcal{EK}_2(\mathbb{C}M^n)=(0,\infty),$$
$$\mathcal{EK}_3(\mathbb{C}M^n)=(0,\infty)^2,\qquad \mathcal{EK}_4(\mathbb{C}M^n)=(0,\infty)^2\times\big(\mathbb{R}\setminus\{0\}\big),$$

set theoretically form a "building structure" like $\mathcal{K}(\mathbb{R}M^n)$. Since the moduli spaces $\mathcal{K}_2(\mathbb{C}M^n)$, $\mathcal{K}_3(\mathbb{C}M^n)$ do not form such structure in canonical sense, we restrict ourselves on moduli spaces of essential Killing helices.

6. Lamination on moduli space of essential Killing helices

We here consider laminations on the moduli spaces of essential Killing helices of proper orders 3 and 4. For $\mathbb{C}P^n(c)$ we have a foliation $\{\mathcal{G}_\mu\}_{\mu\in(-1,1)}$ corresponding to the length spectrum on $\mathcal{EK}_3\big(\mathbb{C}M^n(c)\big)$: It is given as

$$\mathcal{G}_\mu=\left\{\,[\gamma(\kappa_1,\kappa_2)]\;\middle|\;\frac{8(\kappa_1^2+\kappa_2^2)^2+c(9\kappa_1^2-18\kappa_2^2)}{(\kappa_1^2+\kappa_2^2)^{1/2}(4\kappa_1^2+4\kappa_2^2+3c)^{3/2}}=\mu\right\},$$

where $[\gamma(\kappa_1,\kappa_2)]$ denotes the congruence class containing essential Killing helices of proper order 3 with geodesic curvatures κ_1, κ_2 (see Figure 5).

Theorem 6.1. *The length spectrum $\mathcal{L}:\mathcal{EK}_3\big(\mathbb{C}P^n(c)\big)\to(0,\infty]$ is smooth on each leaf $\mathcal{G}_\mu$ with respect to the canonical induced Euclidean differential structure. Each leaf is maximal with respect to this property.*

1) *The leaf $\mathcal{G}_0$ consists of congruence classes of closed helices of proper order 3 satisfying $\mathcal{L}([\gamma(\kappa_1,\kappa_2)])=2\sqrt{3c}\,\pi\big/\sqrt{4\kappa_1^2+4\kappa_2^2+3c}$.*
2) *The leaf $\mathcal{G}_\mu$ ($\mu\neq0$) consists of closed helices of proper order 3 if and only if $\mu=\pm q(9p^2-q^2)(3p^2+q^2)^{-3/2}$ with some relatively prime positive integers p, q satisfying $p>q$.*

By this theorem we find there are embeddings of $\mathcal{K}_2(\mathbb{C}P^n) \setminus \mathcal{E}\mathcal{K}_2(\mathbb{C}P^n)$ into $\mathcal{E}\mathcal{K}_3(\mathbb{C}P^n)$ with respect to the induced Euclidean differential structures which preserve the foliation structure. In other words, there is a two-to-one continuous map $\mathcal{E}\mathcal{K}_3(\mathbb{C}P^n) \to \mathcal{K}_2(\mathbb{C}P^n) \setminus \mathcal{E}\mathcal{K}_2(\mathbb{C}P^n)$ which preserves the foliation structure. The reader should compare Figures 1 and 5. He can find difference between their features near the half line $\{(\kappa, 0) \mid \kappa > 0\}$.

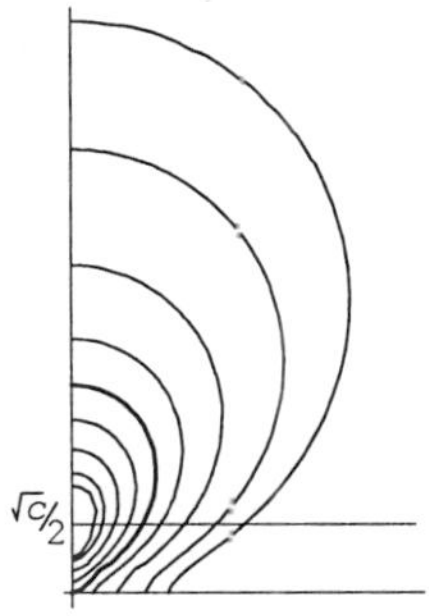

Fig. 5. Folation on $\mathcal{E}\mathcal{K}_3(\mathbb{C}P^n)$

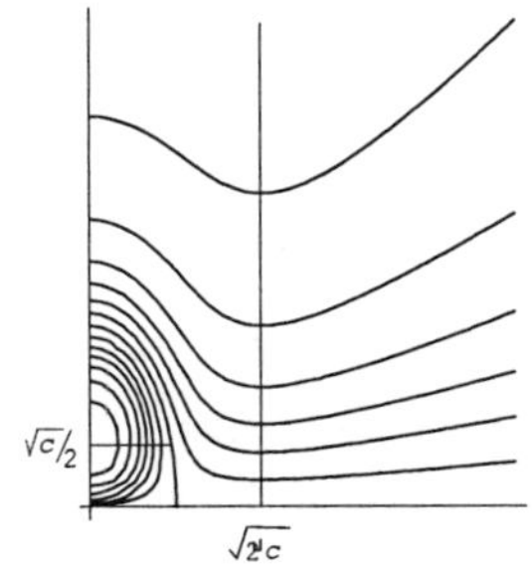

Fig. 6. Foliation on $\mathcal{M}\mathcal{K}_4(\mathbb{C}P^n)$

On $\mathcal{E}\mathcal{K}_4(\mathbb{C}P^n)$ we also have a foliation $\{\mathcal{H}_\mu\}_{\mu\in(-1,1)}$ corresponding to the length spectrum. It is an extension of the foliation on $\mathcal{E}\mathcal{K}_3(\mathbb{C}P^n)$ and we have a projection

$$\mathcal{E}\mathcal{K}_3(\mathbb{C}P^n) \cup \mathcal{E}\mathcal{K}_4(\mathbb{C}P^n) \cong (0,\infty)^2 \times \mathbb{R} \to \mathcal{K}_2(\mathbb{C}P^n) \setminus \mathcal{E}\mathcal{K}_2(\mathbb{C}P^n)$$

which preserves the foliation structure. In order to see this foliation visually we take a plane $\mathcal{M}\mathcal{K}_4(\mathbb{C}P^n(c)) = \{ [\gamma(\kappa_1, \kappa_2, -\kappa_1)] \mid \kappa_1, \kappa_2 > 0 \} \subset \mathcal{E}\mathcal{K}_4(\mathbb{C}P^n(c))$ which consists of congruence classes of Killing helices of proper order 4 with complex torsions $\tau_{12} = -\tau_{34} = 0$, $\tau_{23} = \tau_{14} = \pm 1$. We note that except on this plane absolute values of complex torsions of essential Killing helices of proper order 4 are less than 1. Leaves of the foliation $\{\mathcal{H}\}_\mu$ are transversal to this plane (see Figure 6).

For $\mathbb{C}H^n$ we have a lamination $\{\mathcal{G}_\mu\}_{\mu\in(-\infty,\infty)}$ on $\mathcal{E}\mathcal{K}_3(\mathbb{C}H^n)$ corresponding to the length spectrum which is given by just the same manner as on $\mathcal{E}\mathcal{K}_3(\mathbb{C}P^n)$. It has the same property as of the foliation on $\mathcal{E}\mathcal{K}_3(\mathbb{C}P^n)$ if we restrict ourselves on the moduli space $\mathcal{B}\mathcal{E}\mathcal{K}_3(\mathbb{C}H^n(c))$ of bounded essential Killing helices of proper order 3 which is given by

$$\left\{ [\gamma(\kappa_1, \kappa_2)] \;\middle|\; \begin{array}{l} 4\kappa_1^2 + 4\kappa_2^2 + 3c > 0, \\ \dfrac{|8(\kappa_1^2+\kappa_2^2)^2 + c(9\kappa_1^2 - 18\kappa_2^2)|}{(\kappa_1^2+\kappa_2^2)^{1/2}(4\kappa_1^2+4\kappa_2^2+3c)^{3/2}} < 1 \end{array} \right\}.$$

This moduli space has a cusp at a point $(\sqrt{6c}/3, \sqrt{3c}/6)$ and the lamination on $\mathcal{EK}_3(\mathbb{C}H^n(c))$ has a singularity at this point. We can extend this lamination onto the moduli space $\mathcal{EK}_4(\mathbb{C}H^n(c))$. We should note that some essential Killing helices of proper order 4 are obtained as trajectories for canonical magnetic fields on real hypersurfaces of type A_1 in $\mathbb{C}M^n$ (see [3]).

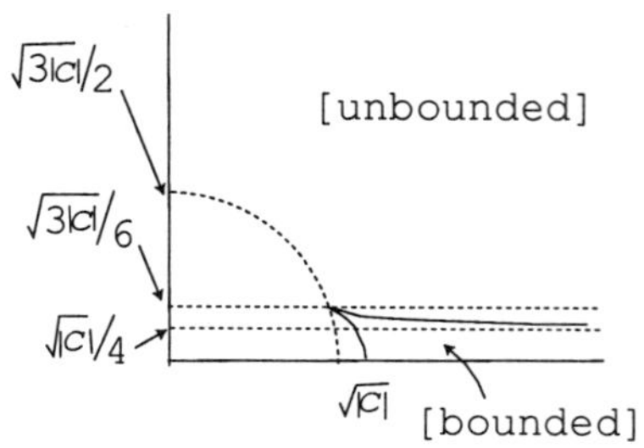

Fig. 7. $\mathcal{K}_3(\mathbb{C}H^n(c))$

Acknowledgments

The author would like to express his hearty thanks to Professors Stancho Dimiev and Milen Hristov and their wives for their hospitality, and also to Ms. Mariana Uzunova in MES for her assistance, during his stay in Bulgaria.

References

1. T. Adachi, *Kähler magnetic flows on a manifold of constant holomorphic sectional curvature*, Tokyo J. Math. **18** (1995), 473-483.
2. ———, *Lamination of the moduli space of circles and their length spectrum for a non-flat complex space form*, Osaka Math. J. **40** (2003), 895–916.
3. ———, *Trajectories on geodesic spheres in a non-flat complex space form.*
4. ———, *Moderate Killing helices of proper order four on a complex projective space.*
5. T. Adachi and S. Maeda, *Holomorphic helix of proper order* 3 *on a complex hyperbolic plane*, Top. Appl. **146–147** (2005), 201–207.
6. ———, *Type two Killing helices of proper order four on a complex projective plane*, in this volume.
7. S. Maeda and T. Adachi, *Holomorphic helices in a complex space form*, Proc. A.M.S. **125** (1997), 1197–1202.
8. S. Maeda and T. Ohnita, *Helical geodesic immersions into complex space forms*, Geom. Dedicata **30** (1983), 93–114.

TYPE TWO KILLING HELICES OF PROPER ORDER FOUR ON A COMPLEX PROJECTIVE PLANE

T. ADACHI*

Department of Mathematics, Nagoya Institute of Technology
Gokiso, Nagoya, 466-8555, Japan
E-mail: adachi@nitech.ac.jp

S. MAEDA

Department of Mathematics, Shimane University,
Matsue, 690-8504, Japan
E-mail: smaeda@riko.shimane-u.ac.jp

We study some geometric properties of helices of proper order 4 on a complex projective space which are of type 2 in the sense of submanifolds in a Euclidean space through the first standard embedding and which are generated by some Killing vector fields.

Keywords: Killing helices of proper order 4; Length; Type 2; Complex torsions; Limit curve.

1. Introduction

In the preceding paper [5] Chen and the second author studied smooth curves on a complex projective space through the first standard embedding into a Euclidean space. They investigated smooth curves from the view point of submanifolds of finite type in a Euclidean space. A submanifold in a Euclidean space is said to be of k-type if the isometric immersion is decomposed as a sum of $(k+1)$-eigenfunctions of the Laplace operator of the submanifold including constant functions (cf. [4]).

Following their result, we see a smooth curve parameterized by its arclength on a complex projective space is of type 1 if and only if it is a geodesic or a circle of positive geodesic curvature with complex torsion ± 1, which lies on some totally geodesic complex line and is interpreted as a

*The first author is partially supported by Grant-in-Aid for Scientific Research (C) (No. 17540072), Japan Society for the Promotion of Science.

trajectory for a Kähler magnetic field. For about smooth curves of type 2, if we pose an additional condition that their tangent vectors and principal normal vectors are $\mathbb{C}$-linearly independent, we find they are either circles lying on some totally geodesic real projective space or special kind of helices of proper order 4. In this paper we focus our mind on these helices of proper order 4 and show that they are always closed.

2. Killing helices of proper order 4

A smooth curve γ parameterized by its arclength on a Riemannian manifold M is said to be a *helix of proper order* 4 if it satisfies the following system of ordinary differential equations:

$$\begin{cases} \nabla_{\dot\gamma}\dot\gamma = \kappa_1 V_2, \\ \nabla_{\dot\gamma}V_2 = -\kappa_1\dot\gamma + \kappa_2 V_3, \\ \nabla_{\dot\gamma}V_3 = -\kappa_2 V_2 + \kappa_3 V_4, \\ \nabla_{\dot\gamma}V_4 = -\kappa_3 V_3 \end{cases}$$

with positive constants κ_1, κ_2, κ_3 and an orthonormal system $\{V_1 = \dot\gamma, V_2, V_3, V_4\}$ of vector fields along γ. Here $\nabla_{\dot\gamma}$ denotes the covariant differentiation along γ with respect to the Riemannian connection ∇ on M. These constants κ_1, κ_2, κ_3 are called the *geodesic curvatures* of γ, and the vector field $\{V_1, V_2, V_3, V_4\}$ are called the Frenet frame of γ.

On a real space form it is well-known that all helices are generated by some Killing vector fields. We are hence interested in helices with such a condition. For a helix γ of proper order 4 on a Kähler manifold M with complex structure J it has important invariants. We define its *complex torsions* by $\tau_{ij} = \langle V_i, JV_j\rangle$ $(1 \leqq i < j \leqq 4)$. On a complex projective space, it is known that a helix is generated by some Killing vector field if and only if all its complex torsions are constant functions (see [9]).

3. Type 2 Killing helices

It is known that the parallel isometric embedding of a complex projective space $\mathbb{C}P^n(c)$ of constant holomorphic sectional curvature c into a Euclidean space $\mathbb{R}^N$ is decomposed as

$$g = f_2 \circ f_1 : \mathbb{C}P^n(c) \xrightarrow{f_1} S^{n(n+2)-1}\left(\frac{(n+1)c}{2n}\right) \xrightarrow{f_2} \mathbb{R}^N, \quad (3.1)$$

where f_1 is the first standard minimal embedding and f_2 is a totally umbilic embedding (cf. [6,7]). This embedding g has nice geometric properties.

A smooth curve γ on a Riemannian manifold M parameterized by its arclength is said to be a *circle* if it satisfies $\nabla_{\dot\gamma}\nabla_{\dot\gamma}\dot\gamma = -\kappa^2\dot\gamma$ with some non-negative constant κ. This constant is called the geodesic curvature of γ. When M is a Kähler manifold with complex structure J, we set its complex torsion by $\tau = \langle\dot\gamma, J\nabla_{\dot\gamma}\dot\gamma\rangle/\|\nabla_{\dot\gamma}\dot\gamma\|$. As we mentioned in Introduction, Chen and the second author [5] studied the shape $g\circ\gamma$ of a smooth curve γ on $\mathbb{C}P^n(c)$ through the parallel embedding g.

Proposition 3.1.

(1) *For each geodesic γ on $\mathbb{C}P^n(c)$, the curve $g\circ\gamma$ is a circle of geodesic curvature $\sqrt{c}$.*

(2) *For each circle of geodesic curvature κ and complex torsion $\tau = \pm 1$ on $\mathbb{C}P^n(c)$, the curve $g\circ\gamma$ is a circle of geodesic curvature $\sqrt{\kappa^2+c}$.*

(3) *If $g\circ\gamma$ is a circle, then γ is either a geodesic or a circle of positive geodesic curvature with complex torsion $\tau = \pm 1$.*

Proposition 3.2. *If a smooth curve γ on $\mathbb{C}P^n(c)$* (with complex structure J) parameterized by its arclength satisfies the conditions that $\langle\dot\gamma, J\nabla_{\dot\gamma}\dot\gamma\rangle/\|\nabla_{\dot\gamma}\dot\gamma\| = 0$ and $g\circ\gamma$ is a helix of proper order 4, then it is one of the following:

1) a circle of null complex torsion, which lies on some totally real totally geodesic $\mathbb{R}P^2(c/4)$,

2) a helix of proper order 4 whose geodesic curvatures satisfy $\kappa_1 = \kappa_3$, $18\kappa_1^2 + 4\kappa_2^2 = 9c$ and whose complex torsions satisfy $\tau_{12} = \tau_{34} = \tau_{13} = \tau_{24} = 0$, $\tau_{14} = -\tau_{23} = \pm 1$.

We here concentrate our attention on helices of proper order 4 in the above Proposition. We shall call them *type* 2 *Killing helices* of proper order 4. It should be noted that their Frenet frames satisfy $V_3 = \pm JV_2$ and $V_4 = \mp J\dot\gamma$.

Theorem 3.1. *On a complex projective space $\mathbb{C}P^n(c)$ of constant holomorphic sectional curvature c, every two type Killing helix is simple and closed. If its first geodesic curvature is κ_1 $(0 < \kappa_1 < \sqrt{c/2})$, then its length is $2\sqrt{2}\pi/\sqrt{2c-\kappa_1^2}$.*

Proof. We consider the case that $c = 4$. We denote by $\varpi : S^{2n+1}(1) \to \mathbb{C}P^n(4)$ the Hopf fibration. We regard vector fields on $\mathbb{C}P^n(4)$ as horizontal vector fields on $S^{2n+1}(1)$ $(\subset \mathbb{C}^{n+1})$. Connections $\overline{\nabla}$ on $\mathbb{C}^{n+1}$ and ∇ on $\mathbb{C}P^n(4)$ are related by the formula

$$\overline{\nabla}_X Y = \nabla_X Y + \langle X, JY\rangle J\mathcal{N} - \langle X, Y\rangle\mathcal{N}$$

for vector fields X, Y on $\mathbb{C}P^n(4)$, where $\mathcal{N}$ is the outward unit normal on S^{2n+1} in $\mathbb{C}^{n+1}$. Thus by the equation

$$\begin{cases} \nabla_{\dot\gamma}\dot\gamma = \kappa_1 V_2, \\ \nabla_{\dot\gamma}V_2 = -\kappa_1\dot\gamma + \kappa_2(\pm J V_2), \end{cases}$$

we find a horizontal lift $\hat\gamma$ of γ satisfies

$$\begin{cases} \overline{\nabla}_{\dot{\hat\gamma}}\dot{\hat\gamma} = \kappa_1 V_2 - \mathcal{N}, \\ \overline{\nabla}_{\dot{\hat\gamma}}V_2 = -\kappa_1\dot{\hat\gamma} + \kappa_2(\pm J V_2), \end{cases}$$

which is equivalent to

$$\frac{d^3\hat\gamma}{dt^3} \mp \sqrt{-1}\,\kappa_2\frac{d^2\hat\gamma}{dt^2} + (\kappa_1^2+1)\frac{d\hat\gamma}{dt} \mp \sqrt{-1}\,\kappa_2\hat\gamma = 0,$$

where the double signs take the same signature. Since its characteristic equation

$$\lambda^3 \mp \sqrt{-1}\,\kappa_2\lambda^2 + (\kappa_1^2+1)\lambda \mp \sqrt{-1}\,\kappa_2 = 0$$

should have pure imaginary solutions, by setting $\lambda = \sqrt{-1}\Lambda$ we have

$$\Lambda^3 \mp \kappa_2\Lambda^2 - (\kappa_1^2+1)\Lambda \pm \kappa_2 = 0,$$

which turns to

$$\left(\Lambda \mp \frac{\kappa_2}{3}\right)^3 - \frac{1}{3}(3\kappa_1^2+\kappa_2^2+3)\left(\Lambda \mp \frac{\kappa_2}{3}\right) \mp \frac{\kappa_2}{27}(9\kappa_1^2+2\kappa_2^2-18) = 0.$$

By the condition $9\kappa_1^2+2\kappa_2^2 = 18$ we see this cubic equation has three distinct real solutions

$$\Lambda = \pm\sqrt{\frac{2-\kappa_1^2}{2}} - \sqrt{\frac{8-\kappa_1^2}{2}},\ \pm\sqrt{\frac{2-\kappa_1^2}{2}},\ \pm\sqrt{\frac{2-\kappa_1^2}{2}} + \sqrt{\frac{8-\kappa_1^2}{2}}.$$

Therefore we find $\hat\gamma$ is of the form

$$\begin{aligned} \hat\gamma(t) = \exp\Big(\pm\sqrt{-(2-\kappa_1^2)/2}\,t\Big)\Big\{A\exp\Big(-\sqrt{-(8-\kappa_1^2)/2}\,t\Big) \\ + B + C\exp\Big(\sqrt{-(8-\kappa_1^2)/2}\,t\Big)\Big\} \end{aligned} \tag{3.2}$$

with some linearly independent $A, B, C \in \mathbb{C}^{n+1}$. As $\hat\gamma$ is a horizontal lift of γ, we see $\gamma(t_0) = \gamma(0)$ if and only if $\hat\gamma(t_0) = e^{\sqrt{-1}\rho}\hat\gamma(0)$ with some real number ρ. One can easily see that such a case occurs only for the case that t_0 is an integer multiple of $2\sqrt{2}\pi/\sqrt{8-\kappa_1^2}$. This guarantees that γ is a simple closed curve with length $2\sqrt{2}\pi/\sqrt{8-\kappa_1^2}$. □

Remark 3.1. Here we give an explicit formula of γ on $\mathbb{C}P^n(4)$. Under the initial condition

$$\gamma(0) = \varpi(z), \quad \dot\gamma(0) = d\varpi((z,u)), \quad \nabla_{\dot\gamma}\dot\gamma(0) = \kappa_1 d\varpi((z,v)),$$

where $z, u, v \in \mathbb{C}^{n+1}$ are linearly independent over $\mathbb{C}$, as we have $\hat\gamma(0) = z$, $\hat\gamma'(0) = u$, $\hat\gamma''(0) = \kappa_1 v - z$, we find those vectors in (3.2) are given as

$$\begin{aligned} A &= \frac{1}{2(8-\kappa_1^2)}\Big\{\Big(4-\kappa_1^2 \pm \sqrt{(2-\kappa_1^2)(8-\kappa_1^2)}\Big)z \\ &\qquad\qquad + \Big(\sqrt{8-\kappa_1^2} \pm 2\sqrt{2-\kappa_1^2}\Big)\sqrt{-2}\,u - 2\kappa_1 v\Big\}, \\ B &= \frac{1}{8-\kappa_1^2}\Big\{4z \mp 2\sqrt{2-\kappa_1^2}\sqrt{-2}\,u + 2\kappa_1 v\Big\}, \\ C &= \frac{1}{2(8-\kappa_1^2)}\Big\{\Big(4-\kappa_1^2 \mp \sqrt{(2-\kappa_1^2)(8-\kappa_1^2)}\Big)z \\ &\qquad\qquad - \Big(\sqrt{8-\kappa_1^2} \mp 2\sqrt{2-\kappa_1^2}\Big)\sqrt{-2}\,u - 2\kappa_1 v\Big\}. \end{aligned}$$

4. Limit curves

In this section we study the phenomena of type 2 Killing helices when we make their geodesic curvatures $(\kappa_1, \kappa_2, \kappa_1)$ with $18\kappa_1^2 + 4\kappa_2^2 = 9c$ tend to $(\sqrt{c/2}, 0, \sqrt{c/2})$ and when make them tend to $(0, 3\sqrt{c}/2, 0)$ on $\mathbb{C}P^n(c)$. Since they are closed with length $2\sqrt{2}\pi/\sqrt{2c-\kappa_1^2}$, we might be able to consider their limit curves (c.f. [2]). On $\mathbb{C}P^n(c)$ we denote by $\gamma_{\kappa_1}^{\pm}$ for $0 < \kappa_1 < \sqrt{c/2}$ a type 2 Killing helix of proper order 4 with geodesic curvatures $\kappa_1 = \kappa_3$, $\kappa_2 = 3\sqrt{c - 2\kappa_1^2}/2$ and complex torsion $\tau_{14} = \pm 1$ which satisfies the initial condition $\gamma_{\kappa_1}^{\pm}(0) = \varpi(z)$, $\dot\gamma_{\kappa_1}^{\pm}(0) = d\varpi((z,u))$, $\nabla_{\dot\gamma_{\kappa_1}^{\pm}}\dot\gamma_{\kappa_1}^{\pm}(0) = \kappa_1 d\varpi((z,v))$.

Theorem 4.1. *On $\mathbb{C}P^n(c)$ limit curves of type 2 Killing helices exist and satisfy the following properties.*

(1) *The limit curves $\lim_{\kappa_1 \uparrow \sqrt{c/2}} \gamma_{\kappa_1}^{+}$, $\lim_{\kappa_1 \uparrow \sqrt{c/2}} \gamma_{\kappa_1}^{-}$ are circles of geodesic curvature $\sqrt{c/2}$ and of null complex torsion.*
(2) *The limit curves $\lim_{\kappa_1 \downarrow 0} \gamma_{\kappa_1}^{+}$, $\lim_{\kappa_1 \downarrow 0} \gamma_{\kappa_1}^{-}$ are geodesics.*

Proof. We consider the case $c = 4$. First we study the case $\kappa_1 \uparrow \sqrt{2}$. By (3.2) and Remark 3.1, we see

$$\begin{aligned} \lim_{\kappa_1 \uparrow \sqrt{2}} \hat\gamma_{\kappa_1}^{+} &= \lim_{\kappa_1 \uparrow \sqrt{2}} \hat\gamma_{\kappa_1}^{-} \\ &= \frac{1}{3}\Big\{(2 + \cos\sqrt{3}t)z + \sqrt{3}\sin\sqrt{3}t\,u + \sqrt{2}(1 - \cos\sqrt{3}t)v\Big\} \end{aligned}$$

$$= \frac{\sqrt{2}}{3}(\sqrt{2}\,z + v) + \frac{1}{3}\cos\sqrt{3}\,t(z - \sqrt{2}\,v) + \frac{1}{\sqrt{3}}\sin\sqrt{3}\,t\,u.$$

As we see in [3], a horizontal lift $\hat{\sigma}$ of a circle σ of geodesic curvature κ and of null complex torsion on $\mathbb{C}P^n(4)$ with initial condition $\sigma(0) = \varpi(z)$, $\dot{\sigma}(0) = d\varpi((z,u))$, $\nabla_{\dot{\sigma}}\dot{\sigma}(0) = \kappa d\varpi((z,v))$ is represented as

$$\hat{\sigma}(t) = \frac{\kappa}{\kappa^2+1}(\kappa z + v) + \frac{\cos\sqrt{\kappa^2+1}\,t}{\kappa^2+1}(z - \kappa v) + \frac{\sin\sqrt{\kappa^2+1}\,t}{\sqrt{\kappa^2+1}}u,$$

and its length is $2\pi/\sqrt{\kappa^2+1}$. Since the limit of lengths of type 2 Killing helices is $\lim_{\kappa_1\uparrow\sqrt{2}} 2\sqrt{2}\pi/\sqrt{8-\kappa_1^2} = 2\pi/\sqrt{3}$, we find the limit curves $\lim_{\kappa_1\uparrow\sqrt{2}}\gamma_{\kappa_1}^+$, $\lim_{\kappa_1\uparrow\sqrt{2}}\gamma_{\kappa_1}^-$ are circles of geodesic curvature $\sqrt{2}$ and of null complex torsion.

Next we study the case $\kappa_1 \downarrow 0$. By (3.2) and Remark 3.1, we have

$$\lim_{\kappa_1\downarrow 0}\hat{\gamma}_{\kappa_1}^+ = \lim_{\kappa_1\downarrow 0}\hat{\gamma}_{\kappa_1}^- = \cos t\,z + \sin t\,u,$$

which is a horizontal lift of a geodesic σ with initial condition $\sigma(0) = \varpi(z)$, $\dot{\sigma}(0) = d\varpi((z,u))$ (see for example [1,3]). As every geodesic on $\mathbb{C}P^n(4)$ is closed of length π and the limit of lengths of type 2 Killing helices is $\lim_{\kappa_1\downarrow 0} 2\sqrt{2}\pi/\sqrt{8-\kappa_1^2} = \pi$, we get the second assertion. □

References

1. T. Adachi, *Kähler magnetic fields on a complex projective space*, Proc. Japan Acad. Sci. **79** Ser. A (1994), 12–13.
2. ———, *Lamination of the moduli space of circles and their length spectrum for a non-flat complex space form*, Osaka J. Math. **40** (2003), 895–916.
3. T. Adachi, S. Maeda and S. Udagawa, *Circles in a complex projective space*, Osaka J. Math. **32** (1995), 709–719.
4. B.Y. Chen, *Total Mean Curvature and Submanifolds of finite type*, World Scientific (1984).
5. B.Y. Chen and S. Maeda, *Extrinsic characterizations of circles in a complex projective space imbedded into a Euclidean space*, Tokyo J. Math. **19** (1996), 169–185.
6. D. Ferus, *Immersions with parallel second fundamental form*, Math. Z. **140** (1974), 87–92.
7. D. Ferus, *Symmetric submanifolds of Euclidean space*, Math. Ann. **247** (1980), 81–93.
8. S. Maeda and T. Adachi, *Holomorphic helices in a complex space form*, Proc. A.M.S. **125** (1997), 1197–1202.
9. S. Maeda and Y. Ohnita, *Helical geodesic immersions into complex space forms*, Geom. Dedicata **30** (1989), 93–114.

REAL ANALYTICITY OF THE HYPERKAHLER ALMOST KÄHLER MANIFOLDS

L.N. APOSTOLOVA

Institute of Mathematics and Informatics,
Bulgarian Academy of Sciences,
Acad. G. Bontchev Street, bl. 8, 1113 Sofia, Bulgaria
E-mail: liliana@math.bas.bg
http://www.math.bas.bg/

Using technique developped in [6] it is proved that a smooth manifold equipped with two symplectic-homotopic symplectic structures ω_0 and ω_1 admits a real analytique structure such that ω_0 and ω_1 are real analytique ones. Then the result is used to prove that each hyperkahler almost Kähler manifolds obtain a real analytic hyperkahler almost Kähler structure C^∞ equivalent to the given one.

Keywords: Real analytic manifold; Hyperkahler manifold; almost Kähler manifold.

1. Preliminaries

Let M be a C^∞-smooth paracompact $4n$-dimensional manifold and J, K be antiinvolutive automorphisms of the tangent bundle on M which anticommute, i.e. $J^2 = K^2 = -E$ and $JK = -KJ$. Let h be a riemannian metric on M satisfying the equalities $h(JX, Y) = h(X, JY)$, $h(KX, Y) = h(X, KY)$ for each two vector fields X, Y defined on an open set in M. Then h is called *an almost hermitian metric* with respect to the almost complex structures J and K and (M, J, h), (M, K, h) are called *almost hermitian manifolds* and $(M; J, K, h)$ is called *a hyperkahler manifold*. The two-form Ω on an almost hermitian manifold (M, J, h) defined by the equality $\Omega(X, Y) = h(X, JY)$ where X, Y are vector fields defined on an open set in M is called *a fundamental form* of the almost hermitian manifold. An almost hermitian manifold is called *an almost Kähler manifold*, if the fundamental form Ω is a closed two-form, i.e. if $d\Omega = 0$.

MS Classification: 53C28, 53D35, 32C05.

Let $(M; J, K, h)$ be an hyperkahler manifold, such that (M, J, h) and (M, K, h) are almost Kähler ones. Then the manifold (M, J, K, h) is called hyperkahler almost Kähler manifold.

Let ω be a closed nondegenerate differential two-form on M. Then the couple (M, ω) is called *a symplectic manifold* and the two-form ω is called *a symplectic form* on the manifold M.

When (M, J, h) is an almost Kähler manifold, the fundamental two-form Ω will be a symplectic form on M and (M, Ω) will be a symplectic manifold too. So each almost Kähler manifold is a symplectic one.

Definition 1.1. We say that an almost complex manifold (M, J) is C^∞-equivalent (or real analytically equivalent) to an almost complex manifold $(\tilde{M}, \tilde{J})$ if there exists a diffeomorphism (or real analytical bijection) $\varphi : M \to \tilde{M}$ which differential $d\varphi$ commute with the authomorphisms J^* on T^*M and $\tilde{J}^*$ on $T^*\tilde{M}$ respectively, i.e. $\tilde{J}^* d\varphi = d\varphi J^*$.

Here the automorphim J^* is defined on T^*M as conjugate automorphism of the automorphism J of the tangent bundle TM of the manifold M. We say that the almost hermitian manifold (M, J, h) is C^∞-equivalent (or real analytically equivalent) to the almost hermitian manifold $(\tilde{M}, \tilde{J}, \tilde{h})$ if the almost complex manifolds (M, J) and $(\tilde{M}, \tilde{J})$ are C^∞- equivalent (or real analytically equivalent) and in addition if the jacobian φ_* maps the almost hermitian metric h on the almost hermitian metric $\tilde{h}$, where φ is the diffeomorphism (or the real analytical bijection) in Definition 1.1.

2. Real analyticity of a smooth manifold with two symplectic-homotopic symplectic forms ω_0 and ω_1

Let M be a smooth paracompact manifold with two symplectic-homotopic symplectic forms ω_0 and ω_1. This means that there exists a homotopy operator $H(p,t) : M \times I \to \wedge^2 T^*M$, I being the unit interval in $\mathbf{R}^1$, such that $H(p,t)$ is a continuous mapping, $H(p,t) = \omega_t(p)$ where ω_t are symplectic forms and $H(p,0) = \omega_0(p)$, $H(p,1) = \omega_1(p)$. With the technique in [6], using systematically the parameter t we shall prove that there exists a C^∞ equivalent to M real analytic manifold $\tilde{M}$ such that the image of the forms ω_0 and ω_1 under the equivalence mapping $f : M \to \tilde{M}$, namely $f^*\omega_0$ and $f^*\omega_1$ are real analytic symplectic forms on $\tilde{M}$.

Theorem 2.1. *Let M be a smooth paracompact manifold with two symplectic-homotopic symplectic forms ω_0 and ω_1. Then there exists a real analytic manifold $\tilde{M}$ and two real analytic symplectic structures $\tilde{\omega}_0$ and $\tilde{\omega}_1$*

on $\tilde{M}$, such that $(\tilde{M}, \tilde{\omega}_0, \tilde{\omega}_1)$ is C^∞-equivalent to (M, ω_0, ω_1). Moreover, the symplectic manifold $(\tilde{M}, \tilde{\omega}_0, \tilde{\omega}_1)$ is unique up to a real analytic isomorphism.

Proof of Theorem 2.1.

Existence: Let us recall first the notion of the Whitney topology for smooth functions and differential forms on a smooth manifold M (c.f. [4]). Let $\mathcal{E}(M)$ denotes the ring of the smooth functions on M and $g \in \mathcal{E}(M)$. A neighbourhood in the Whitney topology for g is given by the data $K = (K_j)_{j\in N}$, $m = (m_j)_{j\in N}$ and $(\varepsilon_j)_{j\in N}$, where $M = \cup K_j$, $K_j \subset \operatorname{int} K_{j+1}$, $(K_j)_{j\in N}$ is a compact exhausting of M, $m_j \in \mathbf{N}$ and $\varepsilon_j > 0$. Then a neighbourhood corresponding to these data are the set of functions $h \in \mathcal{E}(M)$ such that

$$\sum_{|\alpha|\le m_j} \frac{1}{\alpha!} \sup_{x\in K_{j+1}\setminus \operatorname{int} K_j} |D^\alpha (h-g)| < \varepsilon_j.$$

Using the neighbourhoods of the smooth functions on M, it is easy to define a Whitney topology and neighbourhoods for differential forms on M too.

Now let M be a smooth manifold and ω_0 and ω_1 be two smooth two-forms on M, symplectic-homotopic by the homotopy $H(p,t)$, $p \in M$, $t \in [0,1]$ and $H(p,0) = \omega_0$, $H(p,1) = \omega_1$. By Whithney's theorem we may assume that M is real analytic manifold and by theorem of Kutzschebauch and Loose that ω_0 is a real analytic symplectic form on M symplectic-homotopic to ω_1. We must find a diffeomorphism f of M such that $f^*\omega_1 = \omega_1^a$ is real analytic. Then $f^*\omega_1$ will be symplectic form since $d(f^*\omega_1) = f^*(d\omega_1) = 0$ and $(f^*\omega_1)^n = f^*(\omega_1^n) \neq 0$ everywhere on M.

By Grauert's theorem we may assume that M is embedded properly into some euclidean space $\mathbf{R}^N$ and using Whithney's extension theorem we may assume that ω_1 is defined on a fixed neighbourhood $\Omega \subset \mathbf{R}^N$ of M, and there are defined all two-forms from the symplectic homotopy operator $H(t,p)$, $t \in [0,1]$.

Let us give first two lemmas analogous to Lemmas from [6].

Lemma 2.1. *Let M be a real analytic manifold. For every $\varepsilon > 0$ for a given homotopy operator $H(p,t) : I \times M \to \mathcal{E}^{(k)}(M)$ such that $\omega_t = H(p,t)$ are nondegenerate closed differential k-forms $\omega_t(p) \in \mathcal{E}^{(k)}(M)$ $(k \ge 0)$, there exists a homotopy operator $\tilde{H}(p,t) = \tilde{\omega}_t(p)$ such that $D(H(p,t) - \tilde{H}(p,t)) < \varepsilon$, $H(p,0) = \tilde{H}(p,0)$, $H(p,1) = \tilde{H}(p,1)$ and a real analytic family of closed differential k-forms $\tilde{\omega}_t^a$ on M, depending continuously on parameter $t \in I$ and representing the same de Rham cohomology class $[\tilde{\omega}_t^a] = [\tilde{\omega}_t] \in H^k_{dR}(M;\mathbf{R})$.*

Sketch of the proof of Lemma 2.1. Applying the Lemma 1 from § 2 in [6] for every closed nondegenerate k-form ω_t we can find an analytic closed nondegenerate k-form ω_t^a with the same de Rham cohomology class in $H_{dR}(M;\mathbf{R})$. Let $\varepsilon > 0$ and $t_0 = 0 < t_1 < \ldots < t_k$ be points in I such that $D(\omega_{t'} - \omega_{t''}) < \varepsilon/3$ for each $t', t'' \in (t_i, t_{i+1})$ where D denote the distance in the Whitney topology. Then the homotopy $\tilde{H}(p,t)$ can be choosen as follows: $\tilde{H}(p,t_i) = \omega_{t_i}$ for $i = 0, 1, \ldots, k-1$ and in the interval $(t_i, t_{i+1}) \ni t = h.t_i + (1-h).t_{i+1}$, $0 \le h \le 1$, $\tilde{H}(p,t) = h.\tilde{H}(p,t_i) + (1-h).\tilde{H}(p,t_{i+1})$. Then the choosen in this way homotopy operator give the desired homotopic operator in our Lemma 1. We put also $\tilde{\omega}_{t_i}^a := \omega_{t_i}^a$ and $\tilde{\omega}_t^a := h.\tilde{\omega}_{t_i}^a + (1-h).\tilde{\omega}_{t_{i+1}}^a$.

Lemma 2.2. *Let M be real analytic manifold, $\varepsilon > 0$ and $H(p,t) = \omega_t(p)$ homotopy operator for ω_0 and ω_1 consisting of closed differential 2-forms on M. Let $B \subseteq \mathcal{E}^1(M)$ be an arbitrary small convex Whitney neighbourhood of the zero one-form on M. Then there exist a homotopy operator $\tilde{H}(p,t) = \tilde{\omega}_t(p)$ of the two-forms ω_0 and ω_1, $D(H(p,t) - \tilde{H}(p,t)) < \varepsilon$ and a family of one-forms $\beta_t = \tilde{B}(p,t) \in B$, for each $t \in I$, continuous on the parameter t, such that $\tilde{\omega}_t^a := \tilde{\omega}_t + d\beta_t$ is real analytic for each $t \in I$.*

Sketch of the proof of Lemma 2.2. Let us consider the points $t_0, t_1, \ldots, t_k$ and the homotopy operator $\tilde{H}(p,t)$ considered in the proof of Lemma 1. Using Lemma 2 from § 2 in [6] we construct one-forms β_i satisfying the condition of Lemma 2 for each $i = 1, 2, \ldots, k$. Then we consider the one-forms $\beta_t = h.\beta_i + (1-h).\beta_{i+1}$ for $i = 0, 1, \ldots, k-1$ where in the interval $(t_i, t_{i+1}) \ni t = h.t_i + (1-h).t_{i+1}$, $0 \le h \le 1$. As $\tilde{H}(p,t) = h.\tilde{H}(p,t_i) + (1-h).\tilde{H}(p,t_{i+1})$ by the construction, the property $\omega_t^a := \omega_t + d\beta_t$ to be real analytic holds. The Lemma is proved.

Finish of the existing part of proof of Theorem 2.1: Now we have a homotopy operator $\tilde{H}(p,t) = \tilde{\omega}_t(p)$ which is ε-closed to $H(p,t)$ and

(a) a family of real-analytic closed two-forms $\tilde{\omega}_t^a(p)$ with the same de Rham cohomology class with $\tilde{\omega}_t$ according to Lemma 1,
(b) we can choose two-forms $\tilde{\omega}_t^a$ so closed to $\tilde{\omega}_t$ that $\tilde{\omega}_t^h := h.\tilde{\omega}_t + (1-h).\tilde{\omega}_t^a$ to be non-degenerate for all $0 \le h \le 1$, and
(c) we have a family of one-forms β_t such that $\tilde{\omega}_t^a = \tilde{\omega}_t + d\beta_t$ are real-analytic according to Lemma 2 then $d\beta_t = \tilde{\omega}_t - \tilde{\omega}_t^a \in \mathcal{E}^{(1)}(M)$ is such that the induced vector field ξ_t is integrable up to time one.

Now we shall apply the fundamental Moser deformation trick in symplectic geometry to finish the proof of Theorem 2.1.

Suppose that $\tilde{\omega}_t$ and $\tilde{\omega}_t^a$ are symplectic forms on M representing the same de Rhem cohomoly class on M, $[\tilde{\omega}_t] = [\tilde{\omega}_t^a] \in H^2_{dR}(M;\mathbf{R})$, which are closed enough so that the stright line curve $h \to \omega_t^h := (1-h)\tilde{\omega}_t + h\tilde{\omega}_t^a$ has non-degenerate values in $\mathcal{E}^{(2)}(M)$ for all $h \in [0,1]$.

We choose a potential $\beta_t \in \mathcal{E}^{(1)}(M)$ as in Lemma 2 ($-d\beta_t = \tilde{\omega}_t - \tilde{\omega}_t^a$) and find a unique (smooth) curve of vector fields $h \to \xi_h^t$, depending on parameter t, such that $i_{\xi_h^t}\omega_t^h = \beta_t$ as ω_t^h is nondegenerate. Here $i_{\xi_h^t}$ denote the contraction of the form ω_t^h by the vector field ξ_h^t. Taking derivatives one find that $di_{\xi_h^t}\omega_t^h + \sigma_t = 0$, where $\sigma_t := \tilde{\omega}_t - \tilde{\omega}_t^a$.

Finally, in (c) is supposed that non-autonomus vector field ξ_h^t on M is integrable up to time one, i.e. there exists a family of curves $h \to f_h^t$ in the diffeomorphism group of M with $f_0^t = id$ and

$$\frac{d}{dh} f_h^t = \xi_h^t \circ f_h^t.$$

Then since Lie derivative $\mathcal{L}_{\xi_h^t}$ satisfies $\mathcal{L}_{\xi_h^t} = di_{\xi_h^t} + i_{\xi_h^t} d$ and the forms ω_t^h are closed, one concludes that

$$\frac{d}{dh}((f_h^t)^*\omega_t^h) = (f_h^t)^*(\mathcal{L}_{\xi_h^t}\omega_t^h + \frac{d}{dh}\omega_t^h) = (f_h^t)^*(di_{\xi_h^t}\omega_t^h + \sigma_t) = 0$$

for all $h \in [0,1]$. This shows that $f^t := f_h^t$ fulfills the requirement $(f^t)^*\tilde{\omega}_t = \tilde{\omega}_t^a$ since

$$(f^t)^*\tilde{\omega}_t - \tilde{\omega}_t^a = (f_1^t)^*\tilde{\omega}_t^1 - (f_0^t)^*\tilde{\omega}_t^0 = \int_0^1 \frac{d}{dh}((f_h^t)^*\omega_t^h)\, dh = 0.$$

Putting the parameter $t = 1$ and $\varepsilon \to 0$ we obtain the existing part of the theorem.

Uniqueness: Now let us suppose that on the real analytic manifolds M and $\bar{M}$ with two couples of real-analytic symplectic forms ω_0 and ω_1, and $\bar{\omega}_0$ and $\bar{\omega}_1$, respectively, there exists a smooth diffeomorphism f s.t. $f^*\omega_0 = \bar{\omega}_0$ and $f^*\omega_1 = \bar{\omega}_1$. Then f^* will act on the symplectic homotopy operator and produce a symplectic homotopy operator $\bar{H}(p,t)$ for $\bar{\omega}_0$ and $\bar{\omega}_1$ on $\bar{M}$. We would find a real-analytic bijection $\bar{f}$ of M to $\bar{M}$ with properties $\bar{f}^*\omega_0 = \bar{\omega}_0$ and $\bar{f}^*\omega_1 = \bar{\omega}_1$ to prove the uniqueness of the real-analytic structure on M s.t. ω_0 and ω_0 are real analytic symplectic forms.

The homotopy class $\{g \in \mathrm{Diff}\,(M,\bar{M}) : g \sim f\}$ is a Whitney open set and by the Grauert embedding theorem the real analytic diffeomorphisms $\mathrm{Diff}^{\,a}(M,\bar{M})$ are dense in $\mathrm{Diff}\,(M,\bar{M})$. So there exists a real-analytic diffeomorphism $f^a(M,\bar{M})$ arbitrary Whitney near to f and in the same homotopy class. Therefore $\tilde{\omega}_0 := (f^a)^*(\omega_0)$ and $\tilde{\omega}_1 := (f^a)^*(\omega_1)$ are real-analytic, $[\omega_0] = [\tilde{\omega}_0] \in H^2_{dR}(M)$, $[\omega_1] = [\tilde{\omega}_1] \in H^2_{dR}(M)$ and $\tilde{\omega}_0$ is Whitney near to

ω_0, $\tilde{\omega}_1$ is Whitney near to ω_1. Therefore if we find real-analytic potentials $\beta_{0,1} \in \mathcal{E}^1(M)$, $d\beta_0 = \bar{\omega}_0 - \omega_0$, $d\beta_1 = \bar{\omega}_1 - \omega_1$ Whitney near to zero, the Moser's deformation tric would produce a real-analytic diffeomorphism with $g^*\omega_0 = \bar{\omega}_0$, $g^*\omega_1 = \bar{\omega}_1$ and therefore $f^a \circ g \in \text{Diff}\,(M, \bar{M})$ will be the desired symplectomorphism between M and $\bar{M}$.

To do this one need two more Whitney-Poincaré lemmas with parameter — one smooth and one real-analytic. Following the shema in [6] the proof of uniqueness would be finished making the needed changes as in the existence part. □

3. Polarization of a symplectic form

Following § 4 in [1], we consider how to obtain canonical hermitian metric $\tilde{h}$ and almost complex structure $\tilde{J}$ on a manifold $\tilde{M}$, diffeomorphic to a given symplectic manifold (M, ω) in such a way that the symplectic form ω to correspond to the fundamental form $\tilde{\Omega}$ of the almost hermitian manifold $(M, \tilde{J}, \tilde{h})$. Moreover, so obtained almost hermitian manifold will be an almost Kähler one as its fundamental form will be closed two form.

We follow the construction of a polarization of a symplectic form, given in [1], § 4.2. The base of considerations is the procedure of polar decomposition of a non-singular matrix A into product of orthogonal matrix J and a positive definite symmetric matrix G. Let us denote as usual by $O(n)$ the group of orthogonal $n \times n$ matrix and by $H(n)$ the group of positive definite symmetric $n \times n$ matrix.

Let us consider the map $\varphi : O(n) \times H(n) \to GL(n)$ defined by $\varphi(F, G) = FG$ — the product of the matrices F and G. If $O(n)$, $H(n)$ and $GL(n)$ are equipped with their natural real analytic structures, the map φ is a real analytic morphism of $O(n) \times H(n)$ in $GL(n)$. The so written polar decomposition of a non-singular matrix is proved by C. Chevalley in [2; 1946, pp. 14–16] as a continuous mapping and by Y. Hatakayama [5; 1962] as an analytic mapping.

We shall use Theorem 4.2 for real analytycity of the mapping φ^{-1} from [1] which say that the polar decomposition as a map from $GL(n, \mathbf{R}) \to O(n) \times H(n)$ gives an analytic diffeomorphism between these manifolds with respect to the usual analytic structures.

We need also the theorem for existence, Theorem 4.3 from [1] which say that if (M^{2n}, ω) is a symplectic manifold then there exists a Riemannian metric g and an almost complex structure J such that $g(X, JY) = \omega(X, Y)$.

In order to prove this theorem it would be choosen any Riemannian metric k on M and a local k-orthonormal basis $\{X_1, \ldots, X_{2n}\}$. Then to consider the matrix A, where $A_{ij} = \omega(X_i, X_j)$. The matrix A will be a non-singular skew-symmetric $2n \times 2n$ matrix for each point p of an open set U where the vector fields $\{X_1, \ldots, X_{2n}\}$ are defined. This is so because the two-form ω as a symplectic form is a nondegenerate differential two-form. For the matrix A we construct a polar decomposition, $A = FG$ with some orthogonal matrix F and some positive definite symmetric matrix $G \in H(n)$. Then the mertic g is defined locally on U by the equalities $g(X_i, X_j) = H_{ij}$ and the almost complex structure J is defined locally by the equalities $JX_i = F_i^j X_j$. The rest of the proof of the theorem is to check the gluing of the so defined metric and almost complex structure on the common definition set of any two k-orthogonal basis $\{X_1, \ldots, X_{2n}\}$ and $\{Y_1, \ldots, Y_{2n}\}$ which is so because of the construction made above and because of the uniqueness of the polar decomposition.

We remark also that if (J, h) is a starting structure for the polarization of the fundamental form $\omega(X, Y) = h(X, JY)$, then the result of polarization is the same metric h and almost complex structure J.

4. Real analyticity of the hyperkahler almost Kähler manifolds

Theorem 4.1. *Let (M, J, K, h) be a C^∞-smooth hyperkahler almost Kähler manifold. Then there exists a real analytic manifold $\tilde{M}$, real analytic almost complex structures $\tilde{J}$, $\tilde{K}$ and a real analytic almost hermitian metric $\tilde{h}$ on $\tilde{M}$ such that $(\tilde{M}, \tilde{J}, \tilde{K}, \tilde{h})$ is C^∞-equivalent to (M, J, K, h). Moreover, the hyperkahler manifold $(\tilde{M}, \tilde{J}, \tilde{K}, \tilde{h})$ can be choosen to be a hyperkahler almost Kähler manifold. Such a manifold $(\tilde{M}, \tilde{J}, \tilde{K}, \tilde{h})$ is unique up to a real analytic isomorphism.*

Proof of Theorem 4.1.
Existence: Let us consider the fundamental two-forms ω_i, $i = 0, 1$ of the given hyperkahler almost Kähler manifold (M, J, K, h), $\omega_0(X, Y) := h(X, JY)$, $\omega_1(X, Y) := h(X, KY)$ for any two vector fields X, Y on an open set in M. These are a non-degenerate closed two-forms on M and the closed two-forms on M $\omega_t = h(X, tJY) + h(X, (1-t)KY)$ gives a homotopy operator between them by symplectic two-forms, i.e. they are homotopic by symplectic two-forms. So (M, ω_i), $i = 0, 1$ are symplectic manifolds with desired property and the theorem for existence in §2 would be used. According this theorem there exists a diffeomorphim φ from M onto a real analytic manifold $\tilde{M}$ and symplectic real analytic forms $\tilde{\omega}_i$, $i = 0, 1$ on $\tilde{M}$ such that $\varphi_* \tilde{\omega}_i = \omega_i$, $i = 0, 1$.

Now we apply the reasons from §3 above to obtain a Riemannian metric $\tilde{g}$ and almost complex structure $\tilde{J}$ with a fundamental form $g(X, JY)$ coinsiding with $\tilde{\omega}_0$ and once more to obtain an almost complex structure $\tilde{K}$ with a fundamental form $g(X, KY)$ coinsiding with $\tilde{\omega}_1$.

If now we decompose by polarization the fundamental form ω_0 on the given manifold (M, J, h), we obtain the same almost complex structure and metrics h, as is remarked in § 3 above. So, using the conditions in Definition 1 above for the differential of the morphism φ, we obtain that the almost complex structure $\tilde{J}$ and the metric $g = \tilde{h}$ correspond to the initial almost complex structure J and metric h on M. Then we repeat the same for the almost complex structure K. So each hyperkahler almost Kähler manifold (M, J, h) is diffeomorphic to a real analytic hyperkahler almost Kähler manifold $(\tilde{M}, \tilde{J}, \tilde{K}, \tilde{h})$.

Uniqueness: Let now $(\overline{M}, \overline{J}, \overline{K}, \overline{h})$ be another real analytic hyperkahler almost Kähler manifold, diffeomorphic to the given one (M, J, K, h). Now applying the uniqueness part in § 2, we can find a real analytic diffeomorphism from $(\tilde{M}, \tilde{\omega}_i)$ to $(\overline{M}, \overline{\omega}_i)$, $i = 0, 1$. Then reasoning as in the part of existence we obtain that the constructed by polarization almost complex structures $\overline{J}$, $\overline{K}$ and metric $\overline{h}$ correspond to the almost complex structures $\tilde{J}$, $\tilde{K}$ and to the hermitian metric $\tilde{h}$. So the uniqueness of the real analytic hyperkahler almost Kähler manifold up to real analytic isomorphism, corresponding to the given smooth hyperkahler almost Kähler manifold is obtained. □

Acknowledgements are due to Prof. S. Dimiev for his interest on the subject.

References

1. D.E. Blair, *Riemannian Geometry of Contact and Symplectic Manifolds*, Progress in Math., 203, Birkhäuser, Boston, Basel, Berlin, 2002.
2. C. Chevalley, *Theory of Lie Groups*, Princeton Univ. Press, Princeton, 1946.
3. H. Grauert, Ann. of Math. **68**, 460, (1958).
4. M. Hirsch, *Differential Topology*, Springer-Verlag, 1988.
5. Y. Hatakayama, Tohôku Math. J. **14**, 162, (1962).
6. F. Kutzschebauch and F. Loose, PAMS **128** (10), 3009, (2000).
7. J. Moser, Trans. Amer. Math. Soc. **120**, 286, (1965).

KCC AND LINEAR STABILITY FOR THE PARKINSON TREMOR MODEL

V. BALAN

University Politehnica of Bucharest, Faculty of Applied Sciences,
Department Mathematics I, Splaiul Independentei 313,
RO-060042, Bucharest, Romania,
E-mail: vbalan@mathem.pub.ro
http://www.mathem.pub.ro/dept/vbalan.htm

The paper investigates the KCC (structural) stability of the second-order extension of the brain-stimulated Parkinson tremor dynamical system. After a brief presentation of the known linear stability results of the model, are determined and studied the KCC-invariants. It is emphasized that the spectral properties of the second KCC-invariant provide practical information for the behavior of the investigated model. The semispray of the second-order differential system associated to the basic SODE is evidentiated.

Keywords: Dynamical system; Second order extension; KCC-invariants; Jacobi stability; Energy.

1. The neural 3-unit feedback model

The mathematical model which we analise in this paper is a three-dimensional biological system, first introduced by Titcombe [15], which describes in a simplified way the Parkinson tremor model in the presence of deep brain stimulation. It is a three-unit network model of a network with negative feedback to illustrate how an oscillating system interacts with periodic stimulation. Similar to numerous different network models which have been proposed for Parkinsonian tremor, the current model illustrates one way that the Hopf bifurcation might arise in a network context.

The time-evolution of the three units is governed by the following SODE:

$$\begin{cases} \dot{x}^1 = f(x^3) - x^1 \\ \dot{x}^2 = g(x^1) - x^2 \\ \dot{x}^3 = g(x^2) - x^3. \end{cases} \tag{1}$$

Here, f and g are inhibitory and excitatory response functions respectively.

Typically, response functions are sigmoidal, and are monotonically decreasing for inhibition and monotonically increasing for excitation. For our illustration, we use Hill functions, which are sigmoidal in form, to represent the response of the units where the exponent γ controls the slope (gain) of the response:

$$f(u) = \frac{\theta^\gamma}{u^\gamma + \theta^\gamma}, \quad g(u) = \frac{u^\gamma}{u^\gamma + \theta^\gamma}. \tag{2}$$

The effective feedback values of the the units threshold θ range about $\theta_0 = 0.5$ and we note that $g = 1 - f$.

It was proved that for $\gamma = 4$, one obtains a limit cycle and that the SODE exhibits a Hopf bifurcation; it provides a model with negative feedback (inhibition) as shown in the following diagram

$$x^1 \xrightarrow{\textit{excitation}} x^2 \xrightarrow{\textit{excitation}} x^3 \xrightarrow{\textit{inhibition}} x^1.$$

For γ, is commonly used $\gamma(t) = \gamma_0 - z(t)$, where $z(t) = ke^{-t/t_c}$ with $\gamma_0 = 6.0$. In practice are used iterated pulses with initial value z_0 and $z_n = z_0 e^{-n\tau/t_c} + \dfrac{\delta(e^{-n\tau/t_c} - 1)}{1 - e^{\tau/t_c}}$. Then the asymptotic value (for $n \to \infty$) is $z_\infty = \delta/(e^{\tau/t_c} - 1)$.

The response γ or γ_0 and the time constant t_c are refered as *network parameters* — which are inherent properties of the network, and δ and τ are referred as *stimulation parameters* — which are the stimulation pulse amplitude and stimulation period, respectively.

The model described by (1) involves a negative feedback system without time delay and thus, the Hopf bifurcation would be induced by an increase in gain. This network is a special case of an n-unit feedback inhibition network which can display oscillations provided that $n > 3$. The dynamics of systems of this type are a network effect, meaning that the units will not oscillate independently.

2. The covariant extension. Structural stability

In the following we investigate the Jacobi (KCC) structural stability of the second order extension of the dynamical system (1), via the five assocaited KCC-invariants of the SODE.

We shall first introduce several basic elements of KCC-theory [1,2,13], focusing on the second deviation tensor P^i_j and describing the role of the five invariants of the system. Let $x = (x^1, \ldots, x^n)$, $\dot{x} = dx/dt$ and t be the $2n+1$ coordinates of an open connected subset $\Omega \subset \mathbb{R}^n \times \mathbb{R}^n \times \mathbb{R}^1$. The second order system of n second order ODEs has the form

$$\frac{d^2x^i}{dt^2} + g^i(x, \dot{x}, t) = 0, \quad i \in \overline{1, n}, \tag{3}$$

and has been studied by D.D. Kosambi [9], E. Cartan [7] and S.S. Chern [8]. In general, $g^i = \frac{\partial X^i}{\partial x^j}\dot{x}^j$, $i = \overline{1, n}$. In our case, we have

$$\begin{cases} g^1 = .2500z^3\dot{z}/(z^4 + .625)^2 + \dot{x} \\ g^2 = \dot{y} - \left[4x^3/(x^4 + .625) - 4x^7/(x^4 + .625)^2\right]\dot{x} \\ g^3 = \dot{z} - \left[4y^3/(y^4 + .625) - 4y^7/(y^4 + .625)^2\right]\dot{y}. \end{cases} \tag{4}$$

They assumed real analyticity in a neighborhood of some initial conditions $(x_0, \dot{x}_0, t_0) \in \Omega$. Aiming to find the basic differential invariants of the system (3) under the non-singular coordinate transformations

$$\overline{x}^i = f^i(x^1, \ldots, x^n), \quad i \in \overline{1, n}, \quad \bar{t} = t. \tag{5}$$

D.D. Kosambi has introduced a covariant differential operator D, which acts on a contravariant vector field $\xi^i(x)$ as

$$\frac{D\xi^i}{dt} = \frac{d\xi^i}{dt} + \frac{1}{2}g^i_{;r}\xi^r, \tag{6}$$

where we have denoted by semicolon the partial differentiation relative to $\dot{x}$, and where the Eistein summation is used unless contrary stated. Using the operator D from (6), the system (3) rewrites

$$\frac{D\dot{x}^i}{dt} = \frac{1}{2}g^i_{;r}\dot{x}^r - g^i =: \varepsilon^i,$$

which defines a contravariant vector field ε^i on Ω, called *the first KCC-invariant*, and which is interpreted as an external force ([1]). Denoting $(x^1, x^2, x^3) = (x, y, z)$, we obtain the first invariant $\varepsilon = (\varepsilon^1, \varepsilon^2, \varepsilon^3)$,

$$\left(\frac{1}{2}\,\frac{.2500\,z^3\,\dot{z} + \dot{x}\,z^8 + .1250\,\dot{x}\,z^4 + .00390625\,\dot{x}}{(z^4 + .0625)^2} - \frac{.2500\,z^3\,\dot{z}}{(z^4 + .0625)^2} - \dot{x},\right.$$
$$\frac{1}{2}(\dot{y}\,x^8 + .1250\,\dot{y}\,x^4 + .00390625\,\dot{y} - .2500\,x^3\,\dot{x})/(x^4 + .0625)^2 - \dot{y}$$
$$+(4x^3/(x^4 + .0625) - 4x^7/(x^4 + .0625)^2)\dot{x},$$
$$\frac{1}{2}\,\frac{\dot{z}\,y^8 + .1250\,\dot{z}\,y^4 + .00390625\,\dot{z} - .2500\,y^3\,\dot{y}}{(y^4 + .0625)^2} - \dot{z}$$
$$\left. + \left[4y^3/(y^4 + .0625) - 4y^7/(y^4 + .0625)^2\right]\dot{y}\right).$$

The functions $g^i = g^i(x, \dot{x}, t)$ are 2-homogeneous in $\dot{x}$ if and only if $\varepsilon^i = 0$. In other words, $\varepsilon^i = 0$ is a necessary and sufficient condition for a semispray to be a spray. It is obvious that when $\{g^i\}_{i=\overline{1,n}}$ is the geodesic spray of a Riemannian or Finsler manifold, the first invariant vanishes.

Moreover, since the system is of the form $\dot{x} = X(x)$, the first invariant has the components $\varepsilon^i = \frac{1}{2}\frac{\partial X^i}{\partial x^r}\dot{x}^r$, and hence this vanishes for null velocities, i.e., on the stationary points of the field X. We note that for our model, the semispray is 1-homogeneous in velocities, and hence it does not provide a Finsler structure.

For $\Omega = \Omega_x \times \mathbb{R}^n \times \mathbb{R}$, the strongly nonlinear system $\varepsilon^i = 0$, $i = \overline{1,n}$ is governed by $D = \det\left(\frac{\partial X^i}{\partial x^r}\right)_{i,r=\overline{1,n}}$. For $D \neq 0$, the solution is the image of the null section of the vector bundle $\xi = (T\Omega_x, \pi, \Omega_x)$, and for $D = 0$, the solution is a subbundle of rank $k \in \{1, 2\}$ of ξ. In either case, we infer that the first invariant does not identically vanish on the total space of ξ.

As well, it is known that if the trajectories $x^i(t)$ of (3) vary via $\bar{x}^i(t) = x^i(t)+\xi^i(t)\eta$ with small parameter η, one obtains for the first approximation in η, the KCC-equations of variation

$$\frac{d^2\xi^i}{dt^2} + g^i_{;r}\frac{d\xi^r}{dt} + g^i_{,r}\xi^r = 0,$$

where we have denoted by comma the partial differentiation with respect to x. Following Kosambi this can be rewritten using (6) as

$$\frac{D^2\xi^i}{dt^2} = P^i_r\xi^r, \tag{7}$$

where the right side defines *the second KCC-invariant* $\{P^i_r\}_{i,r=\overline{1,n}}$ of the system (3), or *the deviation curvature tensor*

$$P^i_j = -g^i_{,j} - \frac{1}{2}g^r g^i_{;r;j} + \frac{1}{2}\dot{x}^r g^i_{,r;j} + \frac{1}{4}g^i_{;r}g^r_{;j} + \frac{1}{2}\frac{\partial g^i_{;j}}{\partial t}.$$

It is known that the eigenstructure of this (1,1)-tensor field provides an alternative to the Floquet Theory, and the eigenvalues of P^i_j replace the characteristic multipliers — these are also called *Floquet exponents* ([3,12]). In our case, the second invariant has the form

$$P^i_j = \frac{1}{2}\frac{\partial^2 X^i}{\partial x^j \partial x^r}\dot{x}^r + \frac{1}{4}\frac{\partial X^i}{\partial x^r}\frac{\partial X^r}{\partial x^j}.$$

The Jacobi stability is a weaker condition than the stability of periodic orbits given by the sign of eigenvalues of characteristic multipliers. Note that (7) is the Jacobi field equation when the starting system (3) represents geodesic equations in either Finsler or Riemannian geometry. This justifies the usage of the term Jacobi stability for KCC-Theory.

On the other hand, the Jacobi equation (7) of the Finsler manifold $(M = \Omega_x, F)$ can be written in the scalar form $\frac{d^2v}{ds^2} + K \cdot v = 0$, where $\xi^i = v(s)\eta^i$ is a Jacobi field along $\gamma : x^i = x^i(s)$, η^i is the unit normal vector field along γ; K is the flag curvature of (M, F) which describes the shape of the space ([5]). For a Finsler manifold (M, F), the flag curvature is a function of tangent planes and directions, which tells us how curved the space is at a point.

It is also known that the sign of K influences the geodesic rays ([5]). Indeed, if $K > 0$, then the geodesic bunch together (are Jacobi stable), and if $K < 0$, then they disperse (are Jacobi unstable).

Hence negative flag curvature is equivalent to positive eigenvalues of P^i_j, and positive flag curvature is equivalent to negative eigenvalues of P^i_j. It is known the following

Theorem 2.1 **([1,2]).** *The trajectories of* (3) *are Jacobi stable if and only if the real parts of the eigenvalues of the deviation tensor* P^j_i *are strict negative everywhere, and Jacobi unstable, otherwise.*

The notion of Jacobi stability presented until here can be extended to the general case of the SODE (3) using the Theorem above as the definition for the Jacobi stability of the trajectories of a SODE. The third, fourth and fifth invariants of the system (3) are respectively

$$R^i_{jk} = \frac{1}{3}(P^i_{j;k} - P^i_{k;j}), \quad B^i_{jkl} = R^i_{jk;l}, \quad D^i_{jkl} = g^i_{;j;k;l}.$$

It can be easily verified using the Schwartz theorem that in our case these last three invariants are all zero.

A notable result of the KCC-theory which points out the role of the five invariants is the following:

Theorem 2.2 **([1]).** *Two SODE's of form* (3) *on* Ω *can be locally transformed, relative to* (5), *one into another, if and only if their five invariants* $\varepsilon^i, P^i_j, R^i_{jk}, B^i_{jkl}, D^i_{jkl}$ *are equivalent tensors. In particular, there are local coordinates* $(\overline{x})$ *for which* $g^i(x, \dot{x}, t) = 0$ *if and only if all five KCC-tensors vanish.*

Based on Maple computations, we can infer straightforward that for our SODE subject to the requirement of having real (positive) solutions (x, y, z), there exists no coordinate change such that the coefficients of the new second order SODE-semispray do all vanish, i.e., the trajectories of the second-order extended system (including the field lines of the initial SODE) can never be lines, whatever coordinate system one might choose.

3. Linear and KCC stability

The equilibrium points of the system (1) are obtained when the right side of the SODE is set to zero. The resulting nonlinear system has in general several solutions; only the positive ones have significance. In the following we will consider the only real equilibrium point $(x, y, z)_* = (.5; .5; .5)$ — which proves to have positive components, found in the case of the parameter values

$$\theta = .5\,, \quad \gamma = 4. \tag{8}$$

Regarding the linear and Jacobi stability around this equilibrium point, the following results hold true.

Proposition 3.1. *In the case* (8), *the deviation curvature tensor* P^i_j *at the equilibrium point has the associated matrix*

$$[P_*] = \begin{pmatrix} 1.249999999 & -.5000000000 & .5000000000 \\ -.5000000000 & 1.250000000 & -.5000000000 \\ .5000000000 & -.5000000000 & 1.250000000 \end{pmatrix}.$$

This has three positive eigenvalues $\lambda_1 = .7499999993$, $\lambda_2 = .7500000000$ *and* $\lambda_3 = 2.250000000$. *Hence the field lines of the originar system are linear unstable, and the extended system is Jacobi unstable.*

The Jacobi stability depends on the variation of the parameters. We shall discuss further the impact of the change of the parameters γ and θ on the structural behavior of the extended SODE (3)–(4).

Considering the parameter θ variable within the interval (0,1) with γ fixed, we get the following results regarding the linear and the Jacobi stability.

Proposition 3.2. *Let* $\gamma = 4$ *(the limit cycle value). Then the following hold true:*

a) The SODE (1) *exhibits linear stability for* $\theta \in (0, 1)$ *and asymptotic stability for* $\theta \in (0, 1)\backslash\{0.5\}$.

b) For $\theta \in (0.2; 1)$, *the field lines of the extended system* (3)–(4) *are Jacobi unstable.*

Proof. For $\theta \in (0,1)$ the discriminant of the characteristic polyomial of the Jacobi matrix attached to (1) is positive and depends continuously on θ (see Fig. 1).

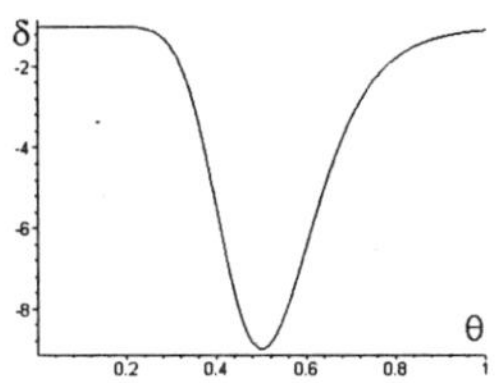

Fig. 1. The discriminant of the cubic characteristic equation in terms of θ, for $\gamma = 4$.

The Jacobian matrix has a real and two complex conjugate eigenvalues, which have all a strictly negative real part except for $\theta = 0.5$ — where the complex roots are pure imaginary ($\lambda_1 = -3$, $\lambda_{2,3} = \pm i\sqrt{3}$; see Fig. 2).

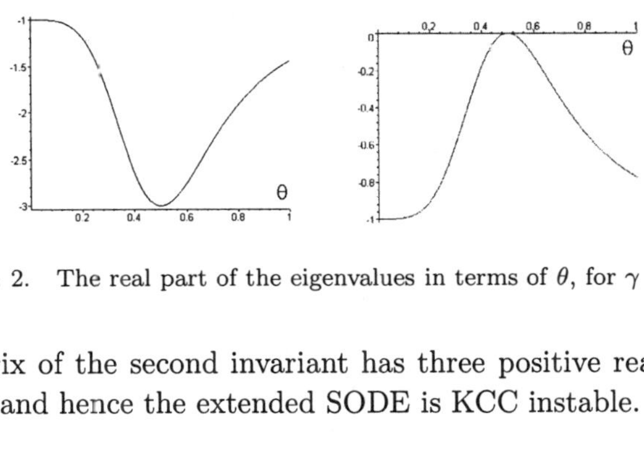

Fig. 2. The real part of the eigenvalues in terms of θ, for $\gamma = 4$.

b) The matrix of the second invariant has three positive real eigenvalues (see Fig. 3), and hence the extended SODE is KCC instable. □

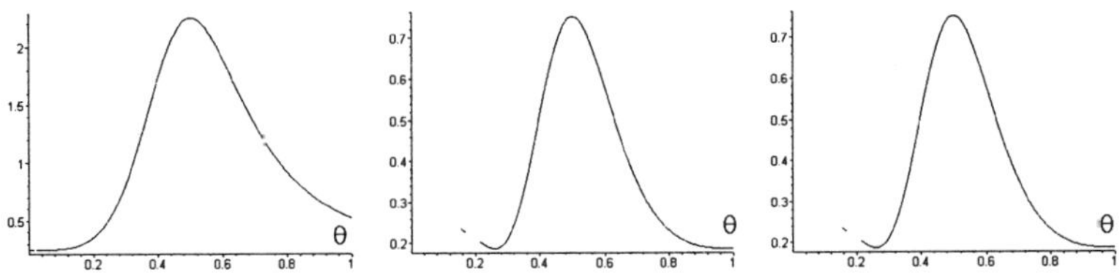

Fig. 3. The Jacobi eigenvalues in terms of θ, for $\gamma = 4$.

Remark. For $\theta = 0.5$ and the parameter γ freely varying in the interval $(0, 10)$, one can easily note that the value of the discriminant which decides the presence of complex eigenvalues of P strongly oscillates (see the subsequent figures).

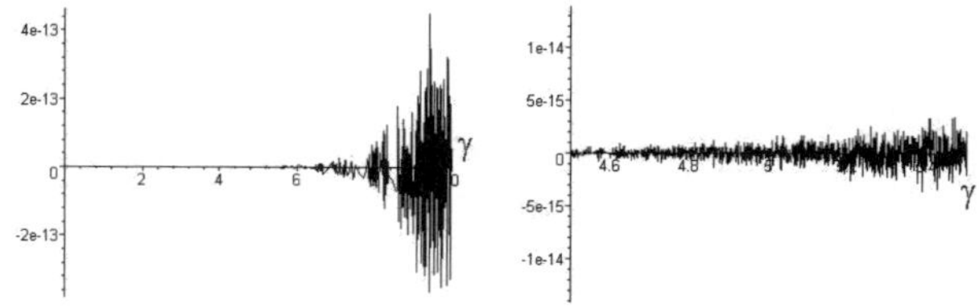

Fig. 4. The discriminant of the KCC characteristic polynomial in terms of γ, for $\theta = 0.5$.

4. The second-order Lagrangian extension

In the following we shall determine the second-order extension of gyroscopic type for the SODE (1). In general, for a given vector field $X \in \mathcal{X}(D)$, $D \subset \mathbb{R}^n$ we associate the corresponding SODE

$$\dot{x}^i = X^i(x), \quad i = \overline{1,n}. \tag{9}$$

Among various alternatives of extending the system (9) to a second order SODE within the framework of second-order jets, a notable one is described in [16]. The resulting dynamical system is in this case of Euler-Lagrange type, being attached to the Lagrangian function defined on the first-order jet space $J^1(\mathbb{R}, \mathbb{R}^n)$,

$$L = \sum_{k=1}^{n} \left(X^k(x) - \dot{x}^k\right)^2, \tag{10}$$

and has the shape

$$\ddot{x}^i = \eta^i(x) + \theta^i_s(x)\dot{x}^s, \tag{11}$$

where we consider on $\mathbb{R}^n$ the canonic Riemannian metric $g = \delta_{ij}dx^i \otimes dx^j$ and we denoted by $\eta^i = g^{is}\frac{\partial f}{\partial x^s}$ the gradient of the energy

$$F = \frac{1}{2}||X(x)||^2_g = g_{ij}X^iX^j = \frac{1}{2}\sum_{k=1}^{n}(X^k)^2 \tag{12}$$

of the field X, $\theta^i_j = g^{is}\omega_{sj}$ and ω is the the curl of X, which entails the gyroscopic character of the extension

$$\omega_{ij} = \frac{\partial(g_{js}X^s)}{\partial x^i} - \frac{\partial(g_{is}X^s)}{\partial x^j}, \quad i,j = \overline{1,n}.$$

In the particular case of a 3D-SODE, the second-order Lagrangian extension $\ddot{x} = \operatorname{grad} F + \operatorname{curl} X \times \dot{x}$ has the detailed form

$$\begin{cases} \ddot{x} = \dfrac{\partial F}{\partial x} + \left(\dfrac{\partial X^1}{\partial y} - \dfrac{\partial X^2}{\partial x}\right)\dot{y} + \left(\dfrac{\partial X^1}{\partial z} - \dfrac{\partial X^3}{\partial x}\right)\dot{z} \\ \ddot{y} = \dfrac{\partial F}{\partial y} + \left(\dfrac{\partial X^2}{\partial z} - \dfrac{\partial X^3}{\partial y}\right)\dot{z} + \left(\dfrac{\partial X^2}{\partial x} - \dfrac{\partial X^1}{\partial y}\right)\dot{x} \\ \ddot{z} = \dfrac{\partial F}{\partial z} + \left(\dfrac{\partial X^3}{\partial x} - \dfrac{\partial X^1}{\partial z}\right)\dot{x} + \left(\dfrac{\partial X^3}{\partial y} - \dfrac{\partial X^2}{\partial z}\right)\dot{y}, \end{cases} \tag{13}$$

where denoting e.g. $g_x = \frac{\partial g}{\partial x}$, and using (12), we get for the extension (13) of (1),

$$F = \tfrac{1}{2}\left[\left(\tfrac{.625}{z^4+.625} - x\right)^2 + \left(\tfrac{x^4}{x^4+.625} - y\right)^2 + \left(\tfrac{y^4}{y^4+.625} - z\right)^2\right]; \tag{14}$$

$$\operatorname{grad} F = -2X + 2(X^2 g_x, X^3 g_y, -X^1 g_z). \tag{15}$$

Within the model, is essential to monitor the energy-level F for $(\theta, \gamma) \in (0,1) \times (2,10)$, at the primary equilibrium point of the field $(x,y,z)_* = (.5,.5,.5)$, computed at $(\theta_0, \gamma_0) = (.5, 4)$ and considered on the null section of the bundle ξ. In the space of variables (θ, γ, F), the energy level attends minimal values at the intersection of the energy surface Σ with the plane $\theta = \theta_0$ (see Fig. 5, including the parallel projection of Σ onto the plane $\gamma = 0$).

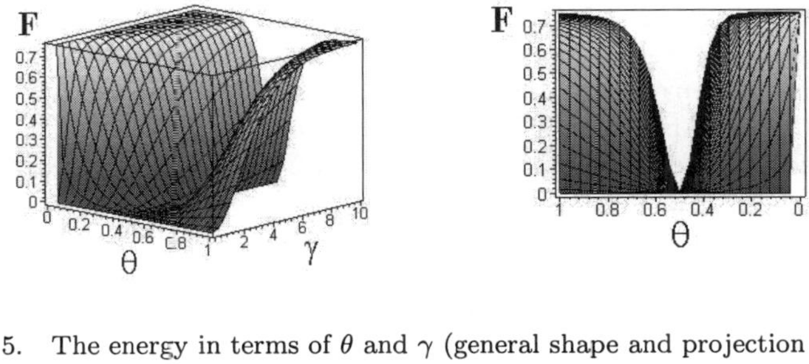

Fig. 5. The energy in terms of θ and γ (general shape and projection onto $\gamma = 0$).

Acknowledgement

The present work was partially supported by the Grant CNCSIS MEN 1478/2006.

References

1. P.L. Antonelli, *Equivalence problem for systems of second order ordinary differential equations*, Encyclopedia of Math., Kluwer Acad. Publ, 2000.
2. P.L. Antonelli, R. Ingarden and M. Matsumoto, *The Theory of Sprays and Finsler Spaces with Applications in Physics and Biology*, Kluwer Acad. Publ, FTPH vol. 58, 1993.
3. P.L. Antonelli, S. Rutz and V.S. Sabău, *A transient-state analysis of Tyson's model for the cell division cycle by means of KCC-theory*, Open System & Information Dynamics, 9 (2002), 3, 223–238.
4. V. Balan and I.R. Nicola, *Linear and structural stability of a cell division process model*, IJMMS, 2006 (2006), Article ID 51848, 1–15.
5. D. Bao, S.S. Chern and Z. Shen, *An Introduction to Riemann-Finsler Geometry*, Springer, GTM 200, 2000.
6. A. Beuter, M.S. Titcombe, D. Guehl and F. Richer, *The effect of deep brain stimulation on amplitude and frequency characteristics of rest tremor in Parkinsons disease*, Thalamus & Related Systems, 1, 3 (2001), 203211.
7. E. Cartan, *Observations sur le mémoir précédent*, Math. Zeitschrift 37 (1933), 619–622.
8. S.S. Chern, *Sur la geometrie d'un système d'equations differentialles du second ordre*, Bull. Sci. Mat. 63 (1939), 206–212.
9. D.D. Kosambi, *Parallelism and path space*, Math. Zeitschrift 37 (1933), 608–618.
10. R. Nicola and V. Balan, *Jacobi stability for dynamical systems with applications to biology*, Proc. of the 3-rd International Colloquium "Mathematics in Engineering and Numerical Physics" (MENP-3), 7–9 October 2004, Bucharest, Romania, BSG Proceedings 12, Geometry Balkan Press, Bucharest, 195–201.
11. E. Pereda, R. Quian Quiroga and J. Bhattacharya, *Nonlinear Multivariate Analysis of Neurophysiological Signals*, to appear in Progress in Neurobiology.
12. C. Robinson, *Dynamical Systems: Stability*, Symbolic Dynamics and Chaos, CRC Press, 1995.
13. V.S. Sabău, *Systems biology and deviation curvature tensor*, Nonlinear Analysis: Real World Applications 6 (2005), 563–587.
14. D.J. Saunders, *The Geometry of Jet Bundles*, Cambridge University Press, 1989.
15. M.S. Titcombe, L. Glass, D. Guehl and A. Beutera, *Dynamics of Parkinsonian tremor during deep brain stimulation*, Chaos, 11, 4 (2001), 766–773.
16. C. Udrişte, *Geometric Dynamics*, Kluwer Academic Publishers, 2000.

THE CAMASSA-HOLM EQUATION AS A GEODESIC FLOW FOR THE H^1 RIGHT-INVARIANT METRIC

A. CONSTANTIN[†] and R.I. IVANOV[‡]

School of Mathematics, Trinity College Dublin,
Dublin 2, Ireland
[†] *E-mail: adrian@maths.tcd.ie*
[‡] *E-mail: ivanovr@tcd.ie*

The fundamental role played by the Lie groups in mechanics, and especially by the dual space of the Lie algebra of the group and the coadjoint action are illustrated through the Camassa-Holm equation (CH). In 1996 Misiołek observed that CH is a geodesic flow equation on the group of diffeomorphisms, preserving the H^1 metric. This example is analogous to the Euler equations in hydrodynamics, which describe geodesic flow for a right-invariant metric on the infinite-dimensional group of diffeomorphisms preserving the volume element of the domain of fluid flow and to the Euler equations of rigid body whith a fixed point, describing geodesics for a left-invariant metric on $SO(3)$.

The momentum map and an explicit parametrization of the Virasoro group, related to recently obtained solutions for the CH equation are presented.

Keywords: Euler top; Sobolev inner product; Coadjoint action; Lie group; Virasoro group; Group of diffeomorphisms.

1. Motion of a rigid body with a fixed point — the $SO(3)$ example

Let us start with a very familiar example — the Euler top. Consider an orthogonal basis $\widetilde{e}_k(t)$, $k = 1, 2, 3$, rotating about a fixed basis ϵ_k. Both bases share the same origin. We can think about the moving frame as a rigid body, moving about the origin.

The relation between the two frames is given by an orthogonal transformation: $\widetilde{e}_k(t) = g_{kj}(t)e_j$, where $g_{kj} = \widetilde{e}_k.e_j$, $g^T = g^{-1}$, i.e. g belongs to the group $G \equiv SO(3)$, the corresponding algebra $\mathfrak{g}$ being

$$\mathfrak{g} \equiv so(3) : \qquad x \in \mathfrak{g} \Leftrightarrow x = -x^T. \tag{1}$$

[‡]On leave from the Institute for Nuclear Research and Nuclear Energy, Bulgarian Academy of Sciences, Sofia, Bulgaria.

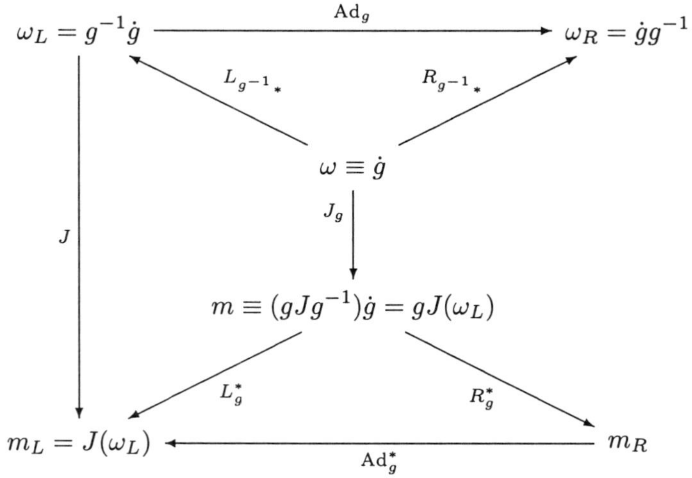

Fig. 1. Quantities and operators related to the so(3) algebra and its dual (Euler top case).

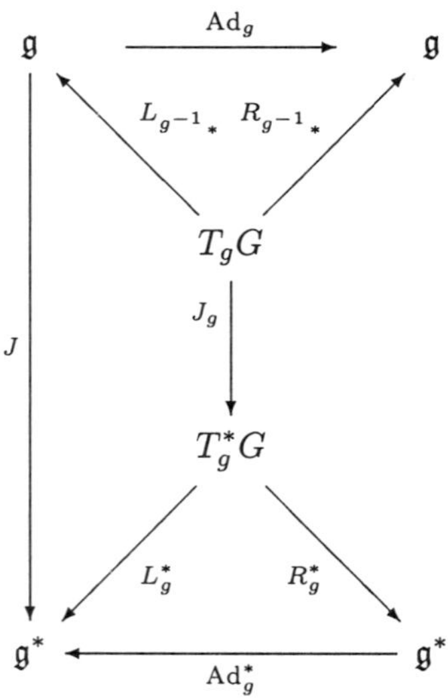

Fig. 2. Mappings between the spaces.

All quantities in the moving frame $\widetilde{e}_k(t)$ (related to the body) will be marked by subscript 'L', the ones in the fixed basis - by subscript 'R'. Let us take the following explicit parametrization for the angular velocity:

$$\omega_L = \begin{pmatrix} 0 & -\omega_3 & \omega_2 \\ \omega_3 & 0 & -\omega_1 \\ -\omega_2 & \omega_1 & 0 \end{pmatrix} \in \mathfrak{g} \quad \leftrightarrow \quad \vec{\omega}_L \equiv \begin{pmatrix} \omega_1 \\ \omega_2 \\ \omega_3 \end{pmatrix} \in \mathbb{R}^3 \tag{2}$$

The quantities related to the Euler top are schematically presented at Fig. 1, (the dot denotes the time derivative) [1,18,21]. The identification between the algebra $\mathfrak{g}$ and its dual is given by the *inertia operator*, see Fig. 2:

$$m_L = J(\omega_L) \equiv A\omega_L + \omega_L A, \tag{3}$$

where $A = \text{diag}(a_1, a_2, a_3)$ is a constant symmetric matrix.

The Hamiltonian is the kinetic energy $H(m_L, \omega_L) = \frac{1}{2}\text{tr}(m_L\omega_L^T)$, given by a left-invariant quadratic form: $H(m_L, \omega_L) = H(m, \omega)$.

This invariance by the virtue of Noether's Theorem leads to the momentum conservation: $\frac{d}{dt}m_R = 0$. This defines a momentum map $TG \to \mathfrak{g}^*$, constant along the geodesics. Furthermore, since $\omega_L = g^{-1}\dot{g}$ we have:

$$m_L = \text{Ad}_g^* m_R = g^{-1} m_R g, \qquad \dot{m}_L = \text{ad}_{\omega_L}^* m_L = -[\omega_L, m_L].$$

Finally we obtain the equations of motion (the Euler top equations):

$$\frac{d}{dt}J(\omega_L) = [J(\omega_L), \omega_L] \quad \text{or} \quad \dot{\omega}_1 = \frac{a_2 - a_3}{a_2 + a_3}\omega_2\omega_3, \quad \text{etc.} \tag{4}$$

2. Camassa-Holm equation – right invariant metric on the diffeomorphism group

The construction described briefly in the previous section can be easily generalized in cases where the Hamiltonian is a left- or right-invariant bilinear form. Such an interesting example is the Camassa-Holm (CH) equation [2,12,17]. This geometric interpretation of CH was noticed firstly by Misiołek [23] and developed further by several other authors [7–9,13,14,18]. Let us introduce the notation $u(g(x)) \equiv u \circ g$ and let us consider the H^1 Sobolev inner product

$$H(u, v) \equiv \frac{1}{2}\int_{\mathcal{M}} (uv + u_x v_x) d\mu(x), \qquad \text{with} \quad \mu(x) = x \tag{5}$$

The manifold $\mathcal{M}$ is $\mathbb{S}^1$ or in the case when the class of smooth functions vanishing rapidly at $\pm\infty$ is considered, we will allow $\mathcal{M} \equiv \mathbb{R}$.

Suppose $g(x) \in G$, where $G \equiv \text{Diff}(\mathcal{M})$. Then $H(u, v) = H(u \circ g, v \circ g)$ is a right-invariant H^1 metric.

Let us define $g(x,t)$ as

$$\dot{g} = u(g(x,t),t), \qquad g(x,0) = x, \quad \text{i.e.} \quad \dot{g} = u \circ g \in T_g G; \tag{6}$$

$u = \dot{g} \circ g^{-1} = R_{g^{-1}*}\dot{g} \in \mathfrak{g}$, where $\mathfrak{g}$ is $\mathrm{Vect}(\mathcal{M})$. Now we recall the following result:

Theorem 2.1 (A. Kirillov, 1980 [19,20]). *The dual space of $\mathfrak{g}$ is a space of distributions but the subspace of local functionals, called the regular dual $\mathfrak{g}^*$, is naturally identified with the space of quadratic differentials $m(x)dx^2$ on $\mathcal{M}$. The pairing is given for any vector field $u\partial_x \in \mathrm{Vect}(\mathcal{M})$ by*

$$\langle m dx^2, u\partial_x \rangle = \int_{\mathcal{M}} m(x)u(x)dx.$$

The coadjoint action coincides with the action of a diffeomorphism on the quadratic differential:

$$\mathrm{Ad}_g^* : \ m dx^2 \mapsto m(g) g_x^2 dx^2.$$

If $m(x) > 0$ for all $x \in \mathcal{M}$, then the square root $\sqrt{m(x)dx^2}$ transforms under G as a 1-form. This means that $C = \int_{\mathcal{M}} \sqrt{m(x)}\, dx$ is a Casimir function, i.e. an invariant of the coadjoint action.

Let us now allow the above pairing to be the H^1 right-invariant metric, mentioned earlier. This is possible by choosing the inertia operator $J = 1-\partial_x^2$, i.e. by taking $m = u - u_{xx}$, see Fig. 3. Again, for the Hamiltonian $H = \frac{1}{2}\int_{\mathcal{M}} mu\, dx$, given by the H^1 right-invariant metric, Noether's Theorem yields [8] the conservation of $m_L \equiv g_x^2 m(g(x,t),t)$, i.e. $g_x^2 m(g(x,t),t) = m(x,0)$.

We have a momentum map $TG \to \mathfrak{g}^*$, constant along the geodesics:

$$0 = \dot{m}_L = g_x^2(2u_x m + u m_x + m_t) \circ g, \tag{7}$$

iff m satisfies the Camassa-Holm equation

$$m_t + 2u_x m + u m_x = 0. \tag{8}$$

Similarly to the Euler top (4), CH can be written also in a Hamiltonian form $\dot{m} = -\mathrm{ad}_u^* m$. Indeed,

$$\langle \mathrm{ad}_{u\partial_x}^* m dx^2, v\partial_x \rangle = \langle m dx^2, [u\partial_x, v\partial_x] \rangle = \int_{\mathcal{M}} m(u_x v - v_x u)dx$$
$$= \int_{\mathcal{M}} v(2mu_x + um_x)dx = \langle (2mu_x + um_x)dx^2, v\partial_x \rangle,$$

i.e. $\mathrm{ad}_u^* m = 2u_x m + u m_x$.

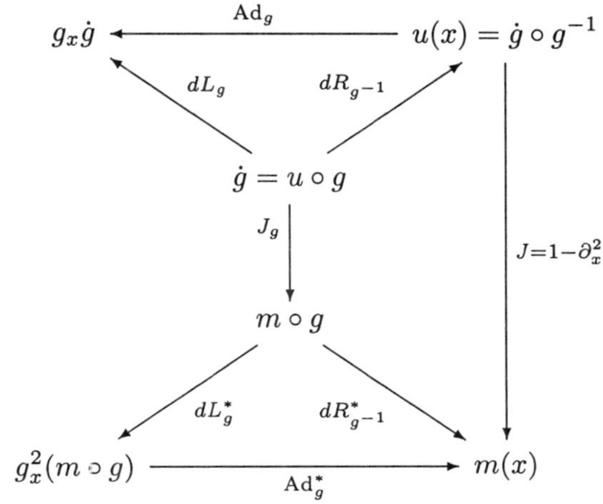

Fig. 3. Quantities related to the Camassa-Holm equation.

3. Inverse Scattering for the CH equation

In this section we consider the Camassa-Holm equation (CH) in the form

$$u_t - u_{xxt} + 2\omega u_x + 3uu_x - 2u_x u_{xx} - uu_{xxx} = 0, \tag{9}$$

which depends on an arbitrary parameter ω (which is not an angular velocity!). The traveling wave solutions of (9) are smooth solitons if $\omega > 0$, and peaked solitons (peakons) if $\omega = 0$ [2,10,11,17,22].

If $\omega \neq 0$ the invariance group of the Hamiltonian is the Virasoro group, Vir $= \mathrm{Diff}(\mathbb{S}_1) \times \mathbb{R}$ and the central extension of the corresponding Virasoro algebra is proportional to ω [7,15]. Thus, for $\omega \neq 0$, CH has various conformal properties [15]. CH is also completely integrable, possesses bi-Hamiltonian form and infinite sequence of conservation laws [2,6,12,16,24]. The Lax pair is

$$\Psi_{xx} = \Big(\frac{1}{4} + \lambda(m+\omega)\Big)\Psi \tag{10}$$

$$\Psi_t = \Big(\frac{1}{2\lambda} - u\Big)\Psi_x + \frac{u_x}{2}\Psi + \gamma\Psi \tag{11}$$

where γ is an arbitrary constant (for a given eigenfunction). CH is obtained from the compatibility condition $\Psi_{xxt} = \Psi_{txx}$. Let us introduce a new spectral parameter k such that $\lambda(k) = -\frac{1}{\omega}\Big(k^2 + \frac{1}{4}\Big)$. From now on we consider the case where m is a Schwartz class function, and $m(x,0)+\omega > 0$. Then $m(x,t) + \omega > 0$ for all t [3]. The spectral picture of (10) is [3]:

continuous spectrum: k — real; discrete spectrum: finitely many points $k_n = \pm i\kappa_n$, $n = 1, \dots, N$ where κ_n is real and $0 < \kappa_n < 1/2$. Eigenfunctions: for all real $k \neq 0$ a basis in the space of solutions can be introduced, fixed by its asymptotic when $x \to \infty$: $\psi(x, k)$ and $\bar{\psi}(x, k)$, such that

$$\psi(x, k) = e^{-ikx} + o(1), \qquad x \to \infty. \tag{12}$$

Another basis can be introduced, fixed by its asymptotic when $x \to -\infty$: $\varphi(x, k)$ and $\bar{\varphi}(x, k)$ such that

$$\varphi(x, k) = e^{-ikx} + o(1), \qquad x \to -\infty. \tag{13}$$

The relation between the two bases is

$$\varphi(x, k) = a(k)\psi(x, k) + b(k)\bar{\psi}(x, k), \tag{14}$$

where [6]

$$|a(k)|^2 - |b(k)|^2 = 1. \tag{15}$$

Further, one can define transmission and reflection coefficients: $\mathcal{T}(k) = a^{-1}(k)$ and $\mathcal{R}(k) = b(k)/a(k)$ correspondingly. According to (15)

$$|\mathcal{T}(k)|^2 + |\mathcal{R}(k)|^2 = 1.$$

The entire information about these two coefficients is provided by $\mathcal{R}(k)$ for $k > 0$. It is sufficient to know $\mathcal{R}(k)$ only on the half line $k > 0$, since $\bar{a}(k) = a(-k)$, $\bar{b}(k) = b(-k)$ and therefore $\mathcal{R}(-k) = \bar{\mathcal{R}}(k)$. At the points of the discrete spectrum, $a(k)$ has simple zeroes, φ and $\bar{\psi}$ are linearly dependent: $\varphi(x, i\kappa_n) = b_n\bar{\psi}(x, -i\kappa_n)$. In other words, the discrete spectrum is simple with eigenfunctions $\varphi^{(n)}(x) \equiv \varphi(x, i\kappa_n)$. The asymptotic behavior of $\varphi^{(n)}$ is

$$\begin{aligned} \varphi^{(n)}(x) &= e^{\kappa_n x} + o(e^{\kappa_n x}), \qquad x \to -\infty; \\ \varphi^{(n)}(x) &= b_n e^{-\kappa_n x} + o(e^{-\kappa_n x}), \qquad x \to \infty. \end{aligned} \tag{16}$$

The sign of b_n obviously depends on the number of the zeroes of $\varphi^{(n)}$. Suppose that $0 < \kappa_1 < \kappa_2 < \ldots < \kappa_N < 1/2$. Then from the oscillation theorem for the Sturm-Liouville problem $\varphi^{(n)}$ has exactly $n-1$ zeroes, i.e. $b_n = (-1)^{n-1}|b_n|$.

The set

$$\mathcal{S} \equiv \left\{ \mathcal{R}(k)\ (k > 0),\ \kappa_n,\ |b_n|,\ n = 1, \dots N \right\} \tag{17}$$

is called *scattering data*. The time evolution of the scattering data can be obtained from (11) with the choice $\gamma = \frac{ik}{2\lambda}$ for the eigenfunction $\varphi(k, x)$

and $x \to \infty$: $\dot{a}(k,t) = 0$, $\dot{b}(k,t) = \frac{ik}{\lambda} b(k,t)$, or

$$a(k,t) = a(k,0), \qquad b(k,t) = b(k,0)\exp\left(\frac{ik}{\lambda}t\right); \tag{18}$$

In other words, $a(k)$ is independent on t and can serve as a generating function of the conservation laws [6].

The time evolution of the data on the discrete spectrum is obtain as follows: $i\kappa_n$ are zeroes of $a(k)$, which does not depend on t, and therefore $\dot{\kappa}_n = 0$. From (11) and (16) in a similar fashion

$$\dot{b}_n = \frac{4\omega\kappa_n}{1-4\kappa_n^2} b_n. \qquad b_n(t) = b_n(0)\exp\left(\frac{4\omega\kappa_n}{1-4\kappa_n^2}t\right). \tag{19}$$

4. Soliton solutions and the diffeomorphisms

The inverse scattering is simplified in the important case of the so-called reflectionless potentials, when the scattering data is confined to the case $\mathcal{R}(k) = 0$ for all real k. This class of potentials corresponds to the N-soliton solutions of the CH equation. In this case $b(k) = 0$ and $|a(k)| = 1$ and $ia'(i\kappa_p)$ is real:

$$ia'(i\kappa_p) = \frac{1}{2\kappa_p} e^{\alpha\kappa_p} \prod_{n\neq p} \frac{\kappa_p - \kappa_n}{\kappa_p + \kappa_n}, \quad \text{where} \quad \alpha = \sum_{n=1}^{N} \ln\left(\frac{1+2\kappa_n}{1-2\kappa_n}\right)^2.$$

Thus, $ia'(i\kappa_p)$ has the same sign as b_n, and therefore $c_n \equiv \frac{b_n}{ia'(i\kappa_p)} > 0$. The time evolution of c_n is $c_n(t) = c_n(0)\exp\left(\frac{4\omega\kappa_n}{1-4\kappa_n^2}t\right)$ in the view of (19).

The N-soliton solution is [5]

$$u(x,t) = \frac{1}{2}\int_0^\infty \exp\Big(-|x - g(\xi,t)|\Big)p(\xi,t)d\xi - \omega, \tag{20}$$

where $g(\xi,t)$, $p(\xi,t)$ can be expressed through the scattering data as :

$$g(\xi,t) \equiv \ln \int_0^\xi \Big(1 - \sum_{n,p} \frac{c_n(t)\underline{\xi}^{-2\kappa_n}}{\kappa_n + 1/2} A_{np}^{-1}[\underline{\xi},t]\Big)^{-2} d\underline{\xi}, \tag{21}$$

$$p(\xi,t) = \omega\xi^{-2} g_\xi^{-1}(\xi,t), \qquad \text{where} \tag{22}$$

$$A_{pn}[\xi,t] \equiv \delta_{pn} + \frac{c_n(t)\xi^{-2\kappa_n}}{\kappa_p + \kappa_n}.$$

Then the computation of $m = u - u_{xx}$ gives

$$m(x,t) = \int_{-\infty}^{\infty} \delta(x - g(\xi,t))p(\xi,t)d\xi - \omega. \tag{23}$$

From the CH equation $m_t + um_x = -2(m+\omega)u_x$, (20) and (23) it follows

$$\dot{g}(\xi,t) = \frac{1}{2}\int_0^\infty e^{-|g(\xi,t)-g(\underline{\xi},t)|}p(\underline{\xi},t)d\underline{\xi} - \omega, \qquad \dot{g}(\xi,t) = u(g(\xi,t),t),$$

therefore $g(x,t)$ in (21) is the diffeomorphism (Virasoro group element) in the purely solitonic case. The situation when the condition $m(x,0) + \omega > 0$ on the initial data does not hold is more complicated and requires separate analysis (if $m(x,0) + \omega$ changes sign there are infinitely many positive eigenvalues accumulating at infinity and singularities might appear in finite time [3,4]).

Acknowledgments

A.C. acknowledges funding from the Science Foundation Ireland, Grant 04/BR6/M0042, R.I.I. acknowledges funding from the Irish Research Council for Science, Engineering and Technology.

References

1. V. Arnold and B. Khesin, *Topological Methods in Hydrodynamics*, Springer Verlag, New York, 1998.
2. R. Camassa and D. Holm, Phys. Rev. Lett. **71**, 1661 (1993).
3. A. Constantin, Proc. R. Soc. Lond. **A457**, 953 (2001).
4. A. Constantin and J. Escher, Acta Math. **181**, 229 (1998).
5. A. Constantin, V. Gerdjikov and R. Ivanov, Inv. Problems **22** 2197 (2006); nlin.SI/0603019.
6. A. Constantin and R. Ivanov, Lett. Math. Phys. **76**, 93 (2006).
7. A. Constantin, T. Kappeler, B. Kolev and P. Topalov, *On geodesic exponential maps of the Virasoro group*, Ann. Global Anal. Geom., to appear (2006).
8. A. Constantin and B. Kolev, Comment. Math. Helv. **78**, 787 (2003).
9. ———, J. Nonlinear Sci. **16**, 109 (2006).
10. A. Constantin and W. Strauss, Comm. Pure Appl. Math. **53**, 603 (2000).
11. ———, J. Nonlinear Sci. **12**, 415 (2002).
12. A. Fokas and B. Fuchssteiner, Physica D **4**, 47 (1981).
13. D. Holm and J. Marsden, in *The breadth of symplectic and Poisson geometry*, Progr. Math. **232**, Birkhäuser, Boston, MA, 2005.
14. D. Holm, J. Marsden and T. Ratiu, Adv. Math. **137**, 1 (1998).
15. R. Ivanov, Phys. Lett. A **345**, 235 (2005); nlin.SI/0507005.
16. ———, Zeitschrift für Naturforschung **61a**, 133 (2006); nlin.SI/0601066.
17. R. Johnson, J. Fluid. Mech. **457**, 63 (2002).
18. B. Khesin and G. Misiołek, Adv. Math. **176**, 116 (2003).
19. A. Kirillov, Funct. Anal. Appl. **15**, 135 (1981).
20. ———, Contemp. Math. **145**, 33 (1993).

21. B. Kolev, J. Nonlinear Math. Phys. **11**, 480 (2004).
22. J. Lenells, J. Diff. Eq. **217**, 393 (2005).
23. G. Misiołek, J. Geom. Phys. **24**, 203 (1998).
24. E. Reyes, Lett. Math. Phys. **59**, 117 (2002).

FERMI-WALKER PARALLEL TRANSPORT, TIME EVOLUTION OF A SPACE CURVE AND THE SCHRÖDINGER EQUATION AS A MOVING CURVE

R. DANDOLOFF

Laboratoire de Physique Théorique et Modélisation,
Université de Cergy-Pontoise, F-95302 Cergy-Pontoise, France
E-mail: rossen.dandoloff@ptm.u-cergy.fr

Based on Fermi-Walker parallel transport along a space curve, we discuss the possible geometric phases that can occur. We give a general condition for the time evolution of a space curve. We identify two types of local geometric phases associated with the evolution of a space curve, the "Fermi-Walker" and "incompatibility" phases, and derive a relationship between them. The time-dependent Schrödinger equation for a particle in a potential $V(s)$ can then be interpreted geometrically as a moving curve whose Fermi-Walker phase density is given by $-(dV/ds)$.

I. Space curves, parallel transport and anholonomy.

A) Static curves

Let us consider a space curve which is given by its curvature $k(s)$ and torsion $\tau(s)$ where s represents the arc length. We will introduce the orthonormal Frenet triad $(\mathbf{t}, \mathbf{n}, \mathbf{b})$ representing the tangent, the normal and the bi-normal to the space curve. The space "evolution" of the Frenet triad is governed by the Frener-Serret equations:[1]

$$\begin{aligned} \mathbf{t}_s &= k\mathbf{n} \\ \mathbf{n}_s &= -k\mathbf{t} + \tau\mathbf{b} \\ \mathbf{b}_s &= -\tau\mathbf{n} \end{aligned} \tag{1}$$

The curvature and torsion are related to the tangent through the following expressions:

$$k^2 = \mathbf{t}_s \cdot \mathbf{t}_s \quad \text{and} \quad \tau = \frac{\mathbf{t} \cdot (\mathbf{t}_s \wedge \mathbf{t}_{ss})}{k^2} \tag{2}$$

PACS numbers: 02.40.Hw, 03.65.Ca, 02.30.Jr, 05.45.+b, 02.40.Dr, 47.32.Cc

There is a more compact way of writing the Frenet-Serret equations which gives a better insight into the structure of these equations. Let us introduce the Darboux vector: $\xi = \tau \mathbf{t} + k\mathbf{b}$. The Darboux vector plays the role of "angular velocity" for the space evolution of the Frenet triad. The Frenet-Serret equations now read:

$$\begin{aligned} \mathbf{t}_s &= \xi \wedge \mathbf{t} \\ \mathbf{n}_s &= \xi \wedge \mathbf{n} \\ \mathbf{b}_s &= \xi \wedge \mathbf{b} \end{aligned} \tag{3}$$

Thus when the triad evolves along the curve the $(\mathbf{n}, \mathbf{b})$-couple rotates around the tangent $\mathbf{t}$ with an "angular velocity" τ and the $(\mathbf{t}, \mathbf{n})$-couple rotates around the bi-normal $\mathbf{b}$ with an "angular velocity" k.

As s increases from s_0 to s_1, a phase develops $\Phi = \int_{s_0}^{s_1} \tau(s)ds$ with respect to a non-rotating frame defined by the Fermi-Walker parallel transport:[2]

$$\frac{DA^i}{ds} = kA^i(t^j n^i - t^i n^j) = (k\mathbf{b} \wedge \mathbf{A})^i \tag{4}$$

Here $\mathbf{A}$ is some vector defined on the space curve.

B) Moving curves

Now the curvature and torsion are time dependent functions $k = k(s, u)$ and $\tau(s, u)$ where u represents the time. For fixed s the triad evolves along a new space curve with u as natural parameter. The equations are as follows:

$$\mathbf{t}_u = g\mathbf{n} + h\mathbf{b}, \quad \mathbf{n}_u = -g\mathbf{t} + \tau^0\mathbf{b}, \quad \mathbf{b}_u = -h\mathbf{t} - \tau^0\mathbf{n}, \tag{5}$$

where the coefficients g, h and τ^0, as well as K and τ in Eq. (2), are functions of s and u.

The corresponding Darboux vector now reads $\xi_0 = \bar{\tau}_0\mathbf{t} + B\mathbf{n} + C\mathbf{b}$. Now, let us consider two neighboring points on a space curve at a time u i.e. $A(s, u)$ and $B(s + \Delta s, u)$. After a time-interval Δu the space curve slightly changes its shape and position in space. The points A and B evolve into $C(s, U + \Delta u)$ and $D(s + \Delta s, u + \Delta u)$ respectively. We follow the triad from A to B and to D and from A to C and to D. We take into consideration only the rotation of the triad around the tangent $\mathbf{t}$. Following the first path from A to D a phase develops $\Phi_1 = \tau(s, u)\Delta s + \bar{\tau}_0(s + \Delta s, u)\Delta u$ and following the second path a different phase develops $\Phi_2 = \bar{\tau}_0(s, u)\Delta u + \tau(s, u + \Delta u)\Delta s$ The phase difference $\delta\Phi = \Phi_1 - \Phi_2 = \left[\frac{\partial \bar{\tau}_0}{\partial s} - \frac{\partial \tau}{\partial u}\right]\Delta s \Delta u$ represents a measure of the anholonomy density.

II. Lamb formalism, geometric phase and gauge potential.

If we introduce complex vectors the Frenet-Serret equations combine to give [3]:

$$(\mathbf{n}+i\mathbf{b})_s + i\tau(\mathbf{n}+i\mathbf{b}) = -k\mathbf{t} \tag{6}$$

Next we introduce the complex vector $\mathbf{N} = (\mathbf{n}+i\mathbf{b})exp(i\int_{-\infty}^{s}(\tau - c_0)ds')$ where $\tau(s \to \pm\infty) \to c_0$ and the complex quantity $q = kexp(i\int_{-\infty}^{s}(\tau - c_0)ds')$. Now instead of the triad $(\mathbf{t},\mathbf{n},\mathbf{b})$ we are going to use the triad $(\mathbf{t},\mathbf{N},\mathbf{N}^*)$. The new triad satisfies the following conditions: $\mathbf{t}\cdot\mathbf{N} = \mathbf{t}\cdot\mathbf{N}^* = \mathbf{N}\cdot\mathbf{N} = 0$ and $\mathbf{N}\cdot\mathbf{N}^* = 2$. Since the tangent is normalized $\mathbf{t}\cdot\mathbf{t}_u = 0$ and therefore we can represent the u derivative in the following form: $\mathbf{t}_u = g\mathbf{n} + h\mathbf{b}$ where the functions g and h are to be determined. Using the compatibility conditions $\mathbf{t}_{su} = \mathbf{t}_{us}$ and $\mathbf{N}_{su} = \mathbf{N}_{us}$ will allow us to compute the u derivatives of $\mathbf{t}$ and $\mathbf{N}$.

$$\mathbf{N}_u = iR\mathbf{N} + \gamma\mathbf{t} \quad \text{and} \quad \mathbf{t}_u = -\frac{1}{2}(\gamma^*\mathbf{N} + \gamma\mathbf{N}^*). \tag{7}$$

where $R_s = \frac{i}{2}(\gamma q^* - \gamma^* q)$ and $q_u + \gamma_s + i(c_0\gamma - Rq) = 0$. The explicit expression for γ is: $\gamma = -(g+ih)\exp(i\int_{-\infty}^{s}(\tau - c_0))ds'$. With this the anholonomy density can be computed directly to give:

$$j(s,u) = \frac{\partial\bar{\tau}_0}{\partial s} - \frac{\partial\tau}{\partial u} = -\frac{\partial R}{\partial s}. \tag{8}$$

Using the s and u derivatives of $\mathbf{t}$ and $\mathbf{N}$ we get $R_s = kh$ and finally we get for the anholonomy density:

$$j(s,u) = -kh. \tag{9}$$

The total phase is then given by the following expression:

$$\Phi = -\int_{u_1}^{u_2} du \int_{-s_0}^{s_0} R_s ds. \tag{10}$$

Using the s and u derivatives of $\mathbf{t}$, we get that $\mathbf{t}_s \wedge \mathbf{t}_u = kh\mathbf{t}$. This allows us to give a geometrical interpretation of the total phase:

$$\Phi = -\int_{u_1}^{u_2} du \int_{-s_0}^{s_0} \mathbf{t}\cdot(\mathbf{t}_u \wedge \mathbf{t}_s)ds. \tag{11}$$

The expression $\mathbf{t}\cdot(\mathbf{t}_u \wedge \mathbf{t}_s)dsdu$ represents the element of area on the unit sphere.

A. Quantum mechanical phase: Berry's phase

Note that the quantity R can be expressed using the complex vector $\mathbf{N}$: $R = -\frac{i}{2}\mathbf{N} * \cdot \mathbf{N}_u$ With this the expression for the geometric phase becomes:

$$\Phi = \frac{i}{2} \int_{-s_0}^{s_0} ds \frac{\partial}{\partial s} \int_{u_1}^{u_2} \mathbf{N} * \cdot \mathbf{N}_u du. \tag{12}$$

If we replace the complex unit vector $\frac{\mathbf{N}}{2}$ by a quantum state $|\mathbf{N}(u)>$, the expression $i \int_{u_1}^{u_2} <\mathbf{N}|\frac{\partial}{\partial u}\mathbf{N}>du$ plays the role of a local Berry phase. The role of the Fermi-Walker parallel transport in the Berry's phase has been discussed in [4].

III. General context: topological terms, Hopf invariant.

Let us now consider a cube with lengths Δx, Δy and Δz and whose lowest left edge is at (x, y, z). Let τ^1, τ^2 and τ^3 represent the three torsions in the three space directions. Now we may consider the three anholonomy densities [5]:

$$\begin{aligned} J^3 &= \frac{\partial \tau^1}{\partial y} - \frac{\partial \tau^2}{\partial x} \\ J^2 &= \frac{\partial \tau^3}{\partial x} - \frac{\partial \tau^1}{\partial z} \\ J^1 &= \frac{\partial \tau^2}{\partial z} - \frac{\partial \tau^3}{\partial y}. \end{aligned} \tag{13}$$

These three quantities represent the components of the topological current $\mathbf{J} = (J^1, J^2, J^3)$. Adding up the anholonomy densities corresponding to the faces of the cube we get:

$$\begin{aligned} \delta H_1 &= [J^1(x+\Delta x, y, z) - J^1(x, y, z)]\Delta y \Delta z \\ &\quad + [J^2(x, y+\Delta y, z) - J^2(x, y, z)]\Delta x \Delta z \\ &\quad + [J^3(x, y, z+\Delta z) - J^3(x, y, z)]\Delta x \Delta y \\ &= \operatorname{div} \mathbf{J} \Delta x \Delta y \Delta z. \end{aligned} \tag{14}$$

Because $\mathbf{J} = curl \mathbf{A}$,where $\mathbf{A} = (\tau^1, \tau^2, \tau^3)$ the divergence of the topological current is identically zero: $\operatorname{div} \mathbf{J} = 0$.

IV. A general constraint on a moving space curve: the Schrdinger equation as a moving space curve.

In view of the relationship between the NLS (Nonlinear Schrödinger equation) and a moving curve, one can ask the following question: Can the time-dependent Schrödinger equation in a potential $V(s)$,

$$i\psi_u + \psi_{ss} - V(s)\psi = 0, \tag{15}$$

itself be associated with a moving space curve [6]? (Here, for convenience we have set $\hbar = 2m = 1$, m being the mass of the quantum particle.)

The conditions under which the moving curve equations (1) and (5) are equivalent to Eq.(15). It is convenient to use the Hasimoto transformation [7]:

$$\psi = K(s,u)\exp[i\int \tau(s,u)ds] \tag{16}$$

in Eq. (15). Equating imaginary and real parts, this leads to the coupled partial differential equations

$$K_u = -(K\tau)_s - K_s\tau, \tag{17}$$

$$\tau_u = [(K_{ss}/K) - \tau^2]_s - V_s. \tag{18}$$

We will establish now appropriate relations between the parameters g, h , τ^0 of Eqs. (1) and the curve parameters K and τ in Eqs. (5), such that Eqs. (17) and (18) follow. The appearance of K_u and τ_u in these latter equations immediately suggests that the direct way to find these is to compare the u-derivatives $\mathbf{t}_{su}$, $\mathbf{n}_{su}$ and $\mathbf{b}_{su}$ of Eqs. (1) with the s-derivatives $\mathbf{t}_{us}$, $\mathbf{n}_{us}$ and $\mathbf{b}_{us}$ of Eqs. (5). Now, smooth curve evolution requires the compatibility condition $\mathbf{t}_{su} = \mathbf{t}_{us}$ for the tangent $\mathbf{t}$. However, we show below that the corresponding conditions on $\mathbf{n}$ and $\mathbf{b}$, $\mathbf{n}_{su} = \mathbf{n}_{us}$ and $\mathbf{b}_{su} = \mathbf{b}_{us}$ are merely sufficiency conditions. Equations (1) and (5) yield the following "incompatibility vectors" [6]:

$$\begin{aligned}\Delta\mathbf{t} &\equiv (\mathbf{t}_{su} - \mathbf{t}_{us}) = \alpha_1\mathbf{n} + \alpha_2\mathbf{b},\\ \Delta\mathbf{n} &\equiv (\mathbf{n}_{su} - \mathbf{n}_{us}) = \alpha_3\mathbf{b} - \alpha_1\mathbf{t},\\ \Delta\mathbf{b} &\equiv (\mathbf{b}_{su} - \mathbf{b}_{us}) = -\alpha_2\mathbf{t} - \alpha_3\mathbf{n}.\end{aligned} \tag{19}$$

Here

$$\alpha_1 = K_u - g_s + h\tau, \quad \alpha_2 = K\tau^0 - h_s - g\tau, \quad \alpha_3 = \tau_u - \tau_s^0 - Kh. \tag{20}$$

It follows from Eqs. (19) that a *general* curve evolution must satisfy the geometric constraint

$$\Delta\mathbf{t}.(\Delta\mathbf{n} \times \Delta\mathbf{b}) = 0, \tag{21}$$

i.e., $\Delta\mathbf{t}, \Delta\mathbf{n}$ and $\Delta\mathbf{b}$ must remain coplanar vectors under time evolution. Further, since Eq. (21) is *automatically* satisfied for $\Delta\mathbf{t} = 0$, we see that

$\Delta\mathbf{n}$ and $\Delta\mathbf{b}$ need not necessarily vanish. In addition, we see from Eqs. (19) that $\Delta\mathbf{t} = 0$ implies $\alpha_1 = \alpha_2 = 0$, so that

$$\Delta\mathbf{n} = \alpha_3\mathbf{b}, \quad \Delta\mathbf{b} = -\alpha_3\mathbf{n}. \tag{22}$$

Thus $\Delta\mathbf{n}$ and $\Delta\mathbf{b}$ are perpendicular to each other and have the same magnitude. Next, we use Eqs. (22) to show that the non-vanishing of $\Delta\mathbf{n}$ and $\Delta\mathbf{b}$ gives rise to a certain geometric phase $\delta\Phi^{inc}$: Let us first consider the incompatibility phase $\delta\Phi^{inc}$: Let the origin of the Frenet-Serret frame $(\mathbf{t}(s,u), \mathbf{n}(s,u), \mathbf{b}(s,u))$ be the center of a unit sphere. Consider the following infinitesimal closed path in space-time: $[s,u] \to [(s+ds),u] \to [(s+ds),(u+du)] \to [(s,(u+du)] \to [s,u]$. Corresponding to this path, let the tips of $\mathbf{t}$, $\mathbf{n}$ and $\mathbf{b}$ create three curves (indicatrices) on the sphere. Now, these curves do not close in general; the non-closure vector connecting the starting point $\mathbf{A}$ and the end point $\mathbf{B}$ for any vector $\mathbf{F}(= \mathbf{t}, \mathbf{n}$ or $\mathbf{b})$ is easily found to be

$$\overrightarrow{AB} = (\mathbf{F}_{su} - \mathbf{F}_{us})\, dsdu \equiv \Delta\mathbf{F}\, dsdu. \tag{23}$$

Thus the non-closure of the indicatrix of $\mathbf{F}$ is a measure of the incompatibility $\Delta\mathbf{F}$. Next, taking $\mathbf{F} = \mathbf{n}$ and $\mathbf{b}$ in succession in Eq. (23) and using Eqs. (22), we see that the two (non-closure) distances for $\mathbf{n}$ and $\mathbf{b}$ are equal, and are given by

$$|\overrightarrow{AB}| = |\Delta\mathbf{n}|\, dsdu = |\Delta\mathbf{b}|\, dsdu = \alpha_3\, dsdu. \tag{24}$$

Since the sphere has unit radius, this distance is also *identical* to a non-closure angle or geometric phase $\delta\Phi^{inc}$ in the $(\mathbf{n}, \mathbf{b})$ plane. This is why we have termed it the "incompatibility phase": The $(\mathbf{n}, \mathbf{b})$ plane does not return to its original orientation but gets rotated by an angle $\delta\Phi^{inc}$ when the triad $(\mathbf{t}, \mathbf{n}, \mathbf{b})$ is taken around an infinitesimal closed path in (s,u) space in such a way that $\Delta\mathbf{t} = 0$. Using this and Eq. (20) for α_3 in Eq. (24), we obtain

$$\delta\Phi^{inc} = [\tau_u - \tau_s^0 - Kh]\, dsdu. \tag{25}$$

Let us now consider the Fermi-Walker phase $\delta\Phi^{FW}$: There is another type of geometric phase [5] that can be associated with curve evolution. Equations (1) and (5) can be written in the form $\mathbf{F}_s = \mathbf{L}^{(1)} \times \mathbf{F}$ and $\mathbf{F}_u = \mathbf{L}^{(0)} \times \mathbf{F}$; here, $\mathbf{L}^{(1)} = \tau\mathbf{t} + K\mathbf{b}$, and $\mathbf{L}^{(0)} = \tau^0\mathbf{t} + h\mathbf{n} + g\mathbf{b}$. If we work in a Fermi-Walker (non-rotating) frame of reference [2] in which $\mathbf{t}$ essentially gets parallel transported, the triad undergoes a rotation τds about $\mathbf{t}$ as one moves along the curve by an amount ds (for fixed u), and a rotation $\tau^0 du$ as one moves along the 'temporal curve' by an interval du (for fixed s). This in turn implies that there is an underlying angle anholonomy or "Fermi-Walker phase" $\delta\Phi^{FW}$ associated with curve evolution, i.e., the $(\mathbf{n}, \mathbf{b})$ plane undergoes a rotation around $\mathbf{t}$ through an angle [5].

$$\delta\Phi^{FW} = (\tau_u - \tau_s^0)\, dsdu \tag{26}$$

with respect to its original orientation, when s and u change along an infinitesimal closed path of area $dsdu$. Further, from Eqs. (2) and (3), we find

$$-Kh\, dsdu = \mathbf{t}.(\mathbf{t}_u \times \mathbf{t}_s)\, dsdu = \delta\Omega, \tag{27}$$

where the area $\mathbf{t}.(\mathbf{t}_u \times \mathbf{t}_s)\, dsdu$ on a unit sphere is just the solid angle $\delta\Omega$. Substituting Eqs. (26) and (27) in Eq. (25), we get

$$\delta\Phi^{inc} = \delta\Phi^{FW} + \delta\Omega. \tag{28}$$

This is the general geometric relationship between the two types of geometric phases, and it is valid for *all* smooth curve evolutions. Returning to Eqs. (19) and (20) with $\Delta\mathbf{t} = 0$, we get

$$K_u = g_s - h\tau, \quad \tau^0 = (h_s + g\tau)/K. \tag{29}$$

So far, our analysis has been quite general. We must now find g, h and τ^0 in such a way that Eqs. (17) and (18) are obtained. In turn, this would imply the equivalence of the Schrödinger equation (15) and the curve evolution equations (1) and (5), as already explained. Comparing Eq. (17) with Eq. (5) immediately identifies g and h as

$$g = -K\tau, \quad h = K_s. \tag{30}$$

Substituting these in the second equation in (29) gives

$$\tau^0 = [(K_{ss}/K) - \tau^2]. \tag{31}$$

Hence Eq. (18) yields

$$(\tau_u - \tau_s^0) = -V_s. \tag{32}$$

Next, setting $h = K_s$ in Eq. (27), and substituting for $\delta\Omega$ thus obtained into Eq. (28), we get $\delta\Phi^{inc} = \delta\Phi^{FW} - KK_s dsdu$. But Eq. (16) gives $\frac{1}{2}|\psi^2|_s = KK_s$. Thus

$$\delta\Phi^{inc} = \delta\Phi^{FW} - \frac{1}{2}|\psi^2|_s\, dsdu. \tag{33}$$

Two possibilities arise, depending on whether $\delta\Phi^{inc} = 0$ or $\delta\Phi^{inc} \neq 0$.
(i) In the special case $\delta\Phi^{inc} = 0$, Eqs. (28) and (33) give $\delta\Phi^{FW} = -\delta\Omega = \frac{1}{2}|\psi^2|_s dsdu$. Using Eq. (26) and substituting for τ^0 from Eq. (31), we get

$\tau_u = [(K_{ss}/K) - \tau^2]_s + \frac{1}{2}|\psi|^2_s$. Noting that Eqs. (17) and (18) follow from Eq. (15) in a formal sense for any V, a comparison of this equation with Eq. (18) yields $V = -\frac{1}{2}|\psi|^2$. Equation (15) then reduces to the NLS equation, which is essentially Hasimoto's result [7].

(ii) More generally, $\delta\Phi^{inc} \neq 0$. Then $\delta\Phi^{FW} \neq \frac{1}{2}|\psi^2|_s dsdu$ any more. Let $\delta\Phi^{FW}$ be of the form $f(s,u)dsdu$, where the Fermi-Walker phase density $f(s,u)$ is some function of s and u. Starting with Eq. (26), we repeat the same steps as in (i) above to find the relationship $V = \int f(s,u)ds$. Hence the identification $f(s,u) = f(s) = -V_s$ leads at once to the time-dependent Schrödinger equation (15) for a particle in the potential $V(s)$. Correspondingly, from Eq. (28), $\delta\Phi^{inc} = (-V_s dsdu + \delta\Omega)$ in this case. The special case $f = 0$, i.e., a vanishing $\delta\Phi^{FW}$, yields the free Schrödinger equation $i\psi_u + \psi_{ss} = 0$. In this case, $\delta\Phi^{inc}$ is identically equal to the solid angle $\delta\Omega$.

References

1. L.P. Eisenhart, *A Treatise on the Differential Geometry of Curves and Surfaces*, Dover, New York, 1960.
2. J.L. Synge, *Relativity: The General Theory* North-Holland, Amsterdam, 1960; R. Dandoloff and W. Zakrzewski J. Phys. A, **22** (1989) L461.
3. G.L. Lamb, J. Math. Phys. **18** (1977) 1654.
4. R. Dandoloff, Phys. Lett. A, **139**, (1989), 19.
5. R. Balakrishnan, A.R. Bishop and R. Dandoloff, Phys. Rev. Lett. **64** (1990) 2107; Phys. Rev. B **47** (1993) 3108; *ibid.* (1993) 5438.
6. R. Balakrishnan and R. Dandoloff, Phys. Lett. A, **260**, (1999), 62.
7. H. Hasimoto, J. Fluid. Mech. **51** (1972) 477.

REMARKS ON BICOMPLEX VARIABLES AND OTHER SIMILAR VARIABLES

S. DIMIEV[†] and R. LAZOV[‡]

Institute of Mathematics and Informatics
Acad. G. Bonchev Str., 8
1113 Sofia, Bulgaria
[†] *E-mail: sdimiev@math.bas.bg*
[‡] *E-mail: lazovr@math.bas.bg*

S. SLAVOVA

Pedagogical College of Dobrich,
Dobrotica, Dobrich, Bulgaria
e-mail: slavova@dobrich.net

In this short report the authors try to clarify the development of different treatments of the algebra of bicomplex numbers and related topics.

Keywords: Bicomplex numbers; Matrix representation; Cauchy-Riemann bicomplex equations; Quadratic geometries.

1. Historical notes

The notion of *bicomplex number* has a long history closely related to the algebra but sporadically with analysis. Hamilton made some remarks about this number system considering it as a commutative alternative of the quaternions, but it was not object of his investigations. H. Hankel in his lectures [3] formulated a general viewpoint considering "complex numbers with n units", which in modern language is "algebra with n generators over $\mathbf{R}$". The special case $n = 4$ is treated in details, but the attention is concentrated over the quaternions, with many geometric applications. The bicomplex number are only remarked, but not studied in details. Berloty [1] considered the case $n = 5$, and from general point of view the problem is studied by Study [15,16] and Scheffers [9]. Scheffers examines in details the

2000 MS Classification: 30G35

problem of construction of an algebra with n generators over $\mathbf{R}$ from algebraic position and presents in details the cases $n = 3, 4, 5$. He gives a complete classification of the irreducible algebras aver $\mathbf{R}$ in dimension 4, but does not use the term "bicomplex". The name "bicomplex" is used by Segre [12] in a paper written entirely in the language of the algebraic geometry and the bicomplex numbers are treated in connections with some geometric constructions. Segre is interested only on some algebraic and geometric properties of this number system and does not consider analytic problems at all. He calculates the zero divisors and the idempotents in the bicomplex algebra, but these subjects remain closely related with the geometry. The algebraic theory is summarized by Study and Cartan [17] with application principally to geometry.

An independent approach to the bicomplex numbers is demonstrated by Em. Ivanov [4], who uses the term "tetranions". His approach is algebraic too: he sets up the problem to construct a 4-dimensional associative, *commutative* algebra over $\mathbf{R}$, which necessary must contain zero divisors. The result is an algebra with 4 generators $\{e_0, e_1, e_2, e_3\}$ and the following multiplication table:

	e_0	e_1	e_2	e_3
e_0	e_0	e_1	e_2	e_3
e_1	e_1	$-e_0$	e_3	$-e_2$
e_2	e_2	e_3	$-e_0$	$-e_1$
e_3	e_3	$-e_2$	$-e_1$	e_0

This composition law is exactly the one introduced by Segre, but Ivanov gives only some geometric and algebraic applications with no relations to analysis. The most valuable contributions in this stage of the development of the theory are made by L. Tchakalov [19], who starts from the paper of Ivanov and concentrates his study on algebraic and analytic aspects. The main results of his work are: determination of the roots of a bicomplex polynomial, the notion of analytic function, the zero divisors and the idempotents of the bicomplex algebra, some questions of arithmetic character: the Euclide division algorithm and unique factorization problem. Having in mind the importance of his results we shall publish in the future a detailed exposition of his work and especially a development of his arithmetic program.

The next step of development of this subject is more closely related with analysis and the starting point is the article of Sheffers [10], which is dedicated to the general problem of the definition of analytic function over general algebra (over $\mathbf{R}$ and $\mathbf{C}$). In this direction are the contributions of Scorza [11], Spampinato [13,14] and Ringleb [7]. The last two authors generalize the notion of analytic function in the case of algebra with (in general) n generators and noncommutative algebras. It is necessary to pay attention to the articles of Ward [23,24], who defines analytic functions in very general context; in some particular cases (especially complex and bicomplex numbers) his definition is equivalent to the Sheffers one. The notion of generalized bicomplex variable was introoduced by Takasu [18]; the system of these number forms an associative commutative linear algebra.

The interest in bicomplex algebra and its relations to analysis is renovated after 1950. Riley [7] treats algebraic as well analytic aspects (power series, analytic functions, Cauchy-Riemann type equations). Vishnevski [21] proves an decomposition theorem for analytic (in sense of Scheffers) functions over general associative, commutative algebras and its connection with analytic structures over Riemannian spaces.

In the last decade the interest to the bicomplex variables and their applications increases and stimulates a range of publications, concerning geometric, analytic as well applied problems. A detailed exposition of different aspects and results is done by Rönn [8]; Pogorui and Rodriguez-Dagnino [6] study the roots of polynomials with bicomplex coefficients, they present a technique for computing zeros of bicomplex polynomials; Turbin [20] is interested on applications of bicomplex analysis in problems of mechanics and physics.

2. Bicomplex and similar variables — a short recall

Bicomplex numbers are defined as follows: $\alpha = a_0+ia_1+ja_2+ea_3$, $a_n \in \mathbf{R}$, $n = 0,1,2,3$, $\alpha \in \mathbf{B}$, $i^2 = j^2 = -e^2 = -1$, where $ij = ji = e$, $ie = ei = -j$, $je = ej = -i$. Each bicomplex number has a complex pair representation

$$\alpha = (a_0 + ia_1) + j(a_2 + ia_3), \quad a_0 + ia_1 \in \mathbf{C}, \quad a_2 + ia_3 \in \mathbf{C}.$$

The bicomplex variable α, and respectively the bicomplex function $f(\alpha)$, admit some complex pair representation. We write $\alpha = z + jw$, with $z = a_0+ia_1$, $w = a_2+ia_3$, $j^2 = -1$, and respectively $f(\alpha) = f_0(z,w)+jf_1(z,w)$. According to the formula for the derivative with respect to α,

$$f'(a_0) = \lim_{\Delta\alpha\to 0} \frac{f(a_0 + \Delta a) - f(a_0)}{\Delta a}, \quad \text{with} \quad \Delta a \neq 0,$$

$\Delta a = \Delta z + j\Delta w = z - z_0 + j(w - w_0)$, we receive the bicomplex Cauchy-Riemann equations

$$\frac{\partial f_0}{\partial z} = \frac{\partial f_1}{\partial w} \quad \text{and} \quad \frac{\partial f_1}{\partial z} = -\frac{\partial f_0}{\partial w}.$$

Anticyclic numbers. Following [22] we introduce the algebra (over $\mathbf{R}$) of the *anticyclic numbers* $\mathbf{R}(i^{m-1})$, $\dim \mathbf{R}(i^{m-1}) = m$. The basis of this algebra so constituated form the powers $i^0 = 1$, $i^1 = i$, $i^2, \ldots, i^{m-1}$ of the symbol i, which satisfies the condition $i^m = -1$.

Double-complex and fourth-complex numbers [2,5]. These kind of numbers appear naturally as a complexification of the 4-dimensional and 8-dimensional anticyclic numbers:

$$a_0 + a_1 f + a_2 f^2 + a_3 f^3, \; f^4 = -1, \; a_k \in \mathbf{R}$$
$$b_0 + b_1 f + b_2 f^2 + b_3 f^3 + \cdots + b_7 f^7, \; f^8 = -1, \; b_k \in \mathbf{R}.$$

The complexification can be obtained as follows: rewriting the above written expressions in the form:

$$a_0 + a_2 f^2 + f(a_1 + a_3 f^2),$$

we can set $f = j$, $f^2 = i$, $i \in \mathbf{C}$. Thus we obtain $\alpha = z + jw$, where $z = a_0 + ia_2$, $w = a_1 + ia_3$.

Analogously, we deduce the complexification of 8-dimensional anticyclic numbers:

$$b_0 + b_4 f^4 + (b_1 + b_5 f^4) f + (b_2 + b_6 f^4) f^2 + (b_3 + b_7 f^4) f^3, \; f^8 = -1.$$

Setting $f = j$, $f^4 = i$, $i \in \mathbf{C}$, we obtain: $\alpha = z_0 + z_1 j + z_2 j^2 + z_3 j^3$ with $j^4 = i$. Addition and multiplication operations are defined by the usual way; in result we receive associative and commutative algebras with units, denoted respectively by $\mathbf{C}(1, j)$ with $j^2 = i$, and $\mathbf{C}(1, j, j^2, j^3)$ with $j^4 = i$.

According to the definition of Scheffers [10] the mapping $f : A \to A$ (A is algebra over $\mathbf{C}$) is called *holomorphic* if there is a *linear* map $f'(\alpha)$, $\alpha \in A$, such that:

$$df(\alpha) = f'(\alpha) d\alpha$$

Applying this definition to the algebra $\mathbf{C}(1, j)$ we receive the Cauchy-Riemann system in a new form:

$$\frac{\partial f_0}{\partial z} = \frac{\partial f_1}{\partial w}, \quad \frac{\partial f_0}{\partial w} = i \frac{\partial f_1}{\partial z},$$
$$f = f_0 + j f_1, \; df = df_0 + j df_1 \quad \text{with} \quad j^2 = i.$$

Here the symbol j can be considered as a hypercomplex number with respect to $\mathbf{C} \times \mathbf{C}$ by an analogy with i which is regarded as complex number (or hyper real) with respect to $\mathbf{R} \times \mathbf{R}$.

In the case of fourth-complex numbers $\mathbf{C}(1, j, j^2, j^3)$, j is considered as a second order hypercomplex number with respect to $\mathbf{C}^2 \times \mathbf{C}^2$. So, the double-complex algebra $\mathbf{C}(1, j)$ is isomorphic to the bicomplex algebra $\mathbf{B}$ introduced by Segre and the fourth-complex algebra $\mathbf{C}(1, j, j^2, j^3)$ is algebraically isomorphic to the algebra $\mathbf{B} \times \mathbf{B}$.

3. Real matrix representation of $\mathbf{C}(1, j)$ and $\mathbf{C}(1, j, j^2, j^3)$

The bi-complex and, respectively, fourth-complex numbers admit real matrix representation by real anticircular matrices. For instance $\mathbf{C}(1, j)$ (or $\mathbf{R}(i^3)$) is represented by the matrices:

$$\alpha = a_0 + a_1 f + a_2 f^2 + a_3 f^3 \mapsto \begin{pmatrix} a_0 & a_1 & a_2 & a_3 \\ -a_3 & a_0 & a_1 & a_2 \\ -a_2 & -a_3 & a_0 & a_1 \\ -a_1 & -a_2 & -a_3 & a_0 \end{pmatrix}.$$

This representation is analogous to the real matrix representation of a complex number:

$$z = a_0 + i a_1 \mapsto \begin{pmatrix} a_0 & a_1 \\ -a_1 & a_0 \end{pmatrix}.$$

In the case of $\mathbf{C}(1, j, j^2, j^3)$ we have real representation with 8×8 anticircular matrices. Anticycles numbers can be considered as a real form of double-complex numbers and, isomorphicaly, of bicomplex numbers. In the book [22] there is nothing about the term bicomplex number.

4. Complex quadratic geometries over $\mathbf{C}(1, j)$ and $\mathbf{C}(1, j, j^2, j^3)$

In the pseudo-Euclidean geometry of the algebra $\mathbf{C}(1, j)$ the scalar product of the basis vectors is defined by the following equalities

$$\langle 1, 1\rangle = 1, \quad \langle 1, j\rangle = \langle j, 1\rangle = 0, \quad \langle j, j\rangle = i.$$

The quadratic form $\langle \alpha, \alpha\rangle = z^2 + iw^2$ determines the isotropic cone $\{\alpha \in \mathbf{C}(1, j) : \langle \alpha, \alpha\rangle = 0\}$ in the introduced geometry (corresponding to the quadratic form $z^2 + iw^2$). In the case $\langle j, j\rangle = -i$ we obtain the quadratic form $\langle \alpha, \alpha\rangle = z^2 - iw^2$, wich corresponds to the analytic set of zero-divisors in the algebra $\mathbf{C}(1, j)$.

Proceeding by analogy, for the algebra $\mathbf{C}(1, j, j^2, j^3)$ we can define the scalar products:

$$\langle 1, 1\rangle = 1, \quad \langle j, j\rangle = a, \quad \langle j^2, j^2\rangle = b,$$
$$\langle j^3, j^3\rangle = c, \quad \langle j^k, j^m\rangle = 0 \quad \text{if} \quad k \neq m, \; a, b, c \in \mathbf{C}.$$

The equation $\langle \alpha, \alpha \rangle = 0$ is the condition for the isotropic cone in the corresponding pseudo-Euclidean geometry. It may be of interest to study the geometric and analytic properties of these isotropic cones.

The authors have no pretences that this brief report on some themes on bicomplex variables can be considered as a complete review of the subject. Our task is to present some aspects, which may be of interest in the future work, but we hope that some of the problems outlined can be developed in different directions. The bicomplex analysis is subject of active interest and every progress in this area may open new nontrivial problems. Let we express our sincere thanks to O. Gerus and A. Pogorui for the useful discussion and some bibliographic indications. Y. Matsushita helps us to complete the list of references published here.

References

1. G. Berloty, *Théorie des quatités complexes à n unités principales*, (Thèse), Paris, 1886.
2. S. Dimiev and S. Slavova, *Bicomplex holomorphicity*, Contemporary Aspects of Complex Analysis, Differential Geometry and Mathematical Physics. Proc. 7th Intern. Workshop on Complex Structures and Vector Fields, St. Dimiev and K. Sekigawa (Eds.), World Scientific, 2005, 41–48.
3. H. Hankel, *Vorlesungen über die complexen Zahlen und ihre Functionen*, Leipzig, 1867.
4. Em. Ivanov, Tetranions, Ann. de l'Université de Sofia, Fac. de Phys.-Math. **2** (1905–1906), 145–154 (in Bulgarian).
5. B. Kiradjiev and P. Stoev, *Fourth-complex holomorphic functions*, Tensor N. S. **64** (2003), 106–111.
6. A.A. Pogorui and R.M. Rodriguez-Dagnino, *On the set of zeros of bicomplex plynomials*, Complex Variables and Elliptic Equations **51** (2006), 725–730.
7. Fr. Ringleb, *Beiträge zur Functionentheorie in hypercomplexen Systemen*, Rendiconti circolo math. Palermo **57** (1933), 311–340.
8. St. Rönn, *Bicomplex algebra and function theory*, arXiv:math.CV/01012001, 2001, 1–71.
9. G. Scheffers, *Zuruckführung complexen Zahlensysteme auf typische Formen*, Math. Annalen **39** (1891), 293–390.
10. ———, *Veralgemeinerung der Grundlagen der gewönlichen komplexen Functionen*, Berichte Sachs. Acad. Wiss. **45** (1893), 828–842.
11. D.G. Scorza, *Sulle funzioni olomorfe di una variabile bicomplessa*, Reale Academia d'Italia, Memorie della classe di scienze fsiche, matematiche e naturali **5** (1934), 597–665.
12. C. Segre, *Le representazioni reali delle forme complesse e gli enti iperalgebrici*, Math. Annalen **40** (1892), 413–467.
13. N. Spampinato, *Sulle funzioni di una variabile in un algebra complessa ad n unita dotata di modulo*, Rendiconti circolo math. Palermo **57** (1933), 235–

272; **59** (1935), 185–227.

14. ———, *Sulle funzioni olomorfe di una variabile bicomplessa*, Annali Math. pura & applicata, Ser. 4, **14** (1936), 305–323.
15. E. Study, *Ueber Syteme von komplexen Zahlen*, Göttingen Nachr., 1889, 237–268.
16. ———, *Ueber Syteme komplexer Zahlen and ihre Anwendung in der Theorie der Transformationsgruppen*, Monatshefte für Math. und Phys. **1** (1890), 283–355.
17. E. Study and E. Cartan, *Nombres complexes*, Encyclopédie des sciences mathématiques pures et appliquées, t. I., vol. 1, 329–468.
18. T. Takasu, *Theorie der Funktionen einer allgemeinen buicomplexen Veranderlichen, I*, Tôhoku Science Reports, **32** (1943), 1.
19. L. Tchakalov, *On the tetranions of E. Ivanov*, Ann. de l'Université de Sofia, Fac.de Phys.-Math., **20** (1924), 89–120 (in Bulgarian, with German abstract). (Cited by the edition: L. Tchakalov, Opera, vol. I., Sofia, 1982, 189–208).
20. A.F. Turbin, *Hypercomplex analysis in Lyusch's algebra*, Problems of analytical mechanics and its applications, Proc. Math. Institute of the NAS of Ukraine **26** (1999), 387–406.
21. V.V. Vishnevski, *On a property of analytic functions over algebras and it application for study of complex structures in Riemannian spaces*, Sci. Reports Kazan Univ. **126** (1966), 5–12.
22. V.V. Vishnevski, A.P. Shirokov and V.V. Shurygin, *Spaces over algebras*, Kazan, 1985 (in Russian).
23. J.A. Ward, *A theory of analytic functions in linear associative algebras*, Duke Math. J. **7** (1940), 233–248.
24. ———, *From generalized Cauchy-Riemann equations to linear algebras*, Proc. Amer. Math. Soc. **4** (1953), 456–461.

INTEGRABILITY, CURVATURE AND DESCRIPTION OF PHOTON-LIKE OBJECTS

S. DONEV* and M. TASHKOVA

Institute for Nuclear Research and Nuclear Energy,
Bulg.Acad.Sci., 1784 Sofia, blvd.Tzarigradsko chaussee 72, Bulgaria
** E-mail: sdonev@inrne.bas.bg*

This paper aims to present a general idea for description of finite physical objects with consistent internal dynamical structure and evolution as a whole making use of the mathematical concepts and structures connected with the Frobenius integrability/nonintegrability theorems and to present an example. The idea consists in to consider some distribution Δ_o of vector fields on a manifold, then to separate some *integrable* subdistribution $\Delta \subset \Delta_o$, representing the integrity of the object considered. The curvatures of all nonintegrable subdistributions of Δ will be interpreted as generators of processes of internal energy-momentum exchange, i.e. of the internal dynamics of the object. The curvatures of distributions including vector fields from Δ_o and Δ will be interpreted as generators of interaction of the physical object with the outside world. Example of photon-like objects is considered in detail.

Keywords: Frobenius theorem; photon-like objects.

1. Introduction

We state the main idea of our approach. Any physical system with a dynamical structure is characterized with some internal energy-momentum redistributions, i.e. energy-momentum fluxes during evolution. Any system of energy-momentum fluxes (as well as fluxes of other interesting for the case physical quantities subject to change during evolution, but we limit ourselfs just to energy-momentum fluxes here) can be considered mathematically as generated by some system of vector fields. A *consistent* and *interelated time-stable* system of energy-momentum fluxes can be considered to correspond to an integrable distribution Δ of vector fields according to the principle *local object generates integral object*. An integrable distribution Δ may contain various *nonintegrable* subdistributions $\Delta_1, \Delta_2, \ldots$ which subdistributions may be interpreted physically as interacting subsytems. Any physical interaction between 2 subsystems is necessarily acompanied with available energy-momentum exchange between them, this could be understood mathematically as nonintegrability of each of the two subdistributions of Δ and could be naturally measured by the corresponding curvatures. For example, if Δ is an integrable 3-dimensional distribution spent by the vector fields (X_1, X_2, X_3) then we may have, in general, three non-integrable 2-dimensional subdistributions $(X_1, X_2), (X_1, X_3), (X_2, X_3)$. Finally, some interaction with the outside world can be described by curvatures of non-integrable distributions in which elements from Δ and vector fields outside Δ are involved (such processes will not be considered in this paper).

We recall the Frobenius theorem on a manifold M^n [1] (further all manifolds are assumed smooth and finite dimensional and all objects defined on M are also assumed smooth). If the system of vector fields $\Delta = [X_1(x), X_2(x), \dots, X_p(x)]$, $x \in M$, $1 < p < n$, satisfies $X_1(x) \wedge X_2(x) \wedge \cdots \wedge X_p(x) \neq 0$, $x \in M$ then Δ is integrable iff the Lie brackets $[X_i, X_j]$, $i, j = 1, 2, \dots, p$ are representable linearly through the very X_i, $i = 1, 2, \dots, p$: $[X_i, X_j] = C^k_{ij} X_k$, where C^k_{ij} are functions. Clearly, an easy way to find out if a distribution is integrable is to check if the exterior products

$$[X_i, X_j] \wedge X_1(x) \wedge X_2(x) \wedge \cdots \wedge X_p(x) \neq 0, \quad x \in M; \quad i, j = 1, 2, \dots, p$$

are identically zero. If this is not the case (which means that at least one such Lie bracket "sticks out" of the distribution Δ) then the corresponding coefficients, which are bilinear combinations of the components of the vector fields and their derivatives, represent the corresponding curvatures [2]. Clearly, if two subdistributions contain at least one common vector field it seems naturally to expect interaction.

In the dual formulation of Frobenius theorem in terms of differential 1-forms (i.e. Pfaff forms) we look for $(n-p)$-Pfaff forms $(\alpha^1, \alpha^2, \dots, \alpha^{n-p})$, i.e. a $(n-p)$-codistribution Δ^*, such that $\alpha^i(X_j) = 0$, and $\alpha^1 \wedge \alpha^2 \wedge \cdots \wedge \alpha^{n-p} \neq 0$, $1 = 1, 2, \dots, n-p$, $j = 1, 2, \dots, p$. Then the integrability of the distribution Δ is equivalent to the requirements

$$\mathbf{d}\alpha^m \wedge \alpha^1 \wedge \alpha^2 \wedge \cdots \wedge \alpha^{n-p} = 0, \quad m = 1, 2, \dots, (n-p),$$

where $\mathbf{d}$ is the exterior derivative. Clearly, if the manifold is 4-dimensional (the usual case in physics) and Δ is 3-dimensional, then to a 2-dimensional nonintegrable subsystem of Δ will correspond a 2-dimensional nonintegrable codistribution of Δ^*.

We are going now to make use of the above general consideration to find appropriate objects and relations in an attempt to describe photon-like objects. Under photon-like object we mean *a spatially finite time-stable massless physical object with a consistent translational-rotational dynamical structure living in 4-dimensional space-time.*

We are going to comment just the requirement for masslessness [3]. The term "*massless*" characterizes the way of propagation: the integral energy E and momentum p of a photon-like object should satisfy the relation $E = cp$, where c is the speed of light in vacuum, and in relativistic terms this means that its integral energy-momentum vector *must be isotropic*, i.e. it must have zero module with respect to Lorentz-Minkowski (pseudo)metric in $\mathbb{R}^4$. Since the translational velocity of every point where the corresponding field functions are different from zero must be equal to c, we have in fact an isotropic vector field $\bar{\zeta}$. The integral trajectories of this vector field are isotropic straight lines. It follows that just the corresponding direction is important, so, canonical coordinates $(x^1, x^2, x^3, x^4) = (x, y, z, \xi = ct)$ on $\mathbb{R}^4$ may be chosen such that $\bar{\zeta}$ may have only two non-zero components of magnitude 1: $\bar{\zeta}^\mu = (0, 0, -\varepsilon, 1)$, where $\varepsilon = \pm 1$ accounts for the two directions along the coordinate z. Further such a coordinate system will be called $\bar{\zeta}$-adapted and will be of main usage.

2. Modelling Photon-like Objects

We consider the space $\mathbb{R}^4$ as a manifold related to standard global coordinates

$$(x^1, x^2, x^3, x^4) = (x, y, z, \xi = ct),$$

and the natural volume form $\omega_o = dx \wedge dy \wedge dz \wedge d\xi$. We introduce the vector field $\bar{\zeta}$, which in appropriate coordinates (throughout used further) is assumed to look as follows:

$$\bar{\zeta} = -\varepsilon \frac{\partial}{\partial z} + \frac{\partial}{\partial \xi}, \quad \varepsilon = \pm 1. \tag{1}$$

Let's denote the corresponding to $\bar{\zeta}$ completely integrable 3-dimensional Pfaff system by $\Delta^*(\bar{\zeta})$. Thus, $\Delta^*(\bar{\zeta})$ is generated by three linearly independent 1-forms $(\alpha_1, \alpha_2, \alpha_3)$ which annihilate $\bar{\zeta}$, i.e.

$$\alpha_1(\bar{\zeta}) = \alpha_2(\bar{\zeta}) = \alpha_3(\bar{\zeta}) = 0; \quad \alpha_1 \wedge \alpha_2 \wedge \alpha_3 \neq 0.$$

We fix $\Delta^*(\bar{\zeta})$ as generated by (A, A^*, ζ), where

$$A = u dx + p dy; \quad A^* = -p dx + u dy; \quad \zeta = \varepsilon dz + d\xi, \tag{2}$$

and (u,p) are two arbitrary functions. This completely integrable 3-dimensional Pfaff system contains three 2-dimensional subsystems: $(A,A^*),(A,\zeta)$ and (A^*,ζ). We have the following

Proposition 1. *The following relations hold:*

$$\mathbf{d}A\wedge A\wedge A^*=0;\quad \mathbf{d}A^*\wedge A^*\wedge A=0;$$
$$\mathbf{d}A\wedge A\wedge\zeta=\varepsilon\big[u(p_\xi-\varepsilon p_z)-p(u_\xi-\varepsilon u_z)\big]\omega_o;$$
$$\mathbf{d}A^*\wedge A^*\wedge\zeta=\varepsilon\big[u(p_\xi-\varepsilon p_z)-p(u_\xi-\varepsilon u_z)\big]\omega_o.$$

Proof. Immediately checked. □

These relations say that the 2-dimensional Pfaff system (A,A^*) is completely integrable for any choice of the two functions (u,p), while the two 2-dimensional Pfaff systems (A,ζ) and (A^*,ζ) are NOT completely integrable in general, and the same curvature factor $\mathbf{R}=u(p_\xi-\varepsilon p_z)-p(u_\xi-\varepsilon u_z)$ determines their nonintegrability.

Since ζ is closed, it determines a 3-dimensional completely integrable distribution (or differential system) $\Delta(\zeta)$, and a basis of this distribution is given by the following vector fields $(\bar{A},\bar{A}^*,\bar{\zeta})$:

$$\bar{A}=-u\frac{\partial}{\partial x}-p\frac{\partial}{\partial y};\quad \bar{A}^*=p\frac{\partial}{\partial x}-u\frac{\partial}{\partial y};\quad \bar{\zeta}=-\varepsilon\frac{\partial}{\partial z}+\frac{\partial}{\partial\xi},\quad \varepsilon=\pm1. \tag{3}$$

We have

$$\zeta(\bar{A})=\zeta(\bar{A}^*)=\zeta(\bar{\zeta})=0;\quad \bar{A}\wedge\bar{A}^*\wedge\bar{\zeta}\neq0.$$

Proposition 2. *The following relations hold:*

$$[\bar{A},\bar{A}^*]\wedge\bar{A}\wedge\bar{A}^*=0 \tag{4}$$
$$[\bar{A},\bar{\zeta}]=(u_\xi-\varepsilon u_z)\frac{\partial}{\partial x}+(p_\xi-\varepsilon p_z)\frac{\partial}{\partial y}; \tag{5}$$
$$[\bar{A}^*,\bar{\zeta}]=-(p_\xi-\varepsilon p_z)\frac{\partial}{\partial x}+(u_\xi-\varepsilon u_z)\frac{\partial}{\partial y}. \tag{6}$$

Proof. Immediately checked. □

From these last relations (4)-(6) and in accordance with Proposition 1 it follows that the distribution $(\bar{A},\bar{A}^*)$ is integrable, and it can be easily shown that the two distributions $(\bar{A},\bar{\zeta})$ and $(\bar{A}^*,\bar{\zeta})$ would be completely integrable only if the same curvature factor

$$\mathbf{R}=u(p_\xi-\varepsilon p_z)-p(u_\xi-\varepsilon u_z) \tag{7}$$

is zero.

We mention also that the projections

$$\langle A, [\bar{A}^*, \bar{\zeta}]\rangle = -\langle A^*, [\bar{A}, \bar{\zeta}]\rangle = u(p_\xi - \varepsilon p_z) - p(u_\xi - \varepsilon u_z)$$

give the same factor **R**. The same curvature factor appears, of course, as coefficient in the exterior products $[\bar{A}^*, \bar{\zeta}] \wedge \bar{A}^* \wedge \bar{\zeta}$ and $[\bar{A}, \bar{\zeta}] \wedge \bar{A} \wedge \bar{\zeta}$. For the other two projections we obtain

$$\langle A, [\bar{A}, \bar{\zeta}]\rangle = -\langle A^*, [\bar{A}^*, \bar{\zeta}]\rangle = \frac{1}{2}\big[(u^2+p^2)_\xi - \varepsilon(u^2+p^2)_z\big]. \tag{8}$$

Clearly, the last relation (8) may be put in terms of the Lie derivative $L_{\bar{\zeta}}$ as

$$\frac{1}{2}L_{\bar{\zeta}}(u^2+p^2) = -\frac{1}{2}L_{\bar{\zeta}}\langle A, \bar{A}\rangle = -\langle A, L_{\bar{\zeta}}\bar{A}\rangle = -\langle A^*, L_{\bar{\zeta}}\bar{A}^*\rangle.$$

Remark. Further in the paper we shall denote $\sqrt{u^2+p^2} \equiv \phi$, and shall assume that ϕ is a *spatially finite* function, so, u and p must also be spatially finite.

Proposition 3. *There is a function $\psi(u,p)$ such, that*

$$L_{\bar{\zeta}}\psi = \frac{u(p_\xi - \varepsilon p_z) - p(u_\xi - \varepsilon u_z)}{\phi^2} = \frac{\mathbf{R}}{\phi^2}.$$

Proof. It is immediately checked that $\psi = \arctan\frac{p}{u}$ is such one. □

We note that the function ψ has a natural interpretation of *phase* because of the easily verified now relations $u = \phi\cos\psi$, $p = \phi\sin\psi$, and ϕ acquires the status of *amplitude*. Since the transformation $(u,p) \to (\phi,\psi)$ is non-degenerate this allows to work with the two functions (ϕ,ψ) instead of (u,p).

From Proposition 3 we have

$$\mathbf{R} = \phi^2 L_{\bar{\zeta}}\psi = \phi^2(\psi_\xi - \varepsilon\psi_z). \tag{9}$$

This last formula (10) shows something very important: at any $\phi \neq 0$ the curvature **R** will NOT be zero only if $L_{\bar{\zeta}}\psi \neq 0$, which admits in principle availability of rotation. In fact, lack of rotation would mean that ϕ and ψ are running waves along $\bar{\zeta}$. The relation $L_{\bar{\zeta}}\psi \neq 0$ means, however, that rotational properties are possible in general, and some of these properties are carried by the phase ψ. It follows that in such a case the translational component of propagation along $\bar{\zeta}$ (which is supposed to be available) must be determined essentially, and most probably entirely, by ϕ. In particular, we could expect the relation $L_{\bar{\zeta}}\phi = 0$ to hold, and if this happens, then the rotational component of propagation will be represented entirely by the phase ψ, and, more specially, by the curvature factor $\mathbf{R} \neq 0$, so, since the objects we are going to describe have consistent translational-rotational dynamical structure, further we assume that, in general, $L_{\bar{\zeta}}\psi \neq 0$.

We consider now the Lie-brackets $[\bar{A}, \bar{\zeta}]$ and $[\bar{A}^*, \bar{\zeta}]$ in terms of ψ:

$$[\bar{A}, \bar{\zeta}] = -L_{\bar{\zeta}}\psi.\bar{A}^*; \quad [\bar{A}^*, \bar{\zeta}] = L_{\bar{\zeta}}\psi.\bar{A}; \quad [\bar{A}, \bar{\zeta}] \wedge [\bar{A}^*, \bar{\zeta}] = -\left(L_{\bar{\zeta}}\psi\right)^2 \bar{A}^* \wedge \bar{A} \neq 0.$$

These relations say that the 2-dimensional frame $(\bar{A}, \bar{A}^*)$ on the (x, y)-plane is transformed to the 2-dimensional frame $([\bar{A}, \bar{\zeta}], [\bar{A}^*, \bar{\zeta}])$ on the same plane by means of the matrix $L_{\bar{\zeta}}\psi.J$, where J is the canonical complex structure in $\mathbb{R}^2$, generating rotation to the angle of $\pi/2$. It is easy now to see that the 4-frame $(\bar{A}, \bar{A}^*, \partial_z, \partial_\xi)$ is transformed to the 4-frame $([\bar{A}, \bar{\zeta}], [\bar{A}^*, \bar{\zeta}], \partial_z, \partial_\xi)$ by means of a linear map with determinant $(L_{\bar{\zeta}}\psi)^2$. Hence, at $\phi \neq 0$ *any of the conditions* $L_{\bar{\zeta}}\psi \neq 0$, $\mathbf{R} \neq 0$ *guarantees availability of rotational component of propagation.*

The two nonintegrable Pfaff systems (A, ζ) and (A^*, ζ) define corresponding 2-forms:

$$G = A \wedge \zeta \quad \text{and} \quad G^* = A^* \wedge \zeta.$$

We have also the 2-vectors

$$\bar{G} = \bar{A} \wedge \bar{\zeta}, \quad \text{and} \quad \bar{G^*} = \bar{A}^* \wedge \bar{\zeta}.$$

We shall need the quantities

$$i(\bar{G})\mathbf{d}G + i(\bar{G^*})\mathbf{d}G^* \quad \text{and} \quad i(\bar{G^*})\mathbf{d}G + i(\bar{G})\mathbf{d}G^*,$$

where

$$i(\bar{G}) = i(\bar{\zeta}) \circ i(\bar{A}), \quad \text{and} \quad i(\bar{G^*}) = i(\bar{\zeta}) \circ i(\bar{A}^*),$$

$\mathbf{d}$ is the exterior derivative, and $i(X)$ is the standard insertion operator in the exterior algebra of differential forms on $\mathbb{R}^4$ defined by the vector field X. Having in view the explicit expressions for $A, A^*, \zeta, \bar{A}, \bar{A}^*$ and $\bar{\zeta}$ we obtain

$$i(\bar{G})\mathbf{d}G = i(\bar{G^*})\mathbf{d}G^* = \frac{1}{2}L_{\bar{\zeta}}\left(\phi^2\right).\zeta. \qquad (*)$$

Also, we have

$$i(\bar{G^*})\mathbf{d}G = -i(\bar{G})\mathbf{d}G^*,$$

$$i(\bar{G^*})\mathbf{d}G + i(\bar{G})\mathbf{d}G^* = \frac{1}{2}\left[(\bar{G^*})^{\alpha\beta}(\mathbf{d}G)_{\alpha\beta\mu} + (\bar{G})^{\alpha\beta}(\mathbf{d}G^*)_{\alpha\beta\mu}\right]dx^\mu = 0.$$

A direct calculation shows that

$$i(\bar{G^*})\mathbf{d}G = -i(\bar{G})\mathbf{d}G^*$$
$$= \varepsilon\Big[u(p_\xi - \varepsilon p_z) - p(u_\xi - \varepsilon u_z)\Big]dz + \Big[u(p_\xi - \varepsilon p_z) - p(u_\xi - \varepsilon u_z)\Big]d\xi = \mathbf{R}.\zeta.$$

3. Physical Interpretation

The relations obtained suggest and allow the following physical interpretation. Let η be the pseudoeuclidean metric with signature $(-,-,-,+)$, then making use of the corresponding Hodge $*$-operator, the following relations can be easily verified:

$$(\varepsilon G^*)_{\mu\nu} = (*G)_{\mu\nu}, \quad \bar{A}^\mu = \eta^{\mu\nu}A_\nu, \quad \bar{\zeta}^\mu = \eta^{\mu\nu}\zeta_\nu,$$
$$\bar{G}^{\mu\nu} = \eta^{\mu\sigma}\eta^{\nu\tau}G_{\sigma\tau}, \quad (\varepsilon\bar{G}^*)^{\mu\nu} = \eta^{\mu\sigma}\eta^{\nu\tau}(*G)_{\sigma\tau}\ .$$

Let now the 2-forms $G = A\wedge\zeta$ and $G^* = A^*\wedge\zeta$ be interpreted as field functions. Then the corresponding analog of the Maxwell stress-energy-momentum tensor is given by

$$\frac{1}{4}G_{\alpha\beta}\bar{G}^{\alpha\beta}\delta^\nu_\mu - G_{\mu\sigma}\bar{G}^{\nu\sigma} = \frac{1}{2}\Big[G_{\mu\sigma}\bar{G}^{\nu\sigma} + (*G)_{\mu\sigma}(\bar{*G})^{\nu\sigma}\Big] = \phi^2\zeta_\mu\bar{\zeta}^\nu\ ,$$

and with respect to the corresponding Levi-Civita covariant derivative ∇ we obtain

$$\nabla_\nu\Big[\frac{1}{4}G_{\alpha\beta}\bar{G}^{\alpha\beta}\delta^\nu_\mu - G_{\mu\sigma}\bar{G}^{\nu\sigma}\Big] = \frac{1}{2}\Big[(\bar{G})^{\alpha\beta}(\mathbf{d}G)_{\alpha\beta\mu} + (\bar{*G})^{\alpha\beta}(\mathbf{d}*G)_{\alpha\beta\mu}\Big].$$

Now, the explicit form of the energy-momentum tensor as a sum of the two fully similar expressions $\frac{1}{2}G_{\mu\sigma}\bar{G}^{\nu\sigma}$ and $\frac{1}{2}(*G)_{\mu\sigma}(\bar{*G})^{\nu\sigma}$, suggests that our photon-like object is described by two vector components, namely G and G^*, so that the correct mathematical modelling object should look like

$$\Omega = G\otimes e_1 + G^*\otimes e_2,$$

where (e_1, e_2) is a basis of an appropriately chosen linear space. The above relations suggest to consider the two expressions

$$i(\bar{G})\mathbf{d}G = *(G\wedge *\mathbf{d}G) = \frac{1}{2}(\bar{G})^{\alpha\beta}(\mathbf{d}G)_{\alpha\beta\sigma}dx^\sigma = \frac{1}{2}L_{\bar{\zeta}}\left(\phi^2\right).\zeta,$$
$$i(\bar{G^*})\mathbf{d}G^* = *(*G^*\wedge *\mathbf{d}*G^*) = \frac{1}{2}(\bar{G^*})^{\alpha\beta}(\mathbf{d}G^*)_{\alpha\beta\sigma}dx^\sigma = \frac{1}{2}L_{\bar{\zeta}}\left(\phi^2\right).\zeta$$

as local changes of the field energy-momentum carried by each vector component. Relation

$$i(\bar{G^*})\mathbf{d}G = -i(\bar{G})\mathbf{d}G^*$$
$$= \varepsilon\Big[u(p_\xi - \varepsilon p_z) - p(u_\xi - \varepsilon u_z)\Big]dz + \Big[u(p_\xi - \varepsilon p_z) - p(u_\xi - \varepsilon u_z)\Big]d\xi = \mathbf{R}.\zeta$$

suggests, in turn, that the two components G and G^* exchange locally energy-momentum simultaniously and in equal quantities. These two expressions are proportional (up to a sign) to the common for the two nonintegrable subsytems $(\bar{A}, \bar{\zeta})$ and $(\bar{A}^*, \bar{\zeta})$ curvature expressions, which in turn are proportional to the energy density $\phi^2 : \mathbf{R} = \phi^2 L_{\bar{\zeta}}\psi = \phi^2(\psi_\xi - \varepsilon\psi_z)$. The proportionality coefficient depends strongly on the change of the phase ψ along the vector field $\bar{\zeta}$. The relation

$$i(\bar{G})\mathbf{d}G = i(\bar{G}^*)\mathbf{d}G^* = \frac{1}{2}L_{\bar{\zeta}}\left(\phi^2\right).\zeta, \qquad (*)$$

together with the zero value of the divergence of the energy-momentum tensor: $\nabla_\nu(\phi^2\zeta_\mu\bar{\zeta}^\nu) = 0$, shows that the changes of the energy-momentum carried by each vector component G and G^* are equal and depend only on ϕ^2 and NOT on the phase ψ, and that if each vector component keeps its energy momentum, then the amplitude ϕ must be running wave along $\bar{\zeta}$.

If l_o is a positive constant (with physical dimension of distance) the two 1-forms A and A^* given by:

$$A = \Big[\phi(x, y, \xi + \varepsilon z)\, \cos\Big(\pm\frac{z}{l_o} + const\Big),$$
$$\phi(x, y, \xi + \varepsilon z)\, \sin\Big(\pm\frac{z}{l_o} + const\Big),\, 0,\, 0\Big];$$
$$A^* = \Big[-\phi(x, y, \xi + \varepsilon z)\, \sin\Big(\pm\frac{z}{l_o} + const\Big),$$
$$\phi(x, y, \xi + \varepsilon z)\, \cos\Big(\pm\frac{z}{l_o} + const\Big),\, 0,\, 0\Big]$$

generate a solution. All the relations obtained suggest to use the corresponding Maxwell stress-energy-momentum tensor defined by any of the 2-forms $G = A \wedge \zeta$ and $*G = A^* \wedge \zeta$ as an unifying quantity characterizing the physical properties of the solutions to the equations

$$i(\bar{G})\mathbf{d}G = 0, \quad i(\bar{G}^*)\mathbf{d}G^* = 0, \quad i(\bar{G}^*)\mathbf{d}G + i(\bar{G})\mathbf{d}G^* = 0. \qquad (10)$$

It is seen from the above expressions for A and A^* that among the solutions of (10) there are spatially finite ones with photon-like nature. A natural measure/characteristic of the rotational properties of these solutions is the 3-form

$$\beta = *\left(2\pi\frac{l_o^2}{c}i(\bar{G})\mathbf{d}G^*\right) = 2\pi\frac{l_o^2}{c}G \wedge *\mathbf{d} * G = 2\pi\frac{l_o^2}{c}G \wedge \delta G = *\left(2\pi\frac{l_o^2}{c}\mathbf{R}.\zeta\right),$$

where $\delta = *\mathbf{d}*$ is the coderivative, because if we choose the phase ψ as mentioned above then this 3-form is closed (for any $\phi(x, y, \xi + \varepsilon z)$): $\mathbf{d}\beta = 0$, so we get a conservative quantity. The corresponding integral of the $\mathbb{R}^3$-restriction of β over the whole $\mathbb{R}^3$ is finite (since ϕ is a spatially finite function) and equal to $2\pi l_o E/c$, where E is the corresponding integral energy, so, denoting $2\pi l_o/c$ by $\tau = 1/\nu$ we obtain an analog $E = \tilde{h}\nu$ of the Planck formula, where $\tilde{h}$ is a conservative quantity equal to the value of the integral $\int_{\mathbb{R}^3} \beta\, dxdydz$, and $\tilde{h} \neq 0$ only if $\mathbf{R} \neq 0$.

4. Conclusion

The above results allow to conclude that equations (10), under suitable choice of the amplitude function ϕ, admit time-stable photon-like solutions, which are localized inside a standard (left or right oriented) finite helical cylinder with height $2\pi l_o$. The evolution of such a solution follows the prolongation of the helical cylinder along some spatial direction (the external axis of the cylinder, chosen for coordinate z in our consideration). The translational and rotational components of propagation exist consistently, the translational velocity is c, the rotational component of propagation is of periodical nature with period $\tau = 2\pi l_o/c$. The dynamical structure of the solution is of intrinsic nature and demonstrates itself through a permanent energy-momentum exchange between the two vector components G and $*G$. A remarkable in our view property of these solutions is that the curvatures obtained are proportional to the energy density of the solution.

The work on this paper was partially supported by contract ϕ1515 with the bulgarian fund "Science Research".

References

1. C. Godbillon, *Géométrie Differentielle et Méchanique Analytique*, Hermann, Paris, 1969.
2. Y. Choquet-Bruat, C. Dewitt-Morette and M. Dillard-Bleick, *Analysis, Manifolds and Physics*, vol. 1, North Holland, 1982.
3. J. Synge. *Relativity: The Special Theory*, North Holland, Amsterdam, 1958.

COMPLEX SUBMANIFOLDS AND LAGRANGIAN SUBMANIFOLDS ASSOCIATE WITH MINIMAL SURFACES IN TORI*

N. EJIRI†

Department of Mathematics,
Meijo University,
Tempaku, Nagoya 466-8555, Japan
E-mail: ejiri@ccmfs.meijo-u.ac.jp

We consider the some energy function on a complex submanifold with parameter in the Siegel upper half space. We proved that the catastrophe set is analytic on an appropriate complex structure. This is a generalization of complex structure of the complex manifolds associated with minimal surfaces in tori. Furthermore, we investigate the catastrophe set as a Lagrangian submanifold.

1. Introduction

Let $L_{n,2\gamma}$ be the space of real $(n, 2\gamma)$ matrices. We consider some energy function E with $L_{n,2\gamma}$ as a parameter space on the Siegel upper half spce H_γ of degree γ and the restriction E_M to a complex manifolod M in H_γ. The catastrophe set of E_M in $M \times L_{n,2\gamma}$ is not analytic with respect to the complex structure induced from the product structure of M and $L_{n,2\gamma}$. Each point τ of H_γ gives a complex structure J_τ on the vector space $L_{n,2\gamma}$ as follows.

$$J_\tau(L_1, L_2) = (-(L_1\mathrm{Re}\tau - L_2)(\mathrm{Im}\tau)^{-1}, -(L_1\mathrm{Re}\tau - L_2)(\mathrm{Im}\tau)^{-1}\mathrm{Re}\tau - L_1\mathrm{Im}\tau),$$

where $L = (L_1, L_2) \in L_{n,2\gamma}$, L_1 and L_2 are (n, γ)-matrices. We can show that J_τ, $\tau \in M$ induces a complex structure on $M \times L_{n,2\gamma}$ such that $M \times L_{n,2\gamma}$ is a holomorphic trivial vector bundle. Thus $M \times L_{n,2\gamma}$ looks like a twistor space of $L_{n,2\gamma}$. With respect to this complex structure, we prove

---

*This work is supported by etc, etc.

†The author is partially supported by Grant-in-Aid for Scientific Research (C) (No. 16540060), Ministry of Education, Science, Sports Culture and Technology.

that a catastrophe set is an analytic set. Moreover an irreducible component containg a non-degenerate critical point gives a Lagrangian cone in the cotangent bundle $T^*L_{n,2\gamma}$ of $L_{n,2\gamma}$. When M is the space of Riemann matrices, we have analytic sets associated with minimal surfaces in flat tori [1], [3]. In the next paper, we shall prove that these Lagrangian cones with complex structure are totally complex cones in the quaternion vector space $T^*L_{n,2\gamma}$.

2. A review on a function with parameter

Let $\mathbf{R^k}$ be the k-dimensional Euclidean space, $(q^1,\ldots,q^k)$ the canonical coordinate system of $\mathbf{R^k}$, and U a neighborhood of the origin 0 of $\mathbf{R^k}$. Similarly, suppose that $(\lambda^1,\ldots,\lambda^n)$ is the canonical coordinate system of $\mathbf{R^n}$ and V is a neighborhood of the origin 0 of $\mathbf{R^n}$. Let F be a real valued function on $U\times V$ such that $F(0,0)=0$ and $0\in U$ is a critical point of $F|_{U\times\{0\}}$. Then we consider that F is a unfolding of $F|_{U\times\{0\}}$ such that $q^1,\ldots,q^k$ are innervariables and $\lambda^1,\ldots,\lambda^n$ are parameter. The catastrophe set $C(F)$ is defined by

$$C(F)=\left\{(q,\lambda)\in U\times V \;\middle|\; \frac{\partial F}{\partial q^1}=\cdots=\frac{\partial F}{\partial q^k}=0\right\}.$$

$0\in U$ is called a non-degenerate critical point if the Hessian of $F|_{U\times\{0\}}$ at $0\in U$ is non-degenerate. We also call $(0,0)\in C(F)$ to be non-degenerate. A sufficient condition such that some neighborhood of $(0,0)\in C(F)$ is a submanifold in $U\times V$ is that $(0,0)$ is a non-degenerate critical point as follows: Since the Jacobi matrix of the map K of $U\times V$ into $\mathbf{R^k}$ defined by

$$\left(\frac{\partial F}{\partial q^1},\ldots,\frac{\partial F}{\partial q^k}\right)$$

is given by

$$\begin{pmatrix} \dfrac{\partial^2 F}{\partial q^1\partial q^1} & \cdots & \dfrac{\partial^2 F}{\partial q^k\partial q^1} & \dfrac{\partial^2 F}{\partial \lambda^1\partial q^1} & \cdots & \dfrac{\partial^2 F}{\partial \lambda^n\partial q^1} \\ \vdots & & \vdots & \vdots & & \vdots \\ \dfrac{\partial^2 F}{\partial q^1\partial q^k} & \cdots & \dfrac{\partial^2 F}{\partial q^k\partial q^k} & \dfrac{\partial^2 F}{\partial \lambda^1\partial q^k} & \cdots & \dfrac{\partial^2 F}{\partial \lambda^n\partial q^k} \end{pmatrix}$$

and

$$\begin{pmatrix} \frac{\partial^2 F}{\partial q^1 \partial q^1} & \cdots & \frac{\partial^2 F}{\partial q^k \partial q^1} \\ \vdots & & \vdots \\ \frac{\partial^2 F}{\partial q^1 \partial q^k} & \cdots & \frac{\partial^2 F}{\partial q^k \partial q^k} \end{pmatrix}$$

is non-degenerate, K is submersive and the implicit function theorem implies that a neighborhood of $(0,0) \in C(F)$ is a graph over a neighborhood of $0 \in V$.

There is a weaker sufficient condition, which is that K is submersive at $(0,0)$. In this case, F is called a Morse family. Then we can construct the Lagrangian embedding of this submanifold $\subset C(F)$ containing $(0,0)$ into $T^*\mathbf{R}^\mathbf{n}$ as

$$(q,\lambda) \mapsto \left(\frac{\partial F}{\partial \lambda^1}(q,\lambda), \dots, \frac{\partial F}{\lambda^n}(q,\lambda), \lambda^1, \dots, \lambda^n \right),$$

where $(p_1, \dots, p_n, x^1, \dots, x^n)$ is the canonical coordinate system of $T^*\mathbf{R}^\mathbf{n}$. The reason is as follows: The Liouville form is $\sum_{j=1}^n p_j dx^j$ and the symplectic form of $T^*\mathbf{R}^\mathbf{n}$ is $\sum_{j=1}^n dp_j \wedge dx^j = d(\sum_{j=1}^n p_j dx^j)$. Since the Liouville form induced on $C(F)$ is given by $\sum_{j=1}^n (\partial F/\partial \lambda^j) d\lambda^j = d(F|_{C(F)})$, the symplectic form induced on $C(F)$ vanishes.

3. A function with parameter on the Siegel upper half space H_γ

Let S^2_C be the space of complex symetric matrices of size γ and H_γ the Siegel upper half space of degree γ defined by $H_\gamma = \{ X + iY \in S^2_C \mid Y > 0 \}$. Then H_γ is an open set of S^2_C. We consider the Hermitian form on S^2_C such that $\langle A, B \rangle = \operatorname{tr} A\overline{B}$, where A, $B \in S^2_C$.

We define $P(\tau)$ for $\tau \in H_\gamma$ by

$$P(\tau) = \begin{pmatrix} (\operatorname{Im}\tau) + (\operatorname{Re}\tau)(\operatorname{Im}\tau)^{-1}(\operatorname{Re}\tau) & -(\operatorname{Re}\tau)(\operatorname{Im}\tau)^{-1} \\ -(\operatorname{Im}\tau)^{-1}(\operatorname{Re}\tau) & (\operatorname{Im}\tau)^{-1} \end{pmatrix}$$

and the energy function E on $H_\gamma \times L_{n,2\gamma}$ by

$$E(\tau, L) = \frac{1}{2} \operatorname{tr} P(\tau)\, {}^tLL.$$

Suppose that M is a k-dimensional complex submanifold in H_γ and τ is the immersion of M into H_γ. Then we have the function E_M restricted to $M \times L_{n,2\gamma}$ and the catastrophe set as

$$C(E_M) = \left\{ (q, L) \in M \times L_{n.2\gamma} \;\middle|\; \frac{\partial E_M}{\partial z^1} = \cdots = \frac{\partial E_M}{\partial z^k} = 0 \right\},$$

where $(z^1, \ldots, z^k)$ be a complex coordinate system of M. We induce a complex structure of $M \times L_{n,2\gamma}$ different from the standard complex structure of $M \times L_{n,2\gamma}$ as the product manifold of complex manifolds M and $L_{n,2\gamma}$ and show the following.

Theorem 3.1. *$C(E_M)$ is an analytic set.*

Using the Hermitian form on S^2_C, we calculate the gradient vector field $\operatorname{grad} E$ on H_γ for a fixed $L \in L_{n,2\gamma}$ [2].

Lemma 3.1. *For $(\tau, L) \in M \times L_{n,2\gamma}$,*

$$\operatorname{grad} E(\tau, L) = 2i \times$$
$$\frac{1}{2}^t\overline{(L_1 + i[L_1 \operatorname{Re}\tau - L_2](\operatorname{Im}\tau)^{-1})}\,\frac{1}{2}\overline{(L_1 + i[L_1 \operatorname{Re}\tau - L_2](\operatorname{Im}\tau)^{-1})}.$$

Proof. Let $K_{n,\gamma}$ be the space of (n, γ) complex matrices. Then we define a diffeomorphism I of $H_\gamma \times L_{n,2\gamma}$ onto $H_\gamma \times K_{n,\gamma}$ by

$$I(\tau, (L_1, L_2)) = \left(\tau, \frac{1}{2}(L_1 + i[L_1 \operatorname{Re}\tau - L_2](\operatorname{Im}\tau)^{-1})\right).$$

We set $K = (1/2)(L_1 + i[L_1 \operatorname{Re}\tau - L_2](\operatorname{Im}\tau)^{-1})$. Then $(q, (L_1, L_2)) \in C(E_M)$ if and only if $\operatorname{grad} E(\tau(q), L)$ is perpendicular to $\partial\tau/\partial z^\ell(q)$, $\ell = 1, \ldots, k$, which is equivalent to

$$\left\langle \frac{\partial \tau}{\partial z^\ell}(q), 2i\overline{({}^tKK)} \right\rangle = 0, \qquad \ell = 1, \ldots, k$$

and hence

$$\operatorname{tr} {}^tKK \frac{\partial \tau}{\partial z^\ell}(q) = 0, \qquad \ell = 1, \ldots, k.$$

Since

$$\operatorname{tr} {}^tKK \frac{\partial \tau}{\partial z^\ell}$$

is holomorphic, $I(C(E_M))$ is an analytic set of $M \times K_{n,\gamma}$. So $C(E_M)$ is an analytic set with respect to the complex structure on $M \times L_{n,2\gamma}$ induced from the complex structure of the product manifold $M \times K_{n,\gamma}$. □

Since the invese mapping of I is given by

$$I^{-1}(\tau, K) = \big(\tau, \big(\mathrm{Re}\,(2K), \mathrm{Re}\,(2K\tau)\big)\big),$$

we get J_τ in the introduction.

Since $C(E_M)$ is an analytic set, we can consider an irreducible component N of $C(E_M)$. If N has a non-degenerate critical point P, then some neighborhood of P is a graph over some neighborhood of $L_{n,2\gamma}$ of real dimension $2n\gamma$. Thus we see $\dim_C N = n\gamma$. By the analyticity of N, N is consists of non-degenerate critical points except proper real analytic subset S. So we obtain the Lagrangian immersion of N with singularity S into $T^*L_{n,2\gamma}$. By a direct calculation, we obtain the following.

Lemma 3.2. *If N has a non-degenerate critical point, then the Lagrangian immersion is given by*

$$\big(q, (L_1, L_2)\big) \in N \mapsto \big((L_1, L_2)P(\tau(q)), (L_1, L_2)\big) \in T^*L_{n,2\gamma}.$$

In particular, the obtained Lagrangian submanifold (with singularity) is a cone.

Let $T_{n,2\gamma}$ be the set of elements $L \in L_{n,2\gamma}$ such that column vectors of L generate a lattice $<L>$. Since $T_{n,2\gamma}$ contains the set of matrices of rank $= n$ with rational components which is dense in $L_{n,2\gamma}$, $T_{n,2\gamma}$ is also dense in $L_{n,2\gamma}$. Let π be the projection of $H_\gamma \times L_{n,2\gamma}$ onto $L_{n,2\gamma}$. Then So $\pi^{-1}(T_{n,2\gamma})$ is dense in $H_\gamma \times L_{n,2\gamma}$. Generally we do not know whether the intersection of N and $\pi^{-1}(T_{n,2\gamma})$ is dense in N. If the density holds, we say that N has the density property. It is easy to see that if N has a non-degenerate critical point, then N has the density property.

Theorem 3.2. *If an irreducible component N of $C(E_M)$ has a non-degenerate critical point, then we obtain*

(1) $\dim_C N = n\gamma$,

(2) *N gives a Lagrangian cone (with singularity) in $T^*L_{n,2\gamma}$,*

(3) *N has the density property.*

4. Minimal surfaces in tori

Let M be a compact Riemann surface of genus γ and $\{A_i, B_i\}$ a canonical homology basis. We define the basis $\{\psi_i\}$ of the space of holomorphic 1-forms on M such that $\int_{A_i} \psi_j = \delta_{ij}$. Then the matrix τ of size γ defined by $(\tau_{ij}) = (\int_{B_j} \psi_i)$ is called the Riemann matrix associated with M and $\{A_i, B_i\}$. It is well-known that $\tau \in H_\gamma$. Suppose that RM is the space of Riemann matrices in H_γ, $RM_{non\text{-}hyper}$ is the space of Riemann matrices for non-hyperelliptic Riemann surfaces, and RM_{hyper} is the space of Riemann matrices for hyperelliptic Riemann surfaces. Then $RM = RM_{non\text{-}hyper} \bigcup RM_{hyper}$ holds. Alfors proved that $RM_{non\text{-}hyper}$ is a $(3\gamma - 3)$-dimensional complex submanifold and RM_{hyper} is a $(2\gamma - 1)$-dimensional complex submanifold in H_γ. We shall consider the geometric meaning of $C(RM_{non\text{-}hyper})$ and $C(RM_{hyper})$. From a Rauch's result, we note the following (see [2], for example).

Lemma 4.1. *Let τ be a Riemann matrix associated with M and $\{A_i, B_i\}$. Then $\overline{A} = \overline{(A_{ij})} \in S^2_C$ is a normal vector at τ if and only if the holomorphic quadratic differential $\sum_{i,j=1}^{\gamma} A_{ij}\psi_i\psi_j$ vanishes.*

By Lemma 3.1, $(\tau, L) \in C(RM_{non\text{-}hyper})$ if and only if

$$\begin{aligned}&\operatorname{grad} E(\tau, L)\\ &= 2i\,\frac{1}{2}\overline{{}^t(L_1 + i[L_1\mathrm{Re}\,\tau - L_2](\mathrm{Im}\,\tau)^{-1})}\,\frac{1}{2}\overline{(L_1 + i[L_1\mathrm{Re}\,\tau - L_2](\mathrm{Im}\,\tau)^{-1})}\end{aligned}$$

is a normal vector of $RM_{non\text{-}hyper}$ at τ, which, together with Lemma 4.1, is equivalent to

$$\sum_{i,j=1}^{\gamma}\left({}^t(L_1{+}i[L_1\mathrm{Re}\,\tau{-}L_2](\mathrm{Im}\,\tau)^{-1})(L_1{+}i[L_1\mathrm{Re}\,\tau{-}L_2](\mathrm{Im}\,\tau)^{-1})\right)_{ji}\psi_i\psi_j = 0.$$

We denote by $\Psi\ {}^t(\psi_1\ \cdots\ \psi_\gamma)$. Then the above condition is

$${}^t\big((L_1 + i[L_1\mathrm{Re}\,\tau - L_2](\mathrm{Im}\,\tau)^{-1})\Psi\big)\big((L_1 + i[L_1\mathrm{Re}\,\tau - L_2](\mathrm{Im}\,\tau)^{-1})\Psi\big) = 0. \qquad (*)$$

We define a $\mathbf{R}^n$-valued harmonic 1-form on M as

$$(L_1, L_2)T_\tau^{-1\,t}(\mathrm{Re}\,\psi_1\ \cdots\ \mathrm{Re}\,\psi_\gamma\ \mathrm{Im}\,\psi_1\ \cdots\ \mathrm{Im}\,\psi_\gamma),$$

where

$$T_\tau = \begin{pmatrix} I & \mathrm{Re}\,\tau \\ 0 & \mathrm{Im}\,\tau \end{pmatrix}$$

and a multivalued harmonic map S of M into $\mathbf{R^n}$ by integrating along a path from a fixed point. Note that the peroid matrix of S is (L_1, L_2) by

$$\left(\int_{A_1} dS, \ldots, \int_{A_\gamma} dS, \int_{B_1} dS, \ldots, \int_{B_\gamma} dS\right) = (L_1, L_2).$$

Let $dS^{1,0}$ be the differential form of type $(1,0)$ for dS. Then

$$dS^{1,0} = \frac{1}{2}(L_1 + i[L_1 \mathrm{Re}\,\tau - L_2](\mathrm{Im}\,\tau)^{-1}) \begin{pmatrix} \psi_1 \\ \vdots \\ \psi_\gamma \end{pmatrix}.$$

So we see that $(*)$ is equivalent to ${}^t dS^{1,0} dS^{1,0} = 0$ which implies that S is weakly conformal. So $C(E_{RM_{non\text{-}hyper}}) \cap \pi^{-1}(L)$ is the space of multivalued branched minimal immersions of a non-hyperelliptic Riemann surface of genus γ into $\mathbf{R}^n$ with period L. Furthermore if $L \in T_{n,2\gamma}$, then S is a branched minimal immersion of M into the torus $\mathbf{R}^n/{<}L{>}$ with period L. We obtain the similar result for RM_{hyper}.

Theorem 4.1. *If an irreducible component N of $C(RM_{non\text{-}hyper})$ admits a non-degenerate critical point, then we see* (1) $\dim_C N = n\gamma$, (2) *N gives a Lagrangian cone in $T^*L_{n,2\gamma}$ and* (3) *N has the density property. Namely there exists a dense det of N whose point gives a branched minimal immersion of a compact Riemann surfaces of genus γ into an n-dimensional torus. We obtain the same result woth respect to a irreducible component of $C(RM_{hyper})$.*

We can give a sufficient criterion that N has a non-degenerate critical point. Suppose that $index_a$ and $nullity_a$ are the index and the nullity of the Jacobi operator associated with the second variation of the area functional at the corresponding minimal surface. Then we obtain the following [2].

Theorem 4.2. *Let S be a full minimal immersion of a Riemann surface M into a torus $\mathbf{R}^n/{<}L{>}$, where L is the period matrix associated with a canonical homology basis of M. If M is not hyperelliptic, then*

(1) *$index_a = index$ of $E_{RM_{non\text{-}hyper}}$,*
(2) *$nullity_a = n + nullity$ of $E_{RM_{non\text{-}hyper}}$.*

If M is hyperelliptic

(1) *$index_a = index$ of $E_{RM_{hyper}} + \alpha$,*
(2) *$nullity_a = n + nullity$ of $E_{RM_{hyper}} + 2\gamma - 4 - 2\alpha$,*

where α is an integer satisfying $0 \leq \alpha \leq \gamma - 2$.

We see that if S has only trivial Jacobi fields (Killing Jacobi fields), which is equivalent to $nullity_a = n$, then the corresponding critical point is non-degenerate.

Corollary 4.1. *An irreducible component containing a critical point corresponding a minimal surface with only trivial Jacobi fields satisfies* (1), (2) *and* (3) *in Theorem 4.1.*

References

1. C. Arezzo and G.P. Pirola, *On the existence of periodic minimal surfaces*, J. Alg. Geom. **8** (1999), 765–785.
2. N. Ejiri, *A Differential-Geometric Schottoky Problem, and Minimal Surfaces in Tori*, Contemporary Mathematics **308** (2002), 101–144.
3. G.P. Pirola, *The infinitesimal variation of the spin abelian differentials and periodic minimal surfaces*, Communications in Analysis and Geometry **6** (1998), 393–426.

ABOUT RESONANCES FOR SCHRÖDINGER OPERATORS WITH SHORT RANGE SINGULAR PERTURBATION

V. GEORGIEV* and N. VISCIGLIA†

Department of Mathematics, University of Pisa,
Largo Bruno Pontecorvo 5, 56127 Pisa, Italy
**E-mail: georgiev@dm.unipi.it*
†E-mail: viscigli@dm.unipi.it

We prove that there are no resonances embedded in the positive real line for the family of operators

$$-\Delta + \mathbf{i}a_j(x)\frac{\partial}{\partial x_j} + b(x),$$

under suitable short - range assumptions at infinity and possible local singularities of the coefficients $a_j(x)$ and $b(x)$.

Keywords: Resonances; Schrödinger equation; Resolvent estimates.

1. Introduction

In this work we study the existence of non-trivial solutions to the equation

$$-\Delta u + \mathbf{i}a_j(x)\frac{\partial u}{\partial x_j} + b(x)u = \lambda u, \quad x \in \mathbf{R}^n, \ \lambda > 0, \tag{1}$$

is a typical obstacle to establish dispersive properties of the time evolution group associated with the perturbed Laplace operator

$$-\Delta + \mathbf{i}a_j(x)\frac{\partial}{\partial x_j} + b(x).$$

In this work we shall obtain a simple sufficient condition for the non - existence of (non-trivial) solutions to the eigenvalue problem (1) under the additional assumption that $u(x)$ satisfies appropriate radiation conditions at infinity.

The case of vanishing magnetic field (i.e. $a_j \equiv 0$) is treated by Kato [1], Agmon [2], Alsholm-Schmidt [3], under the assumption $u \in H^2_{loc}$, $b \in L^2_{loc}$. The result in [1] concerns the solutions $u \in L^2$ and therefore triviality of solutions to (1) can be interpreted as absence of embedded eigenvalues in this case. The phenomena of nonexistence of solutions in weaker L^2 weighted spaces is known as absence of resonances and it is treated in the works [2,3]. Recently, Ionescu-Jerison [4], and Koch-Tataru [5] considered the case of rough potentials (not necessarily L^2_{loc}) and established a unique continuation principle for (1). Typical decay assumption imposed there is

$$\int (1+|x|)^{-1+\delta}|u(x)|^2 dx < \infty, \quad \delta > 0,$$

without any additional radiation condition. The results in [5] imply also the unique continuation principle for more general magnetic type perturbation.

To give a precise definition of the radiation conditions let us recall that due to the self-adjointness of the operator

$$\Delta : L^2 \supset H^2 \to L^2,$$

it is meaningful to consider the operators $(-\Delta - \lambda \pm \mathbf{i}\epsilon)^{-1} \in \mathcal{L}(L^2, H^2)$ for any $\epsilon > 0$ and hence also the existence of the following limit in a suitable sense

$$R(\lambda \pm \mathbf{i}0) = \lim_{\epsilon \to 0} (-\Delta - \lambda \mp \mathbf{i}\epsilon)^{-1}, \tag{2}$$

(this is called in the literature limiting absorption principle for the free resolvent, see [6]). It is well-known (see chapter 14, [6]) that the operators $R(\lambda \pm \mathbf{i}0)$, $\lambda > 0$, act from the Banach space $\mathcal{B}$, that consists of measurable functions $f \in L^2_{loc}$ with finite norm

$$\|f\|_{\mathcal{B}} = \left(\int_{|x|<1} |f(x)|^2 dx\right)^{\frac{1}{2}} + \sum_{k=0}^{\infty} 2^{\frac{k}{2}} \left(\int_{2^k<|x|<2^{k+1}} |f(x)|^2 dx\right)^{\frac{1}{2}} \tag{3}$$

into its dual space $\mathcal{B}^*$ (see (12) for the definition of the norm in this space).

One can verify (see Propositions 2.1 and 2.2) that for any $\lambda > 0$, we have

$$R(\lambda \pm \mathbf{i}0) : \mathcal{B} + L^{2n/(n+2)}_{comp} \longrightarrow \mathcal{W}^*, \tag{4}$$

where $\mathcal{W}^*$ is the space of all functions in $\mathcal{B}^*$ having gradient in $\mathcal{B}^*$, endowed with the following norm:

$$\|f\|^2_{\mathcal{W}^*} = \|f\|^2_{\mathcal{B}^*} + \|\nabla f(x)\|^2_{\mathcal{B}^*}. \tag{5}$$

Our goal is to show that any solution to (1) that satisfies the additional outgoing (respectively incoming) radiation condition

$$u \in \operatorname{Im} R(\lambda + \mathbf{i}0) \quad (\text{respectively } u \in \operatorname{Im} R(\lambda - \mathbf{i}0)) \tag{6}$$

is trivial, i.e. $u \equiv 0$.

To be more precise, we shall say that $u \in \operatorname{Im} R(\lambda + \mathbf{i}0)$ if there exists an element $f \in \mathcal{B} + L^{2n/(n+2)}_{comp}$, so that

$$u = R(\lambda + \mathbf{i}0) f.$$

Let us describe next the main assumptions that we require on $a_j(x)$ and $b(x)$:

$$b(x), a_j(x) \in \mathbf{R}, \tag{7}$$
$$|a_j(x)| \le M|x|^{-1-\epsilon_0}, |b(x)| \le M|x|^{-1-\epsilon_0} \quad \forall x \in \mathbf{R}^n \text{ s.t. } |x| > D,$$

where $M, D, \epsilon_0 > 0$ are suitable constants,

$$b(x) \in L^{\frac{n}{2}}_{loc}(\mathbf{R}^n), \quad a_j(x) \in W^{1,n}_{loc}(\mathbf{R}^n). \tag{8}$$

Finally, we fix a Coulomb gauge for the magnetic potential a_j, i.e.

$$\sum_{j=1}^{n} \frac{\partial a_j}{\partial x_j}(x) = 0. \tag{9}$$

It is easy to verify that if $u \in \mathcal{W}^*$, then

$$p(x, D_x)u = -\mathbf{i}a_j(x)\frac{\partial u}{\partial x_j} - b(x)u \in \mathcal{B} + L^{2n/(n+2)}_{comp}, \tag{10}$$

so if $u \in \mathcal{W}^*$ is a solution to

$$R(\lambda \pm \mathbf{i}0)\, p(x, D_x)u = u, \quad \lambda > 0, \tag{11}$$

then u satisfies also (1) in distribution sense.

Adding the outgoing radiation condition in (6), one can see that any solution to (1) is also a solution to (11) (with sign +). Similarly, the incoming radiation condition in (6) and (1) yield (11) (with sign −). For this, since now on our purpose is to study the non - existence of non-trivial solutions to (11), taking $u \in \mathcal{W}^*$.

We can now state the main result of this paper.

Theorem 1.1. *Assume $n \ge 3$, and $a_j(x)$, $b(x)$ satisfy* (7), (8) *and* (9). *Then for any $\lambda > 0$, the operator $R(\lambda \pm \mathbf{i}0)\, p(x, D_x)$ is a linear endomorphism of the Banach space $\mathcal{W}^*$. If moreover $u \in \mathcal{W}^*$ satisfies* (11), *then $u \equiv 0$.*

Remark 1.1. The solutions to (11) that belong to $\mathcal{W}^*$ are called often resonances. Hence theorem 1.1 can be interpreted as a result on the absence of resonances under short range assumptions on $a_j(x)$ and $b(x)$.

Remark 1.2. Let us notice that theorem 1.1 is false if we consider general solutions to (1). In fact the nontrivial function

$$u(x) = \int_{|\xi|=\sqrt{\lambda}} e^{\mathbf{i}x\cdot\xi}\, d\sigma_{\sqrt{\lambda}},$$

where $d\sigma_{\sqrt{\lambda}}$ denotes the natural measure on the sphere $\{|\xi| = \sqrt{\lambda}\}$, satisfies (1) in the free case, and moreover a stationary phase argument shows that $u \in \mathcal{W}^*$.

Remark 1.3. The assertion of theorem 1.1 is a key point in order to extend the limiting absorption principle to the first order perturbations of the free laplacian. In particular theorem 1.1 guarantees that the operator $Id - R(\lambda \pm \mathbf{i}0)\, p(x, D_x) \in \mathcal{L}(\mathcal{W}^*, \mathcal{W}^*)$ is injective and hence, via the Fredholm theory, it is surjective.

Next we fix some notations that will be used in the sequel.

Notations. If $A \subset \mathbf{R}^n$ is any measurable set then we shall denote by $I_A(x)$ the function defined as follows:

$$I_A(x) = 1 \quad \forall x \in A \quad \text{and} \quad I_A(x) = 0 \quad \forall x \in \mathbf{R}^n \setminus A;$$

for any $\rho > 0$ we shall write

$$S_\rho = \{\, x \in \mathbf{R}^n \mid |x| = \rho \,\}$$

and

$$B_\rho = \{\, x \in \mathbf{R}^n \mid |x| < \rho \,\};$$

the measure $d\sigma_\rho$ is the natural measure on the sphere S_ρ; for any function $f(x)$ we shall denote by *supp* f its support; we shall denote by $W^{k,p}_{loc}$ the usual localized Sobolev spaces and in the case $k = 0$ we shall denote by L^p_{loc} the localized Lebesgue spaces, we shall also write $W^{k,2} = H^k$; for any $\lambda, \epsilon > 0$ we shall use the notation

$$R(\lambda \pm \mathbf{i}\epsilon) = (-\Delta - \lambda \mp \mathbf{i}\epsilon)^{-1} \in \mathcal{L}(L^2, H^2);$$

if X, Y are Banach spaces then $\mathcal{L}(X, Y)$ shall denote the space of the linear and continuous operators between X and Y; the space $\mathcal{B}$ is defined by (3); the space $\mathcal{B}^*$, dual to the space $\mathcal{B}$, is a Banach space with norm

$$\|f\|^2_{\mathcal{B}^*} = \sup_{\rho \geq 1} \rho^{-1} \int_{|x|<\rho} |f(x)|^2\, dx; \tag{12}$$

along the paper we shall make use also of the space $\mathcal{B}_0^*$ defined as the functions $f \in \mathcal{B}^*$ such that

$$\lim_{\rho \to +\infty} \rho^{-1} \int_{|x|<\rho} |f(x)|^2 \, dx = 0; \tag{13}$$

finally, the space $\mathcal{W}^*$ is defined as the set of functions in $\mathcal{B}^*$ having finite norm (5).

2. On the continuity of the operator $R(\lambda \pm \mathrm{i}0)$

The space

$$\mathcal{X} = \mathcal{B} + L^{2n/(n+2)}_{comp},$$

appeared in the introduction seems to be very natural since the following property

$$p(x, D) : \mathcal{W}^* \to \mathcal{B} + L^{2n/(n+2)}_{comp}$$

is guaranteed by the assumptions (7), (8). Hence in order to justify the composition operator $R(\lambda \pm \mathrm{i}0)\, p(x, D_x)$ in the basic equation (11), we have to verify the property (4).

Typically, the elements $f \in \mathcal{X}$ are measurable functions $f : \mathbf{R}^n \to \mathbf{C}$, such that

$$I_D(x) f \in L^{\frac{2n}{n+2}} \quad \text{and} \quad (1 - I_D(x)) f \in \mathcal{B}$$

where $I_D(x)$ denotes the characteristic function of the set $\{|x| < D\}$ and $D > 0$ is a suitable number. Notice that for any fixed $D > 0$ we can introduce the norm

$$\|f\|_{\mathcal{X}} = \|I_D(x) f\|_{L^{\frac{2n}{n+2}}} + \|(1 - I_D(x)) f\|_{\mathcal{B}}, \tag{14}$$

that depends on $D > 0$. In the sequel, when it is not better specified, it will be assumed that the norm $\|\cdot\|_{\mathcal{X}}$ is computed with respect to a fixed parameter $D > 0$ that in general will be clear from the context.

Let us recall the following basic result that includes in himself the definition of the operator $R(\lambda \pm \mathrm{i}0)$.

Proposition 2.1. *The following limit is well-defined*

$$\lim_{\epsilon \to 0} R(\lambda \pm \mathrm{i}\epsilon) f = R(\lambda \pm \mathrm{i}0) f \text{ in } \mathcal{W}^*$$

for any $f \in \mathcal{B}$ and for any $\lambda > 0$. Moreover the following estimate holds

$$\|R(\lambda \pm \mathrm{i}0) f\|_{\mathcal{W}^*} \le C \|f\|_{\mathcal{B}} \quad \forall f \in \mathcal{B},$$

where $C = C(\lambda) > 0$.

The aim of next proposition is to study some other continuity properties of the operators $R(\lambda \pm \mathrm{i}0)$.

Proposition 2.2. *Let D be a positive number, $\lambda > 0$, $f \in L^{\frac{2n}{n+2}}$ and $\operatorname{supp} f \subset \{|x| < D\}$. Then there exists a constant $C = C(\lambda) > 0$, such that*

$$\|R(\lambda \pm \mathrm{i}0)f\|_{\mathcal{B}^*} + \|\nabla R(\lambda \pm \mathrm{i}0)f\|_{\mathcal{B}^*} \leq C\|f\|_{L^{\frac{2n}{n+2}}}.$$

As a consequence of the previous propositions we get the following

Corollary 2.1. *For any $\lambda > 0$ the following estimate holds:*

$$\|R(\lambda \pm \mathrm{i}0)f\|_{\mathcal{W}^*} \leq C\|f\|_{\mathcal{X}} \quad \forall f \in \mathcal{X},$$

where $C = C(\lambda)$ and the norm in $\mathcal{X}$ is the one defined in (14).

3. Some preliminary computations

In the sequel the trace operator will play a relevant role, hence we recall the following basic result.

Proposition 3.1. *For any $1 \leq p < n$ and $\rho > 0$ there exists $C(\rho, p) > 0$ such that*

$$\left(\int_{S_\rho} |f|^{\frac{p(n-1)}{n-p}}\, d\sigma_\rho\right)^{\frac{n-p}{p(n-1)}} \leq C(\rho, p)\|f\|_{W^{1,p}(B_\rho)} \quad \forall f \in C_0^\infty(\mathbf{R}^n). \tag{15}$$

In the limit case $p = n$ we get the following estimate:

$$\left(\int_{S_\rho} |f|^q\, d\sigma_\rho\right)^{\frac{1}{q}} \leq C(\rho, q)\|f\|_{W^{1,n}(B_\rho)} \quad \forall f \in C_0^\infty(\mathbf{R}^n), \tag{16}$$

for any $1 \leq q < \infty$. In particular by a density argument it is meaningful to define restriction on S_ρ for any $f \in \dot{W}^{1,p}(B_\rho)$ with $1 \leq p \leq n$.

In next proposition the restriction of functions on spheres will be intended in the trace sense.

Proposition 3.2. *Assume that $a_j(x)$ and $b(x)$ are as in theorem 1.1 and $u \in \mathcal{W}^*$ satisfies* (1). *Then we have $u \in W^{2,\frac{2n}{n+2}}_{loc}$ and moreover*

$$\mathcal{I}m \int_{S_\rho} (\bar{u}\, \partial_{|x|} u)\, d\sigma_\rho = \frac{1}{2}\int_{S_\rho} |u|^2 (a_j(x)\nu_j(x))\, d\sigma_\rho \quad \forall \rho > 0. \tag{17}$$

Proof. *Regularity of* u: By assumption u satisfies the following partial differential equation

$$-\Delta u - \lambda u + \mathbf{i} a_j(x) \frac{\partial u}{\partial x_j} + b(x) u = 0, \quad x \in \mathbf{R}^n, \ \lambda > 0,$$

that can be rewritten as follows:

$$-\Delta u = f,$$

where

$$f = \lambda u - \mathbf{i} a_j(x) \frac{\partial u}{\partial x_j} - b(x) u. \tag{18}$$

Notice that the assumptions done imply

$$u \in W^{1,2}_{loc}(\mathbf{R}^n), \quad a_j(x) \in L^n_{loc}(\mathbf{R}^n) \quad \text{and} \quad b(x) \in L^{\frac{n}{2}}_{loc}(\mathbf{R}^n).$$

By using the Sobolev embedding we get $u \in L^{\frac{2n}{n-2}}_{loc}$, that can be combined with the Hölder inequality in order to deduce $b(x)u \in L^{\frac{2n}{n+2}}_{loc}$.

Due again to the Hölder inequality, and recalling that $\nabla u \in L^2_{loc}$ we have that $a_j(x)\frac{\partial u}{\partial x_j} \in L^{\frac{2n}{n+2}}_{loc}$. In particular, looking at (18) and recalling that $u \in L^2_{loc} \subset L^{\frac{2n}{n+2}}_{loc}$ we get

$$-\Delta u = f \in L^{\frac{2n}{n+2}}_{loc}$$

that in turn, due to classical regularity results, implies

$$u \in W^{2,\frac{2n}{n+2}}_{loc}. \tag{19}$$

Proof of (17): Let us fix $\rho > 0$ and let u_k be a sequence such that

$$u_k \in C^\infty_0(\mathbf{R}^n) \quad \text{and} \quad u_k \to u \text{ in } W^{2,\frac{2n}{n+2}}(B_\rho). \tag{20}$$

Notice that the following identities are satisfied:

$$-\int_{B_\rho} \bar{u}_k \, \Delta u_k \, dx = \int_{B_\rho} |\nabla u_k|^2 dx - \int_{S_\rho} (\bar{u}_k \, \partial_{|x|} u_k) \, d\sigma_\rho \tag{21}$$

and

$$\mathcal{I}m \int_{B_\rho} \mathbf{i} a_j(x) \bar{u}_k \frac{\partial u_k}{\partial x_j} \, dx = \frac{1}{2} \int_{B_\rho} a_j(x) \frac{\partial |u_k|^2}{\partial x_j} \, dx \tag{22}$$

$$= -\frac{1}{2} \int_{B_\rho} |u_k|^2 \frac{\partial a_j}{\partial x_j} \, dx + \frac{1}{2} \int_{S_\rho} |u_k|^2 a_j(x) \nu_j(x) \, d\sigma_\rho$$

$$= \frac{1}{2} \int_{S_\rho} |u_k|^2 a_j(x) \nu_j(x) \, d\sigma_\rho$$

where $\nu_j(x) = \frac{x_j}{|x|}$ and where we have exploited the assumption $\sum_{j=1}^n \frac{\partial a_j}{\partial x_j} = 0$. On the other hand, due to (20) we get

$$\Delta u_k \to \Delta u \text{ in } L^{\frac{2n}{n+2}}(B_\rho)$$

that in turn, due to the Sobolev embedding, implies

$$u_k \to u \text{ in } L^{\frac{2n}{n-2}}(B_\rho).$$

By combining these facts we get:

$$\lim_{k\to\infty} \int_{B_\rho} u_k\, \Delta u_k\, dx = \int_{B_\rho} u\, \Delta u\, dx. \tag{23}$$

Due again to the Sobolev embedding and (20) we deduce $\nabla u_k \to \nabla u$ in $L^2(B_\rho)$ that gives

$$\lim_{k\to\infty} \int_{B_\rho} |\nabla u_k|^2 dx = \int_{B_\rho} |\nabla u|^2 dx. \tag{24}$$

Finally by combining the trace inequality (15) with (20), we deduce that the restrictions of u_k and $\partial_{|x|} u_k$ on the spheres S_ρ converge to the restrictions (in trace sense) of u and $\partial_{|x|} u$ respectively in the spaces $L^{\frac{2(n-1)}{n-2}}(S_\rho)$ and $L^{\frac{2(n-1)}{n}}(S_\rho)$ and hence

$$\lim_{k\to\infty} \int_{S_\rho} (\bar{u}_k\, \partial_{|x|} u_k)\, d\sigma_\rho = \int_{S_\rho} (\bar{u}\, \partial_{|x|} u)\, d\sigma_\rho. \tag{25}$$

By combining (21), (23), (24) and (25) we get:

$$-\int_{B_\rho} \bar{u}\, \Delta u\, dx = \int_{B_\rho} |\nabla u|^2 dx - \int_{S_\rho} (\bar{u}\, \partial_{|x|} u)\, d\sigma_\rho. \tag{26}$$

In similar way we can take the limit in (22) in order to get

$$\mathcal{I}m \int_{B_\rho} \mathbf{i} a_j(x) \bar{u} \frac{\partial u}{\partial x_j}\, dx = \frac{1}{2} \int_{S_\rho} |u|^2 (a_j(x) \nu_j(x))\, d\sigma_\rho \tag{27}$$

where $\nu_j(x) = \frac{x_j}{|x|}$. Hence if we multiply (1) by $\bar{u}$, we integrate on B_ρ and we take the imaginary part, then due to (26) and (27) we get

$$\mathcal{I}m \int_{S_\rho} (\bar{u}\, \partial_{|x|} u)\, d\sigma_\rho = \frac{1}{2} \int_{S_\rho} |u|^2 (a_j(x) \nu_j(x))\, d\sigma_\rho. \tag{28}$$

□

4. Proof of theorem 1.1

The basic result that we shall need is the following

Proposition 4.1. *Assume that u, $a_j(x)$ and $b(x)$ are as in theorem 1.1, then there exists $\delta > 0$ such that*

$$\int_{\mathbf{R}^n} |u|^2 \langle x\rangle^{-1+\delta} dx < \infty. \tag{29}$$

The idea of the proof of proposition 4.1 is based on the following proposition 4.2, where

$$\widehat{f}(\xi) = \int_{\mathbf{R}^n} e^{-\mathbf{i}x\cdot\xi} f(x)\, dx.$$

Let us recall that a possible proof of next proposition can be found in [7], however in order to make the paper self - contained we present a simplified argument.

Proposition 4.2. *The following asymptotic formulas are satisfied for any $f \in \mathcal{B}$:*

$$R(1 \pm \mathbf{i}0) f(x) = \mp \sqrt{\frac{\pi}{2}} \frac{e^{\mp \mathbf{i}\frac{(n+1)\pi}{4}}}{(2\pi)^{\frac{n}{2}}} \frac{e^{\pm \mathbf{i}|x|}}{|x|^{(n-1)/2}} \widehat{f}\left(\pm \frac{x}{|x|}\right) + \epsilon(x), \tag{30}$$

$$\partial_{|x|} R(1 \pm \mathbf{i}0) f(x) = \mp \mathbf{i} \sqrt{\frac{\pi}{2}} \frac{e^{\mp \mathbf{i}\frac{(n+1)\pi}{4}}}{(2\pi)^{\frac{n}{2}}} \frac{e^{\pm \mathbf{i}|x|}}{|x|^{(n-1)/2}} \widehat{f}\left(\pm \frac{x}{|x|}\right) + \delta(x). \tag{31}$$

where $\epsilon(x)$, $\delta(x) \in \mathcal{B}_0^*$.

Proof. Notice that the space $\mathcal{B}_0^*$ is a closed subspace of $\mathcal{B}^*$. Due to this fact, to the density of C_0^∞ in $\mathcal{B}$ and to proposition 2.1 it is enough to show (30) and (31) for any $f \in C_0^\infty$.

Next we shall focus on the proof of (30) since the proof of (31) is similar and we shall treat the operator $R(1+\mathbf{i}0)$ (in fact $R(1-\mathbf{i}0)$ can be treated in similar way). Let us recall that $R(1+\mathbf{i}0)$ can be expressed by convolution with the appropriate kernel.

Due to the decay property it is easy to prove that

$$\left(\frac{e^{\mathbf{i}|x|} a_+(|x|)}{|x|^{\frac{n-1}{2}}}\right) * f \in \mathcal{B}_0^* \quad \forall f \in C_0^\infty,$$

and hence it is sufficient to show that

$$\int_{\mathbf{R}^n} \frac{e^{\mathbf{i}|x-y|}}{|x-y|^{\frac{n-1}{2}}} f(y)\, dy - \int_{\mathbf{R}^n} \frac{e^{\mathbf{i}|x|} e^{-\mathbf{i}\frac{x}{|x|}\cdot y}}{|x|^{\frac{n-1}{2}}} f(y)\, dy \in \mathcal{B}_0^* \quad \forall f \in C_0^\infty, \tag{32}$$

in order to deduce (30) for $f \in C_0^\infty$. Let us fix $R > 0$ such that

$$\operatorname{supp} f \subset B_R \tag{33}$$

and let us introduce the function $\phi(x, y)$ defined as follows:

$$e^{\mathrm{i}(|x| - \frac{x}{|x|}\cdot y)} = e^{\mathrm{i}|x-y|} + \phi(x, y) \quad \forall x \neq 0,\ y \in \mathbf{R}^n. \tag{34}$$

Then the proof of (32) will follow from the following facts:

$$\int_{\mathbf{R}^n} e^{\mathrm{i}|x-y|} \Big(\frac{1}{|x-y|^{\frac{n-1}{2}}} - \frac{1}{|x|^{\frac{n-1}{2}}} \Big) f(y)\, dy \in \mathcal{B}_0^* \tag{35}$$

and

$$\int_{\mathbf{R}^n} \phi(x, y) f(y) \frac{dy}{|x|^{\frac{n-1}{2}}} \in \mathcal{B}_0^* \tag{36}$$

where ϕ is defined as in (34).

Notice that the following inequality follows from the mean value theorem applied to the one variable function $t \rightarrow \frac{1}{t^{\frac{n-1}{2}}}$:

$$\left| \frac{1}{|x-y|^{\frac{n-1}{2}}} - \frac{1}{|x|^{\frac{n-1}{2}}} \right|$$
$$\leq C \max \left\{ \frac{1}{|x|^{\frac{n+1}{2}}}, \frac{1}{|x-y|^{\frac{n+1}{2}}} \right\} ||x-y| - |x|| \quad \forall x, y \in \mathbf{R}^n$$

and hence by changing in a suitable the constant $C > 0$, we get

$$\left| \frac{1}{|x-y|^{\frac{n-1}{2}}} - \frac{1}{|x|^{\frac{n-1}{2}}} \right| \leq \frac{C}{|x|^{\frac{n+1}{2}}} \quad \forall y \in B_R \text{ and } |x| > 2R$$

where $R > 0$ is the constant given in (33). It is now easy to deduce (35).

In order to show (36) let us notice that due to (34) we get:

$$\begin{aligned} |\phi(x, y)| &= \left| \int_{|x-y|}^{|x| - \frac{x}{|x|}\cdot y} e^{\mathrm{i}t}\, dt \right| \\ &\leq \left| |x-y| - |x| + \frac{x}{|x|} \cdot y \right| \quad \forall x, y \in \mathbf{R}^n, x \neq 0. \end{aligned} \tag{37}$$

In particular if we prove

$$\left| |x-y| - |x| + \frac{x}{|x|} \cdot y \right| \rightarrow 0 \quad \text{as} \quad |x| \rightarrow \infty, \tag{38}$$

where the limit is uniform for $y \in B_R$, then by using (37) we get

$$\int_{\mathbf{R}^n} \phi(x, y) f(y) \frac{dy}{|x|^{\frac{n-1}{2}}} = \frac{o(1)}{|x|^{\frac{n-1}{2}}},$$

where $o(1) \to 0$ as $|x| \to \infty$, that in turn implies (36).

Next we shall prove (38). Let us consider the family of functions

$$F_x : B_R \ni y \to |x - y| \in \mathbf{R}$$

where $|x| > 2R$. Since we are assuming $|x| > 2R$, the function F_x are of class C^∞ on the domain B_R, then we can write its Taylor development at the first order in the origin $y = 0$ for any fixed x such that $|x| > 2R$:

$$|x - y| = |x| - \frac{x}{|x|} \cdot y + r(x, y). \tag{39}$$

Let us recall that the reminder $r(x, y)$ can be expressed in terms of the second derivatives of F_x. An explicit computation of these derivatives allows us to deduce

$$|r(x, y)| \to 0 \quad \text{as} \quad |x| \to \infty \text{ and } y \in B_R.$$

The proof of (38) follows then by (39). □

Acknowledgments

The authors are partially supported by Contract PRIN-2005010482 (2005-2007) with Italian Ministry of Education, University and Research and the INDAM Contract "Mathematical modelling and numerical analysis of quantum systems with applications to nanosciences".

References

1. T. Kato, Comm. Pure Appl. Math. **12** (1959), 403.
2. S. Agmon, Ann. Scuola Norm. Sup. Pisa Cl. Sci. **2** (1975), 151.
3. P. Alsholm and G. Schmidt, Arch. Rational Mech. Anal. **40** (1970/1971), 281.
4. A. Ionescu and D. Jerison, GAFA **13** (2003), 1029.
5. H. Koch and D. Tataru, Comm. Math. Phys. **267** (2006), 419.
6. L. Hormander, *The Analysis of Linear Partial Differential Operators II; Differential Operators with Constant coefficients*, Springer-Verlag, Berlin, Heidelberg, Tokyo, New York, 1983.
7. Y. Gatel and D. Yafaev, Ann. Inst. Fourier, Grenoble, **45** (1999), 1581.

SOLITON EQUATIONS WITH DEEP REDUCTIONS. GENERALIZED FOURIER TRANSFORMS

V. GERDJIKOV[†], N. KOSTOV[‡] and T. VALCHEV[†]

[†]*Institute for Nuclear Research and Nuclear Energy,*
Bulgarian Academy of Sciences,
72 Tzarigradsko chaussèe, 1784 Sofia, Bulgaria

[‡]*Institute of Electronics,*
Bulgarian Academy of Sciences,
72 Tzarigradsko chaussèe, 1784 Sofia, Bulgaria

We consider the application of the inverse scattering method to a special nonlinear equation related to $\mathfrak{so}(5)$ with a $\mathbb{Z}_4$ reduction imposed on it. Typical representative of these equations is the affine Toda field theory (ATFT). Here we outline the spectral properties of the Lax operator and prove the completeness relation for its 'squared solutions'. They are used to construct the action-angle variables of the corresponding ATFT and demonstrate the special properties of their hierarchy of Hamiltonian structures.

Keywords: Soliton equations; Deep reductions; Hamiltonian formalism.

1. Introduction

The reduction group technique introduced by Mikhailov [1] gives an effective tool to derive new integrable equations from generic ones. Integrable equations obtained by imposing certain symmetry conditions (or reductions) on generic equations form a class of significant importance. A common case is the one related to a $\mathbb{Z}_2$-reduced equation: nonlinear Schrödinger equation [2] (NLS), N-wave equation [2,3], chiral field models [4] etc. Deep reductions, i.e. reductions connected with more complicated groups — $\mathbb{Z}_n$, $n > 2$ or the dihedral group $\mathbb{D}_n$, are of practical interest too. They appear to be natural for some models such as the affine Toda field theory (ATFT), NLS etc.

In this paper we consider a special case of an integable nonlinear evolution equation (NLEE), an affine Toda field model, connected with $\mathfrak{so}(5)$ algebra and obeying a $\mathbb{Z}_4$ symmetry condition. Affine Toda field models are widely investigated, see for example [5–7]. Our aim in this paper is to discuss the application of the inverse scattering method (IST) to it. We stress on the squared solutions and derive their completeness relations.

2. Preliminaries

In this section we are going to outline the basic features of the scattering theory in the case when a $\mathbb{Z}_4$ reduction is imposed on the Lax operators. The scattering theory of $\mathbb{Z}_h$-reduced Lax operators connected with a generalized Zakharov-Shabat system was developed by Beals and Coifman [8] for $\mathfrak{sl}(n)$ and it was later generalized by Gerdjikov and Yanovski [9] for an arbitrary simple Lie algebra.

It is said that a $1+1$ differential equation is S-integrable if it admits a zero curvature representation

$$[L(\lambda), M(\lambda)] = 0, \tag{1}$$

holding identically with respect to the spectral parameter λ. We will restrict ourselves with considering Lax operators $L(\lambda)$ and $M(\lambda)$ in the form

$$L(\lambda)\psi(x,t,\lambda) \equiv (i\partial_x - i\partial_x q(x,t) - \lambda J)\psi(x,t,\lambda) = 0, \tag{2}$$

$$M(\lambda)\psi(x,t,\lambda) \equiv (i\partial_t - I(x,t)/\lambda)\psi(x,t,\lambda) = 0, \tag{3}$$

where J is a constant element of the Cartan subalgebra $\mathfrak{h}$ of $\mathfrak{so}(5)$. The potential functions $q_x(x,t)$ (the subscript means differentiation with respect to x) and $I(x,t)$ are linear combinations of its Weyl generators, see eqs. (9), (10) below. The corresponding ATFT model looks as follows

$$\partial^2_{xt} q_1 = e^{q_2 - q_1} - e^{q_1 + q_2}, \tag{4}$$

$$\partial^2_{xt} q_2 = -e^{q_2 - q_1} + e^{-q_2} - e^{q_1 + q_2}. \tag{5}$$

Reminder. We shall use a basis in $\mathbb{C}^5$ such that the matrix elements of the metric S that is involved in the definition of $\mathfrak{so}(5)$

$$X \in \mathfrak{so}(5) \quad \Leftrightarrow \quad X^T S + SX = 0,$$

are given by

$$S_{ij} = (-1)^{i+1}\delta_{i\,6-j}, \qquad i,j = 1,\ldots,5.$$

Thus the Cartan subalgebra $\mathfrak{h}$ consists of diagonal matrices of the form

$$\mathfrak{a} \in \mathfrak{h} \Rightarrow \mathfrak{a} = a^1 H_1 + a^2 H_2,$$

where $H_1 = \mathrm{diag}\,(1,0,0,0,-1)$ and $H_2 = \mathrm{diag}\,(0,1,0,-1,0)$. The root system Δ of $\mathfrak{so}(5)$ is formed by the roots

$$\Delta = \{\pm e_1 \pm e_2,\ \pm e_1,\ \pm e_2\},$$

where $\{e_1, e_2\}$ is an orthonormal basis in $\mathbb{R}^2$. There exist many ways to introduce an ordering in Δ. We shall use a "canonical" ordering which leads to the following set of all positive roots

$$\Delta^+ = \{e_1 - e_2,\ e_1 + e_2,\ e_1,\ e_2\}.$$

The Lie algebra $\mathfrak{so}(5)$ has 2 simple roots: $\alpha_1 = e_1 - e_2$ and $\alpha_2 = e_2$. The maximal root is $\alpha_0 = e_1 + e_2$, the Coxeter number is $h = 4$. We require that the Cartan-Weyl generators H_j and E_α are normalized with respect to Killing form, i.e.

$$\langle E_\alpha, E_\beta \rangle \equiv \frac{1}{2}\mathrm{tr}\,(E_\alpha E_\beta) = \delta_{\alpha,-\beta}, \qquad \langle H_j, H_k \rangle = \delta_{jk}, \tag{6}$$

where $\delta_{\alpha,\beta}$ denotes the Kronecker symbol.

The second Casimir operator P which has the important property

$$P(A \otimes B) = (B \otimes A)P, \qquad \forall\, A, B \in SO(5) \tag{7}$$

looks as follows

$$P = \sum_{j=1,2} H_j \otimes H_j + \sum_{\alpha\in\Delta} E_\alpha \otimes E_{-\alpha}. \tag{8}$$

We recommend the classical book by Helgasson[10] for more details on Lie algebra and Lie group theory. □

Remark 2.1. If we choose $L^{(0)}$ and $M^{(0)}$ with $q_x \equiv q_x^{(0)} = q_{1,x}H_1 + q_{2,x}H_2$, $J \equiv J^{(0)} = E_{\alpha_1} + E_{\alpha_2} + E_{-\alpha_0}$ and

$$I(x,t) \equiv I^{(0)}(x,t) = e^{-(\alpha_1,\vec{q})}E_{-\alpha_1} + e^{-(\alpha_2,\vec{q})}E_{-\alpha_2} + e^{(\alpha_0,\vec{q})}E_{-\alpha_0},$$

where $\vec{q} = q_1 e_1 + q_2 e_2$, we recover one of the well known Lax representations for the $\mathfrak{so}(5)$ ATFT.

However, analyzing the spectral properties of L it will be more convenient to consider an equivalent Lax pair:

$$L = U_0^{-1} L^{(0)} U_0, \qquad M = U_0^{-1} M^{(0)} U_0,$$

where $U_0 \in SO(5)$ is a constant matrix diagonalizing $J^{(0)}$. Thus we get:

$$J \equiv U_0^{-1} J^{(0)} U_0 = H_1 + iH_2, \qquad I(x,t) = U_0^{-1} I^{(0)}(x,t) U_0, \tag{9}$$

$$\begin{aligned} q_x \equiv U_0^{-1}(q_{1,x}H_1 + q_{2,x}H_2)U_0 &= -\frac{1}{2} q_{1,x}(iE_{e_1} + E_{e_2} - iE_{-e_1} + E_{-e_2}) \\ &- \frac{1}{2} q_{2,x}(E_{e_1-e_2} - E_{e_1+e_2} + E_{-e_1+e_2} - E_{-e_1-e_2}). \end{aligned} \tag{10}$$

Both sets of Lax operators possess $\mathbb{Z}_4$ reduction group in the sense of Mikhailov.[1]In the first case with $J^{(0)}$ the automorphism C is realized by the element $C_0 = \exp(-i\pi(H_1 + H_2/2))$ of the Cartan subgroup. In the second case with $J = H_1 + iH_2$, C is the Coxeter element of the Weyl group: $C = S_{\alpha_1} S_{\alpha_2}$.

Let us consider the action of $\mathbb{Z}_4$ in the set of fundamental solutions $\{\psi(x,t,\lambda)\}$ given by

$$C\psi(x,t,\kappa(\lambda))C^{-1} = \tilde{\psi}(x,t,\lambda), \tag{11}$$

where $C = C_0$ satisfies $C_0^4 = \mathbb{1}$ and $\kappa : \lambda \to \omega\lambda$, $\omega = e^{i\pi/2}$. We require that the Lax representation $[L(\lambda), M(\lambda)] = 0$ remains invariant under the action of G_R. As a result the potentials $U(x,t,\lambda)$ and $V(x,t,\lambda)$ must satisfy the following symmetry conditions [1]:

$$C(U(x,t,\kappa(\lambda)))C^{-1} = U(x,t,\lambda), \;\; C(V(x,t,\kappa(\lambda)))C^{-1} = V(x,t,\lambda). \tag{12}$$

The continuous spectrum of L consists of a bunch of 8 rays l_ν closing equal angles $\pi/4$ (see the figure below). With each ray l_ν it is associated a $\mathfrak{sl}(2)$ subalgebra. It is the algebra $\{E_\alpha, E_{-\alpha}, H_\alpha\}$ generated by the root α to fulfill the condition [9]:

$$\delta_\nu \equiv \{\alpha \in \Delta \mid \operatorname{Im} \lambda\alpha(J) = 0, \; \forall \lambda \in l_\nu\}, \quad \nu = 1, \dots, 8. \tag{13}$$

Obviously, the following relations between these root subsystems hold true

$$\delta_\nu = \delta_{\nu+4}, \qquad \Delta = \bigcup_{\nu=1}^{4} \delta_\nu .$$

The list of roots β_ν and the correspondent rays l_ν are shown in the table below

ray l_ν	$l_1,\ l_5$	$l_2,\ l_6$	$l_3,\ l_7$	$l_4,\ l_8$
root β_ν	$\pm e_1$	$\pm(e_1 - e_2)$	$\pm e_2$	$\pm(e_1 + e_2)$

.

The condition (13) allows one to introduce ordering in Ω_ν

$$\Delta_\nu^\pm \equiv \{\alpha \in \Delta \mid \operatorname{Im} \lambda\alpha(J) \lessgtr 0, \; \forall \lambda \in \Omega_\nu\}. \tag{14}$$

We shall use the notation $\delta_\nu^\pm = \Delta_\nu^\pm \cap \delta_\nu$ as well. One can easily see that the following symmetries hold

$$\Delta_{\nu+4}^\pm = \Delta_\nu^\mp, \qquad \delta_{\nu+4}^\pm = \delta_\nu^\mp . \tag{15}$$

Following the approach in [9] we are able to construct fundamental solutions $\eta^\nu(x,t,\lambda)$ of the auxiliary linear problem

$$i\partial_x \eta^\nu(x,t,\lambda) + q(x,t)\eta^\nu(x,t,\lambda) - \lambda[J, \eta^\nu(x,t,\lambda)] = 0 \tag{16}$$

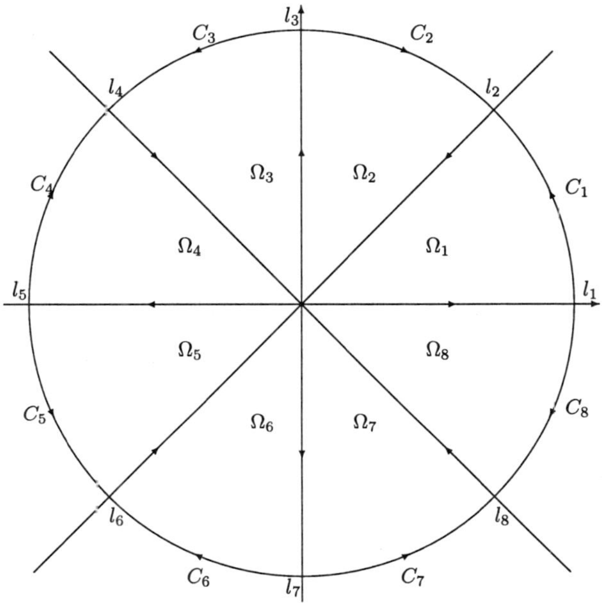

Fig. 1. The contours of integration $\gamma_\nu = l_\nu \cup C_\nu \cup l_{\nu+1}$.

to possess analytic properties in each sector Ω_ν. By analogy with the nonreduced situation when the FAS $\eta^\pm(x,t,\lambda)$ are solutions of a local Riemann-Hilbert problem on $\mathbb{R}$ the FAS $\eta^\nu(x,t,\lambda)$ and $\eta^{\nu-1}(x,t,\lambda)$ are interrelated via a sewing function $G^\nu(\lambda)$

$$\eta^\nu(x,t,\lambda) = \eta^{\nu-1}(x,t,\lambda)e^{-i\lambda Jx}G^\nu(\lambda)e^{i\lambda Jx}, \qquad \lambda \in l_\nu, \tag{17}$$

$$G^\nu(\lambda) = \hat{S}_\nu^-(\lambda,0)S_\nu^+(\lambda,0) = \hat{D}_\nu^-(\lambda)\hat{T}_\nu^+(\lambda,0)T_\nu^-(\lambda,0)D_\nu^+(\lambda). \tag{18}$$

The matrices $S_\nu^\pm(t,\lambda)$, $T_\nu^\pm(t,\lambda)$ and $D_\nu^\pm(\lambda)$ are Gauss factors of the scattering matrices $T_\nu(t,\lambda)$ which can be related to each of the rays l_ν, i.e.

$$T_\nu(t,\lambda) = T_\nu^\pm(t,\lambda)D_\nu^\mp(\lambda)\hat{S}_\nu^\mp(t,\lambda). \tag{19}$$

Note that $T_\nu(t,\lambda)$ take values in the subgroup $\mathcal{G}_\nu \subset SO(5)$ corresponding to the subalgebra $\mathfrak{g}_\nu$ with a root system δ_ν [9].

It can be proven that the Gauss factors have the form

$$S_\nu^\pm(\lambda) = \exp(s_\nu^\pm(\lambda)E_{\pm\beta_\nu}), \quad T_\nu^\pm(\lambda) = \exp(t_\nu^\pm(\lambda)E_{\pm\beta_\nu}) \tag{20}$$

$$D_\nu^\pm(\lambda) = \exp(\pm d_\nu^\pm(\lambda)H_{\beta_\nu}), \quad \beta_\nu \in \delta_\nu^+. \tag{21}$$

$D_\nu^\pm(\lambda)$ are piece-wise analytic functions of λ which generate the integrals of motion of the corresponding NLEE. The reduction leads to certain symmetries of the scattering data which is related to different sectors, namely:

$$S^{\pm}_{\nu+2}(\lambda\omega) = C\left(S^{\pm}_{\nu}(\lambda)\right)C^{-1}, \quad T^{\pm}_{\nu+2}(\lambda\omega) = C\left(T^{\pm}_{\nu}(\lambda)\right)C^{-1}, \tag{22}$$

$$D^{\pm}_{\nu+2}(\lambda\omega) = C\left(D^{\pm}_{\nu}(\lambda)\right)C^{-1}. \tag{23}$$

We refer the reader who is interested to the paper [11] where this topic is considered in great detail in the case of a reduced generalized Zakharov-Shabat system.

3. Wronskian relations and squared solutions

The IST can be viewed as a generalization of the Fourier transform. In order to demonstrate this statement one usually derives the so-called Wronskian relations. Starting from the equality

$$(\hat{\chi}^{\nu} J \chi^{\nu}(x,\lambda) - J)|_{-\infty}^{\infty} = \int_{-\infty}^{\infty} dx \hat{\chi}^{\nu}[J, q_x]\chi^{\nu}(x,\lambda) \tag{24}$$

where $\chi^{\nu}(x,\lambda)$ is a FAS that is related to $\eta^{\nu}(x,\lambda)$ via

$$\eta^{\nu}(x,\lambda) = \chi^{\nu}(x,\lambda)e^{i\lambda J x},$$

we obtain a Wronskian relations which connect the scattering data and the potential $q_x(x,t)$

$$s^{+}_{\nu}(\lambda) = -\frac{1}{\beta_{\nu}(J)}\int_{-\infty}^{\infty} dx \left\langle q_x(x), [J, \chi^{\nu} E_{-\beta_{\nu}} \hat{\chi}^{\nu}(x,\lambda)]\right\rangle \tag{25}$$

$$s^{-}_{\nu}(\lambda) = \frac{1}{\beta_{\nu}(J)}\int_{-\infty}^{\infty} dx \left\langle q_x(x), [J, \chi^{\nu-1} E_{\beta_{\nu}} \hat{\chi}^{\nu-1}(x,\lambda)]\right\rangle \tag{26}$$

Another type of Wronskian relation occurs when we consider the following equality

$$\hat{\chi}^{\nu}\delta\chi^{\nu}(x,\lambda)|_{-\infty}^{\infty} = \int_{-\infty}^{\infty} dx \hat{\chi}^{\nu}\delta q_x \chi^{\nu}(x,\lambda). \tag{27}$$

The result reads

$$\delta s^{+}_{\nu}(\lambda) = -\int_{-\infty}^{\infty} dx \left\langle \delta q_x, \chi^{\nu} E_{-\beta_{\nu}} \hat{\chi}^{\nu}(x,\lambda)\right\rangle, \tag{28}$$

$$\delta s^{-}_{\nu}(\lambda) = -\int_{-\infty}^{\infty} dx \left\langle \delta q_x, \chi^{\nu-1} E_{\beta_{\nu}} \hat{\chi}^{\nu-1}(x,\lambda)\right\rangle. \tag{29}$$

Introduce squared solutions in the following manner

$$e^{(\nu)}_{\alpha}(x,\lambda) = P_J(\chi^{\nu} E_{\alpha} \hat{\chi}^{\nu}(x,\lambda)), \quad h^{(\nu)}_{j}(x,\lambda) = P_J(\chi^{\nu} H_j \hat{\chi}^{\nu}(x,\lambda)), \tag{30}$$

where P_J stands for the projection which maps onto $\mathfrak{so}(5)/\mathfrak{h}$.

4. Completeness of the squared solutions

The next theorem represents the main result in our paper.

Theorem 4.1. *The squared solutions form a complete set with the following completeness relations*

$$\begin{aligned}\delta(x-y)\Pi = \frac{1}{2\pi}\sum_{\nu=1}^{8}(-1)^{\nu+1}\int_{l_\nu} d\lambda \left(G_{\beta_\nu}^{(\nu)}(x,y,\lambda) - G_{-\beta_\nu}^{(\nu-1)}(x,y,\lambda)\right) \\ - i\sum_{\nu=1}^{8}\sum_{n_\nu} \operatorname*{Res}_{\lambda=\lambda_{n_\nu}} G^{(\nu)}(x,y,\lambda), \qquad (31)\end{aligned}$$

where

$$G_{\beta_\nu}^{(\nu)}(x,y,\lambda) = e_{\beta_\nu}^{(\nu)}(x,\lambda)\otimes e_{-\beta_\nu}^{(\nu)}(y,\lambda),\ \Pi = \sum_{\alpha\in\Delta+}\frac{E_\alpha\otimes E_{-\alpha} - E_{-\alpha}\otimes E_\alpha}{\alpha(J)}.$$

Proof. We will derive the completeness relations (31) starting from the expression

$$\mathcal{J}(x,y) = \sum_{\nu=1}^{8}(-1)^{\nu+1}\oint_{\gamma_\nu} G^{(\nu)}(x,y,\lambda)d\lambda, \qquad (32)$$

where the contours $\gamma_{\nu 1}$ are shown on figure 1. The Green functions $G^{(\nu)}(x,y,\lambda)$ have the form

$$\begin{aligned}G^{(\nu)}(x,y,\lambda) = \theta(y-x)\sum_{\alpha\in\Delta_\nu^+} e_\alpha^{(\nu)}(x,\lambda)\otimes e_{-\alpha}^{(\nu)}(y,\lambda) - \theta(x-y) \\ \times\left[\sum_{\alpha\in\Delta_\nu^-} e_\alpha^{(\nu)}(x,\lambda)\otimes e_{-\alpha}^{(\nu)}(y,\lambda) + \sum_{j=1}^{2} h_j^{(\nu)}(x,\lambda)\otimes h_j^{(\nu)}(y,\lambda)\right].\end{aligned}$$

The next important result underlies the proof of the theorem

Lemma 4.1. *The following equality holds for $\lambda\in l_\nu$*

$$\begin{aligned}\sum_{\alpha\in\Delta} e_\alpha^{(\nu-1)}(x.\lambda)\otimes e_{-\alpha}^{(\nu-1)}(y,\lambda) + \sum_{j=1,2} h_j^{(\nu-1)}(x,\lambda)\otimes h_j^{(\nu-1)}(y,\lambda) \\ = \sum_{\alpha\in\Delta} e_\alpha^{(\nu)}(x,\lambda)\otimes e_{-\alpha}^{(\nu)}(y,\lambda) + \sum_{j=1,2} h_j^{(\nu)}(x,\lambda)\otimes h_j^{(\nu)}(y,\lambda). \qquad (33)\end{aligned}$$

Proof. The proof is based on the interrelation (17), the definition of $\chi^\nu(x,\lambda)$ and the properties of the Casimir operator P. □

According to Cauchy's residue theorem we have

$$\mathcal{J}(x,y) = 2\pi i \sum_{\nu=1}^{8} \sum_{n_\nu} \mathrm{Res}_{\lambda=\lambda_{n_\nu}} G^{(\nu)}(x,y,\lambda). \tag{34}$$

In the simplest case when $e_\alpha^{(\nu)}(x,\lambda)$ and $h_j^{(\nu)}(x,\lambda)$ possess only a single simple pole $\lambda_\nu \in \Omega_\nu$, i.e.

$$e_\alpha^{(\nu)}(x,\lambda) \approx \frac{e_{\alpha,n_\nu}^{(\nu)}(x)}{\lambda - \lambda_{n_\nu}} + \dot{e}_\alpha^{(\nu)}(x) + O(\lambda - \lambda_{n_\nu})$$

$$h_j^{(\nu)}(x,\lambda) \approx \frac{h_j^{(\nu)}(x)}{\lambda - \lambda_{n_\nu}} + \dot{h}_j^{(\nu)}(x) + O(\lambda - \lambda_{n_\nu}).$$

The following lemma holds true

Lemma 4.2. *Residues of $G^{(\nu)}(x,y,\lambda)$ are given by*

$$\mathrm{Res}_{\lambda=\lambda_{n_\nu}} G^{(\nu)}(x,y,\lambda) = \dot{e}_{\beta_\nu,n}^{(\nu)}(x) \otimes e_{-\beta_\nu,n}^{(\nu)}(y) + e_{\beta_\nu,n}^{(\nu)}(x) \otimes \dot{e}_{-\beta_\nu,n}^{(\nu)}(y).$$

Proof. The statement of that lemma follows from the result formulated in lemma 4.1 and the observation that Jost solutions (for potentials on compact support) do not possess any pole singularities. □

On the other hand taking into account the orientation of the contours γ_ν the expression (32) turns into

$$\begin{aligned}\mathcal{J}(x,y) = &\sum_{\nu=1}^{8}(-1)^{\nu+1}\int_{l_\nu}(G^{(\nu)}(x,y,\lambda) - G^{(\nu-1)}(x,y,\lambda))d\lambda \\ &+ \sum_{\nu=1}^{8}(-1)^{\nu+1}\int_{C_\nu} G^{(\nu)}(x,y,\lambda)d\lambda.\end{aligned} \tag{35}$$

It turns out that the sum above can be simplified, namely the following lemma holds

Lemma 4.3. *In the integrals along the rays contribute only terms related to the roots that belong to δ_ν^+ and δ_ν^- respectively, i.e.*

$$\begin{aligned}G^{(\nu)}(x,y,\lambda) - G^{(\nu-1)}(x,y,\lambda) = &\, e_{\beta_\nu}^{(\nu)}(x,\lambda) \otimes e_{-\beta_\ni}^{(\nu)}(y,\lambda) \\ &- e_{-\beta_\nu}^{(\nu-1)}(x,\lambda) \otimes e_{\beta_\nu}^{(\nu-1)}(y,\lambda).\end{aligned} \tag{36}$$

Proof. As a consequence of lemma 4.1 one can verify that

$$G^{(\nu)}(x,y,\lambda) - G^{(\nu-1)}(x,y,\lambda) = \sum_{\alpha\in\Delta_\nu^+} e_\alpha^{(\nu)}(x,\lambda)\otimes e_{-\alpha}^{(\nu)}(y,\lambda)$$
$$- \sum_{\alpha\in\Delta_{\nu-1}^+} e_\alpha^{(\nu-1)}(x,\lambda)\otimes e_{-\alpha}^{(\nu-1)}(y,\lambda). \tag{37}$$

At this point we make use of the property $\Delta_\nu^+\backslash\delta_\nu^+ = \Delta_{\nu-1}^+\backslash\delta_\nu^-$ and the fact that the sewing function $G^\nu(\lambda)$ is an element of $SL(2)$ group related to l_ν. Then the sums in $G^{(\nu)}(x,y,\lambda)$ and in $G^{(\nu-1)}(x,y,\lambda)$ over these subsets annihilate each other. □

In order to evaluate the integrals along the arcs C_ν we need the asymptotic behavior of $G^{(\nu)}(x,y,\lambda)$ as $\lambda\to\infty$. It is given by the expression

$$G^{(\nu)}(x,y,\lambda) \underset{\lambda\to\infty}{\approx} \sum_{\alpha\in\Delta_\nu^+} e^{i\lambda\alpha(J)(y-x)}E_\alpha\otimes E_{-\alpha} - \theta(x-y)$$
$$\times\left(\sum_{\alpha\in\Delta} e^{i\lambda\alpha(J)(y-x)}E_\alpha\otimes E_{-\alpha} + \sum_{j=1,2} H_j\otimes H_j\right). \tag{38}$$

Asymptotically $G^{(\nu)}(x,y,\lambda)$ is an entire function hence we are allowed to deform the arcs C_ν into $\bar{l}_\nu\cup\bar{l}_{\nu+1}$. Hence the integrals along the arcs C_ν we rewrite as follows

$$\sum_{\nu=1}^{8}(-1)^{\nu+1}\int_{C_\nu} G^{(\nu)}(x,y,\lambda)d\lambda = \sum_{\nu=1}^{8}(-1)^{\nu+1}\int_{l_\nu} d\lambda$$
$$\times\left(e^{-i\lambda\beta_\nu(J)(y-x)}E_{-\beta_\nu}\otimes E_{\beta_\nu} - e^{i\lambda\beta_\nu(J)(y-x)}E_{\beta_\nu}\otimes E_{-\beta_\nu}\right). \tag{39}$$

After we combine the term associated with l_ν and that one associated with $l_{\nu+4}$ and perform an integration, the result reads

$$2\pi\delta(x-y)\sum_{\alpha\in\Delta^+}\frac{(E_\alpha\otimes E_{-\alpha} - E_{-\alpha}\otimes E_\alpha)}{\alpha(J)}. \tag{40}$$

That way we reach the completeness relations (31). □

5. Expansions over the squared solutions

The completeness relations (31) allows one to expand a function $X(x,t)$ over the squared solutions $e_\alpha^{(\nu)}(x,t,\lambda)$ as displayed below in the case when

L does not have discrete eigenvalues

$$X(x,t) = \sum_{\nu=1}^{8} \frac{(-1)^{\nu+1}}{2\pi} \int_{l_\nu} d\mu \left(\xi_\nu^+(t,\mu) e_{\beta_\nu}^{(\nu)}(x,t,\mu) - \xi_{\nu-1}^-(t,\mu) e_{-\beta_\nu}^{(\nu)}(x,t,\mu) \right), \tag{41}$$

where the coefficients $\xi_\nu^+(x,t)$ and $\xi_{\nu-1}^-(t,\mu)$ are given by

$$\xi_\nu^+(t,\mu) = \int_{-\infty}^{\infty} dy \, \mathrm{tr}\left([J, X(y,t)] e_{\beta_\nu}^{(\nu)}(y,t,\mu)\right) \tag{42}$$

$$\xi_\nu^-(t,\mu) = \int_{-\infty}^{\infty} dy \, \mathrm{tr}\left([J, X(y,t)] e_{-\beta_\nu}^{(\nu-1)}(y,t,\mu)\right).$$

For example, if we consider $X(x,t) = [J, q_x(x,t)]$ and $X(x,t) = \delta q_x(x,t)$ then using the Wronskian relations (25) and (28) respectively, we obtain

$$q_x(x,t) = \sum_{\nu=1}^{8} \frac{(-1)^{\nu}}{2\pi} \int_{l_\nu} d\mu \, \beta_\nu(J) \left(s_\nu^+ e_{\beta_\nu}^{(\nu)} + s_\nu^- e_{-\beta_\nu}^{(\nu-1)} \right)(x,t,\mu), \tag{43}$$

$$\mathrm{ad}_J^{-1} \delta q_x(x,t) = \sum_{\nu=1}^{8} \frac{(-1)^{\nu}}{2\pi} \int_{l_\nu} d\mu \left(\delta s_\nu^+ e_{\beta_\nu}^{(\nu)} - \delta s_\nu^- e_{-\beta_\nu}^{(\nu-1)} \right)(x,t,\mu), \tag{44}$$

$$\delta q(x,t) = i \sum_{\nu=1}^{8} \frac{(-1)^{\nu+1}}{2\pi} \int_{l_\nu} \frac{d\mu}{\mu} \left(\delta s_\nu^+ e_{\beta_\nu}^{(\nu)} - \delta s_\nu^- e_{-\beta_\nu}^{(\nu-1)} \right)(x,t,\mu). \tag{45}$$

These formulae make obvious the fact that the IST for the Lax operator (2) can be interpreted as generalized Fourier transform. The squared solutions are analogues of the usual plane waves $e^{i\lambda x}$. In order to complete the analogy we need also the recursion operator Λ defined by

$$\Lambda_+ e_{\pm\alpha}^{(\nu)}(x,\lambda) = \lambda^4 e_{\pm\alpha}^{(\nu)}(x,\lambda), \qquad \Lambda_- e_{\mp\alpha}^{(\nu)}(x,\lambda) = \lambda^4 e_{\mp\alpha}^{(\nu-1)}(x,\lambda), \tag{46}$$

where $\alpha \in \Delta_\nu^+$. These integro-differential operators, whose explicit form will be published elsewhere, generalize the ones obtained in [9,12–14] for the case of $\mathfrak{sl}(3,\mathbb{R})$. It can be used to derive all fundamental properties of the relevant NLEE.

6. Hamiltonian formalism

The ATFT model has obvious Hamiltonian formulation

$$H_{\mathrm{ATFT}}^{(0)} = \int_{-\infty}^{\infty} dx \left((\vec{q}_x, \vec{q}_t) + e^{-(\vec{\alpha}_1,\vec{q})} + e^{-(\vec{\alpha}_2,\vec{q})} + e^{-(\vec{\alpha}_0,\vec{q})} \right) \tag{47}$$

$$\Omega_0 = \int_{-\infty}^{\infty} dx \left\langle \mathrm{ad}_J^{-1} \delta q_x \wedge [J, \delta q] \right\rangle. \tag{48}$$

Using the expansions (44) and (45) after some calculations we are able to cast Ω_0 in the canonical form

$$\Omega_0 = \sum_{\nu=1}^{8} \int_{l_\nu} \frac{d\lambda}{\lambda}\, \delta\pi_\nu(\lambda) \wedge \delta\kappa_\nu(\lambda), \tag{49}$$

where the action-angle variables $\pi_\nu(\lambda)$ and $\kappa_\nu(\lambda)$ are defined by

$$\pi_\nu(\lambda) = \frac{1}{\pi} \ln\left(1 - s_\nu^+ s_\nu^-(\lambda)\right), \qquad \kappa_\nu(\lambda) = \frac{1}{2} \ln\left(\frac{b_\nu^+}{b_\nu^-}\right). \tag{50}$$

Along with Ω_0 there exists a hierarchy Ω_m, $m = 0, 1, \dots$ of symplectic forms

$$\Omega_m = \int_{-\infty}^{\infty} dx \, \left\langle \operatorname{ad}_J^{-1} \delta q_x \wedge \Lambda^m [J, \delta q] \right\rangle \tag{51}$$

Skipping the calculational details we formulate the most important results concerning the Hamiltonian hierarchies of the ATFT, namely their pairwise compatibility. Indeed, calculating Ω_m in terms of action-angle variables due to equation (46) we obtain:

$$\Omega_m = \sum_{\nu=1}^{8} \int_{l_\nu} d\lambda\, \lambda^{4m-1} \delta\pi_\nu(\lambda) \wedge \delta\kappa_\nu(\lambda), \qquad m = 0, 1, \dots. \tag{52}$$

7. Conclusions

We have demonstrated how IST can be applied to $\mathbb{Z}_4$-reduced two dimensional affine Toda model and have derived completeness relations for squared solutions. This allows one to interpret IST as a generalized Fourier transform. A more detailed analysis of these topics will be published elsewhere.

Acknowledgement

This work is partially supported by a grant 1410 with the National Science Foundation of Bulgaria.

References

1. A. Mikhailov, Physica D **3**, 73–117 (1981).
2. V. Zakharov , S. Manakov, S. Novikov and L. Pitaevskii, *Theory of Solitons: The Inverse Scattering Method*, Plenum, New York, 1984.
3. V. Gerdjikov, G. Grahovski and N. Kostov, J. Phys. A **34**, 9425–9461 (2001).
4. V. Zakharov and A. Mikhailov, Commun. Math. Phys. **74**, 21–40 (1980).

5. A. Mikhailov, M. Olshanetsky and A. Perelomov, Commun. Math. Phys. **79**, 473–488 (1981).
6. D. Olive and N.Turok, Nucl. Phys. B **257**, 277–301 (1985).
7. M. Kneipp and D. Olive, preprint (hep-th/9305154).
8. R. Beals and R. Coifman, Commun. Pure and Appl. Math. **38**, n. 29 (1985).
9. V. Gerdjikov and A. Yanovski, J. Math. Phys. **35**, 3687–3725 (1994).
10. S. Helgasson, *Differential Geometry, Lie Groups and Symmetric Spaces*, Academic Press, 1978.
11. G. Grahovski, Proc. of Third International Conference on Geometry, Integrability and Quantization, June 14–23, 2001, Varna, Bulgaria, Eds. Ivailo Mladenov and Gregory Naber.
12. A.Fordy and J. Gibbons, J. Math. Phys. **22**, 1170–1175 (1981).
13. V. Gerdjikov, Inv. Problems **2**, 51–74 (1986).
14. ———, Contemp. Math. **301**, 35–67 (2002).

ON SOME DEFORMATIONS OF ALMOST COMPLEX STRUCTURES OF 6-DIMENSIONAL SUBMANIFOLDS IN THE OCTONIONS

H. HASHIMOTO

Department of Mathematics,
Meijo University
Tempaku, Nagoya 468-8502, Japan
E-mail: hhashi@ccmfs.meijo-u.ac.jp

We give some examples of deformations of the induced almost complex structures of 6-dimensional submanifolds in the octonions.

Keywords: *Spin*(7)-congruence; Almost complex structure; Octonions.

1. Introduction

Let $\mathfrak{C}$ be the octonions (or Cayley algebra). In ([1]), R.L. Bryant showed that any oriented 6-dimensional submanifold $\varphi : M^6 \to \mathfrak{C}$ of the octonions admits the almost complex (Hermitian) structure is defined by

$$\varphi_*(JX) = \varphi_*(X)(\eta \times \xi)$$

where $\{\xi, \eta\}$ is the oriented orthonormal frame field of the normal bundle of φ. The frame field $\{\xi, \eta\}$ is defined locally, however, since the exterior product is skew symmetric, $\eta \times \xi$ is a $S^6(\subset \operatorname{Im} \mathfrak{C})$-valued function whole on M^6. Also, this almost complex structure J is compatible with the induced metric, automatically, therefore M^6 admits the almost Hermitian structure.

We note that the induced almost complex (Hermitian) structure is a *Spin*(7) invariant in the following sense. Let $\varphi_1, \varphi_2 : M^6 \to \mathfrak{C}$ be two isometric immersions from the same manifold to the octonions. If there exist an element $(g, b) \in Spin(7) \times \mathfrak{C}$ such that $g \circ \varphi_1 + b = \varphi_2$, then the two maps are said to be *Spin*(7)-congruent. If two maps are *Spin*(7)-congruent, then the induced almost Hermitian structures coincide.

For any $h \in SO(8)$, then $h \circ \varphi$ induce the almost complex structure

$J^{h\circ\varphi}$. If there exists an isometry $\phi \in \mathrm{Iso}\,(M^6, \langle\, , \,\rangle)$ such that

$$J^{h\circ\varphi} \circ \phi_* = \phi_* \circ J^{\varphi},$$

then $J^{h\circ\varphi}$ and J^{φ} are said to be equivalent. In this paper, we consider the equivalence classes of some simple immersions.

2. Clifford algebras, spinor groups and octonions

First we review the definition of the Clifford algebras and spinors.

2.1. *Clifford algebras and spinor groups*

Let V be a n-dimensional vector space over the real field $\mathbf{R}$, with positive definite metric $\langle\, , \,\rangle$. The Clifford algebra $Cl(V, \langle\, , \,\rangle)$ associated to V and $\langle\, , \,\rangle$ is an associated algebra with unit defined as follows. Let

$$\mathcal{T}(V) = \sum_{r=0}^{\infty} \bigotimes^{r} V$$

denote the tensor algebra of V, and define $\mathcal{I}$ to be the ideal in $\mathcal{T}(V)$ generated by all elements of the form $v \otimes v + \langle v, v\rangle 1$ for any $v \in V$. The Clifford algebra is defined to be the quotient

$$Cl(V, \langle\, , \,\rangle) = \mathcal{T}(V)/\mathcal{I}.$$

We recall the universal property of the Clifford algebra. Let A be an associative algebra with unit and $f : V \to A$ be a linear map satisfying

$$f(v)f(v) = -\langle v, v\rangle 1$$

for any $v \in V$. Then f extends uniquely to an algebra homomorphism

$$F : Cl(V, \langle\, , \,\rangle) \to A,$$

such that $F|_V = f$. Furthermore, $Cl(V, \langle\, , \,\rangle)$ is the unique associative algebra with this property.

Next, we shall give the relationship the representation of *Spin* (7) and the Clifford group. To do this, we define the octonions as follows. Let $\mathbf{H}$ be the skew field of all quaternions with canonical basis $\{1, i, j, k\}$, which satisfy

$$i^2 = j^2 = k^2 = -1, \quad ij = -ji = k, \quad jk = -kj = i, \quad ki = -ik = j.$$

The octonions $\mathfrak{C}$ over $\mathbf{R}$ can be considered as a direct sum $\mathbf{H} \oplus \mathbf{H} = \mathfrak{C}$ with the following multiplication

$$(a + b\varepsilon)(c + d\varepsilon) = ac - \bar{d}b + (da + b\bar{c})\varepsilon,$$

where $\varepsilon = (0,1) \in \mathbf{H} \oplus \mathbf{H}$ and $a, b, c, d \in \mathbf{H}$, the symbol "$^-$" denote the conjugation of the quaternion. For any $x, y \in \mathfrak{C}$, we have

$$\langle xy, xy \rangle = \langle x, x \rangle \langle y, y \rangle$$

which is called "normed algebra" in ([2]). The octonions is a non-commutative, non-associative, alternative, division algebra. It is known that the group of automrphisms of the octonions is the exceptional simple Lie Group

$$G_2 = \{ g \in SO(8) \mid g(uv) = g(u)g(v) \text{ for any } u, v \in \mathfrak{C} \}.$$

For any $v \in \operatorname{Im} \mathfrak{C}$, we define the linear endomorphism R_v of $\mathbf{R}^8 \simeq \mathfrak{C}$ by setting

$$R_v(x) = xv, \quad (f(v) = R_v),$$

for any $x \in \mathfrak{C}$ and $f : \operatorname{Im} \mathfrak{C} \to \operatorname{End}_{\mathbf{R}}(\mathfrak{C})$. By altenativity, we see that

$$R_v{}^2 = -\langle v, v \rangle \, id_{\mathfrak{C}}, \quad (= f(v)^2).$$

From the universal property of the Clifford algebra, f extends to the a representation

$$F : Cl(V, \langle\, , \rangle) \to \operatorname{End}_{\mathbf{R}}(\mathfrak{C}).$$

In order to define the Clifford groups, we set the automorphism $\alpha : Cl(V, \langle\, , \rangle) \to Cl(V, \langle\, , \rangle)$, such that $\alpha(v) = -v$ for any $v \in V$. Then we have the splitting of the Clifford group $Cl(V, \langle\, , \rangle) = Cl^+(V, \langle\, , \rangle) \oplus Cl^-(V, \langle\, , \rangle)$, where $Cl^\pm(V, \langle\, , \rangle) = \{ g \in Cl(V, \langle\, , \rangle) \mid \alpha(g) = \pm g \}$. We denote by $Cl^\times(V, \langle\, , \rangle)$ the multiplicative group of units in the Clifford algebra. The Clifford group $\Gamma^+$is defined by

$$\Gamma^+ = \{ g \in Cl^+(V, \langle\, , \rangle) \cap Cl^\times(V, \langle\, , \rangle) \mid gvg^{-1} \in V \text{ for any } v \in V \}.$$

Then the Spinor group is given by

$$Spin\,(n) = \{ g \in \Gamma^+ \mid g\overline{g} = 1 \},$$

where $\overline{g}$ is the conjugation of the Clifford algebra defined by

$$\overline{e_{i_1} e_{i_2} \cdots e_{i_k}} = (-1)^k e_{i_k} e_{i_{(k-1)}} \cdots e_{i_1}$$

whenever $1 \leq i_1 < i_2 < \cdots < i_k \leq n$, and $g = e_{i_1} e_{i_2} \cdots e_{i_k}$. If we identify V with $\{ R_v \mid v \in \operatorname{Im} \mathfrak{C} \} \simeq \mathbf{R}^7$, then we see that (from the definition as above)

$$Spin\,(7) = \{ g \in SO(8) \mid g \circ R_v \circ g^{-1} \in V \text{ for any } v \in \mathfrak{C} \}.$$

Therefore there exists $u \in \operatorname{Im}\mathfrak{C}$ such that

$$g \circ R_v \circ g^{-1} = R_u.$$

From this, we have $u = g(g^{-1}(1)v)$. We set $\chi_g(v) = g(g^{-1}(1)v)$. Then, we obtain

$$g \circ R_v \circ g^{-1} = R_{\chi_g(v)}, \quad \text{or} \quad g \circ R_v = R_{\chi_g(v)} \circ g.$$

From this we get the following definition of *Spin* (7)

$$Spin\,(7) = \{\, g \in SO(8) \mid g(uv) = g(u)\chi_g(v) \text{ for any } u, v \in \mathfrak{C}\,\}.$$

We see that G_2 is a Lie subgroup of *Spin* (7); $G_2 = \{\, g \in Spin(7) \mid g(1) = 1\,\}$. The map χ defines the double covering map from *Spin* (7) to $SO(7)$, which satisfy the following equivariance

$$g(u) \times g(v) = \chi_g(u \times v)$$

for any $u, v \in \mathfrak{C}$ and $u \times v = (1/2)(\bar{v}u - \bar{u}v)$ (which is called the "exterior product") where $\bar{v} = 2\langle v, 1\rangle - v$ is the conjugation of $v \in \mathfrak{C}$. We note that $u \times v$ is an element of pure-imaginary part of $\mathfrak{C}$.

2.2. *Spin* (7)*-structure equations*

In this section, we shall recall the structure equation of *Spin* (7) which is established by R. Bryant ([1]). To construct this, we fix a basis of the complexification of the octonions $\mathbf{C}\otimes_{\mathbf{R}}\mathfrak{C}$ as follows

$$N = (1/2)(1 - \sqrt{-1}\,\varepsilon), \quad \bar{N} = (1/2)(1 + \sqrt{-1}\,\varepsilon)$$

$$E_1 = iN,\ E_2 = jN,\ E_3 = -kN,\ \bar{E}_1 = i\bar{N},\ \bar{E}_2 = j\bar{N},\ \bar{E}_3 = -k\bar{N}.$$

We extend the multiplication of the octonions complex linearly. We define a $\mathfrak{C} \rtimes Spin\,(7)$ (semi-direct product) admissible frame field as follows. Let o be the origin of the octonions. The Lie group $\mathfrak{C} \rtimes Spin\,(7)$ acts on $\mathfrak{C} \oplus \operatorname{End}(\mathbf{C}\otimes_{\mathbf{R}}\mathfrak{C})$ such that

$$\begin{aligned}(x, g)(o; N, E, \bar{N}, \bar{E}) &= (g \cdot o + x, g(N), g(E), g(\bar{N}), g(\bar{E}))\\ &= (x, g(N), g(E), g(\bar{N}), g(\bar{E}))\\ &= (o; N, E, \bar{N}, \bar{E})\begin{pmatrix} 1 & 0_{1\times 8} \\ \rho(x) & \rho(g)\end{pmatrix}\end{aligned}$$

where $(x, g) \in \mathfrak{C} \rtimes Spin\,(7)$ and $\begin{pmatrix} 1 & 0_{1\times 8} \\ \rho(x) & \rho(g)\end{pmatrix}$ is an its matrix representation. A frame $(x; n, f, \bar{n}, \bar{f})$ is said to be a $\mathfrak{C} \rtimes Spin\,(7)$ admissible one if there exists a $(x, g) \in \mathfrak{C} \rtimes Spin\,(7)$ such that

$$(x; n, f, \bar{n}, \bar{f}) = (x, g)(o; N, E, \bar{N}, \bar{E}).$$

Proposition 2.1 ([1]). *The Maurer-Cartan form of* $\mathfrak{C} \rtimes Spin(7)$ *is given by*

$$d(x;n,f,\bar{n},\bar{f}) = (x;n,f,\bar{n},\bar{f}) \left(\begin{array}{c|cc|cc} 0 & 0 & 0_{1\times 3} & 0 & 0_{1\times 3} \\ \nu & \sqrt{-1}\rho & -{}^t\bar{\mathfrak{h}} & 0 & -{}^t\theta \\ \omega & \mathfrak{h} & \kappa & \theta & [\bar{\theta}] \\ \hline \bar{\nu} & 0 & -{}^t\bar{\theta} & -\sqrt{-1}\rho & -{}^t\mathfrak{h} \\ \bar{\omega} & \bar{\theta} & [\theta] & \bar{\mathfrak{h}} & \bar{\kappa} \end{array}\right)$$

$$= (x;n,f,\bar{n},\bar{f})\psi$$

where ψ *is the* $Spin(7) \oplus \mathfrak{C}(\subset M_{9\times 9}(\mathbf{C}))$*-valued 1-form,* ρ *is a real-valued 1-form,* ν *is a complex valued 1-form,* $\omega, \mathfrak{h}, \theta$ *are* $M_{3\times 1}$*-valued 1-form,* κ *is a* $\mathfrak{u}3$*-valued 1-form which satisfy* $\sqrt{-1}\rho + tr\kappa = 0$*, and*

$$[\theta] = \begin{pmatrix} 0 & \theta^3 & -\theta^2 \\ -\theta^3 & 0 & \theta^1 \\ \theta^2 & -\theta^1 & 0 \end{pmatrix}$$

for $\theta = {}^t(\theta^1, \theta^2, \theta^3)$*. The* ψ *satisfy the following integrability condition* $d\psi + \psi \wedge \psi = 0$.

3. Gram-Schmidt process of *Spin*(7)

To construct the *Spin*(7)-frame field, we recall the Gram-Schmidt process of G_2-frame. Let $\operatorname{Im}\mathfrak{C} = \{u \in \mathfrak{C} \mid \langle u, 1\rangle = 0\}$ be the subspace of purely imaginary octonions.

Lemma 3.1. *For a pair of mutually orthogonal unit vectors* e_1, e_4 *in* $\operatorname{Im}\mathfrak{C}$ *put* $e_5 = e_1e_4$*. Take a unit vector* e_2*, which is perpendicular to* e_1, e_4 *and* e_5*. If we put* $e_3 = e_1e_2$, $e_6 = e_2e_4$ *and* $e_7 = e_3e_4$ *then the matrix*

$$g = [e_1, e_2, e_3, e_4, e_5, e_6, e_7] \in SO(7)$$

is an element of G_2.

By Lemma 3.1, we can take $e_4 = \eta \times \xi$, we can get the G_2-frame field as follows. We set

$$\begin{aligned} N^* &= (1/2)(1 - \sqrt{-1}\,e_4), & \bar{N}^* &= (1/2)(1 + \sqrt{-1}\,e_4), \\ E_1^* &= (1/2)(e_1 - \sqrt{-1}\,e_5), & \bar{E}_1^* &= (1/2)(e_1 + \sqrt{-1}\,e_5), \\ E_2^* &= (1/2)(e_2 - \sqrt{-1}\,e_6), & \bar{E}_2^* &= (1/2)(e_2 + \sqrt{-1}\,e_6), \\ E_3^* &= -(1/2)(e_3 - \sqrt{-1}\,e_7), & \bar{E}_3^* &= -(1/2)(e_3 + \sqrt{-1}\,e_7). \end{aligned}$$

Then $\mathrm{span}_{\mathbf{C}}\{N^*, E_1^*, E_2^*, E_3^*\}$ is a $\sqrt{-1}$-eigen space $T_p^{(1,0)}\mathfrak{C}(\subset \mathfrak{C}\otimes \mathbf{C})$ with respect to the almost complex structure $J = R_{\eta\times\xi}$ at $p \in \mathfrak{C}$. On the other hand, $n = (1/2)(\xi - \sqrt{-1}\,\eta)$ is a local orthonormal frame field of the complexified normal bundle $T^{\perp(1,0)}M$. Since $T^{\perp(1,0)}_{\varphi(m)}M \subset T^{(1,0)}_{\varphi(m)}\mathfrak{C}$, there exists a $M_{4\times 1}(\mathbf{C})$-valued function $a_1 = {}^t(a_{11}, a_{21}, a_{31}, a_{41})$, such that

$$n = (1/2)(\xi - \sqrt{-1}\,\eta) = (N^*, E_1^*, E_2^*, E_3^*)a_1.$$

By the Gram-Schmidt orthonormalization with respect to the Hermitian inner product of $T^{(1,0)}_{\varphi(m)}\mathfrak{C}$, there exist three $M_{4\times 1}(\mathbf{C})$-valued functions $\{a_2, a_3, a_4\}$ such that $\{a_1, a_2, a_3, a_4\}$ is a special unitary frame. We set

$$f_i = (N^*, E_1^*, E_2^*, E_3^*)a_{i+1}$$

for $i = 1, 2, 3$, then

$$(n, f, \bar{n}, \bar{f}) = (n, f_1, f_2, f_3, \bar{n}, \bar{f}_1, \bar{f}_2, \bar{f}_3)$$

is a (local) *Spin* (7)-frame field on M.

4. Invariants of *Spin*(7)

We shall recall the invariants of Spin(7)-congruence classes. By Proposition 2.1, we have

Proposition 4.1 ([1]). *Let $\varphi : M^6 \to \mathfrak{C}$ be an isometric immersion from an oriented 6-dimensional manifold to the octonions. Then*

$$\begin{aligned} d\varphi &= f\omega + \bar{f}\bar{\omega}, \\ \nu &= 0, \\ dn &= n\sqrt{-1}\rho + f\mathfrak{h} + \bar{f}\bar{\theta}, \\ df &= n(-{}^t\bar{\mathfrak{h}}) + f\kappa + n(-{}^t\bar{\theta}) + \bar{f}[\theta]. \end{aligned}$$

The second fundamental form II is given by

$$\mathrm{II} = -2\mathrm{Re}\left\{({}^t\bar{\mathfrak{h}} \odot \omega + {}^t\bar{\theta}\odot\bar{\omega})\otimes n\right\}$$

where the symbol "$\odot$" is the symmetric tensor product. By Cartan's Lemma (since $\nu = 0$), there exist $M_{3\times 3}$-valued matrices A, B, C such that

$$\begin{pmatrix}\mathfrak{h}\\ \theta\end{pmatrix} = \begin{pmatrix}\overline{B} & \overline{A}\\ {}^tB & \overline{C}\end{pmatrix}\begin{pmatrix}\omega\\ \bar{\omega}\end{pmatrix} \tag{1}$$

where ${}^tA = A$ and ${}^tC = C$. We have the following decomposition

$$\begin{aligned} \mathrm{II}^{(2,0)} &= (-{}^t\omega \odot A\omega) \otimes n \\ \mathrm{II}^{(1,1)} &= (-{}^t\bar{\omega} \odot {}^tB\omega - {}^t\omega \odot B\bar{\omega}) \otimes n \\ \mathrm{II}^{(0,2)} &= (-{}^t\bar{\omega} \odot \bar{C}\bar{\omega}) \otimes n. \end{aligned}$$

We shall write each elements more explicitly. There exists a unitary frame $\{e_i, Je_i\}$ for i=1,2,3, such that

$$n = (1/2)(\xi - \sqrt{-1}\eta),\, f_i = (1/2)(e_i - \sqrt{-1}Je_i).$$

Thus elements of second fundamental form are given by

$$\begin{aligned} A_{ij} &= -2\langle \mathrm{II}(f_i, f_j), \bar{n}\rangle, \\ B_{ij} &= -2\langle \mathrm{II}(f_i, \bar{f}_j), \bar{n}\rangle, \\ C_{ij} &= -2\langle \mathrm{II}(\bar{f}_i, \bar{f}_j), \bar{n}\rangle. \end{aligned}$$

We shall recall the relation of Ricci $*$-curvature ρ^* and $*$-scalar curvature τ^* which are fundamental invariants on almost Hermitian manifolds. Generically, these curvatures of an almost Hermitian manifold $M = (M, J, \langle\, ,\, \rangle)$ with even dimension $2n$, are defined by

$$\rho^*(x,y) = (1/2)\sum_{i=1}^{2n}\langle R(e_i, Je_i)Jy, x\rangle \quad \text{and} \quad \tau^* = \sum_{i=1}^{2n}\rho^*(e_i, e_i),$$

respectively. We note that Ricci $*$-curvature tensor is neither symmetric nor skew-symmetric.

Proposition 4.2 ([4])**.** *The Ricci $*$-curvature and $*$-scalar curvature of oriented 6-dimensional submanifolds in $\mathfrak{C}$ are given by*

$$\begin{aligned} \rho^*(x,y) &= {}^t\alpha(A\bar{B} - BC - {}^t(A\bar{B} - BC))\beta \\ &\quad - {}^t\alpha(A\bar{A} - B\,{}^t\bar{B} - {}^t\bar{B}B + C\bar{C})\bar{\beta} + \textit{its conjugation} \\ \tau^* &= -4(\operatorname{tr} A\bar{A} - 2\operatorname{tr}{}^t\bar{B}B + \operatorname{tr} C\bar{C}), \end{aligned}$$

where $x = f\alpha + \bar{f}\bar{\alpha}$, $y = f\beta + \bar{f}\bar{\beta}$ and $\alpha, \beta \in M_{3\times 1}(\mathbf{C})$.

We give the equivalent condition for *Spin*(7)-congruence.

Proposition 4.3. *Let M^6 be a connected 6-dimensional manifold and $\varphi_1, \varphi_2 : M^6 \to \mathfrak{C}$ be two isometric immersions with same induced metrics and almost complex structures. Let $\mathrm{II}^{(2,0)}_{\varphi_1}, \mathrm{II}^{(2,0)}_{\varphi_2}$ be the corresponding $(2,0)$ part of the 2nd fundamental forms. Then there exists an element $g \in Spin(7)$ such that $g \circ \varphi_1 = \varphi_2$ if and only if $\mathrm{II}^{(2,0)}_{\varphi_1} \cong \mathrm{II}^{(2,0)}_{\varphi_2}$*

5. Homogeneous spaces related to *Spin*(7)

The Lie group *Spin* (7) acts on some homogeneous spaces, transitively. From the above argument, a 7-dimensional sphere S^7 has an exceptional representation $S^7 = Spin\,(7)/G_2$. If we identify $\mathbf{R}^7$ with the tangent space T_pS^7 for some point $p \in S^7$, the isotropy group G_2 acts transitively on a 6-dimensional sphere $S^6 \subset T_pS^7$, so we have $S^6 = G_2/SU(3)$. In the same way, we identify $\mathbf{R}^6$ with the tangent space T_qS^6 for some point $q \in S^6$, we see that $S^5 = SU(3)/SU(2)$. In this case, $SU(2)$ can not acts transitively on S^4.

Let $V_k(\mathbf{R}^8)$ be the Stiefel manifold of orthonormal k-frames in $\mathbf{R}^8$ and $G_k^+(\mathbf{R}^8)$ the Grassmann manifold of oriented k-planes in $\mathbf{R}^8$.

From these arguments, *Spin* (7) can not acts transitively on the Stiefel manifold $V_4(\mathbf{R}^8)$ of orthonormal 4-frames in $\mathbf{R}^8$ and the Grassmann manifold $G_4^+(\mathbf{R}^8)$ of oriented 4-planes in $\mathbf{R}^8$. However, we have the exceptional representation of the homogeneous manifolds as follows

$$V_2(\mathbf{R}^8) \simeq Spin\,(7)/SU(3), \quad V_3(\mathbf{R}^8) \simeq Spin\,(7)/SU(2),$$

$$G_2^+(\mathbf{R}^8) \simeq Spin\,(7)/U(3), \quad G_3^+(\mathbf{R}^8) \simeq Spin\,(7)/SO(4).$$

The Grassmann manifold of the quaternionic plane in the octonions can be represented by $Spin\,(7)/(Sp\,(1) \times Sp\,(1) \times Sp\,(1)/Z_2)$. Note that $\rho : Sp\,(1) \times Sp\,(1) \times Sp\,(1) \to Spin\,(7)$ is given by

$$\rho(q_1, q_2, q_3)(a + b\varepsilon) = q_2 a\overline{q_1} + (q_3 a\overline{q_1})\varepsilon$$

where $(q_1, q_2, q_3) \in Sp\,(1) \times Sp\,(1) \times Sp\,(1)$ and $(a, b) \in \mathbf{H} \oplus \mathbf{H}$.

6. Deformations of some examples

In this section, we consider the deformations of almost complex structure of some simple examples.

6.1. *On* $S^2 \times S^4$

In this section, we give the $*$-scalar curvature $*\tau$ of the immersion

$$\varphi_{\alpha_0}(p, y_0, y) = \cos(\alpha_0)p + \sin(\alpha_0)(y_0 \cdot 1 + y\varepsilon),$$

where $\alpha_0 \in (0, \pi/2)$ is a constant, $p \in S^2 \subset \operatorname{Im}\mathbf{H}$ ($|p| = 1$), and $y_0 \cdot 1 + y\varepsilon \in S^4 \subset \mathbf{R} \oplus \mathbf{H}\varepsilon$ ($y_0^2 + |y|^2 = 1$). Then the oriented orthonormal basis $\{\xi, \eta\}$ of the normal bundle $T^{\perp}M$ is given by $\xi = y_0 \cdot 1 + y\varepsilon$, $\eta = p$. The almost complex structure is given by the right multiplication of the vector field

$u = \eta \times \xi = y_0\, p + (y\, p)\varepsilon$. Let $\{e_1, e_2\}$ be the oriented orthonormal basis at p (i.e., $e_2 = e_1 p$), then $\{e_1, e_2, p\}$ is an associated plane in $\mathrm{Im}\,\mathfrak{C}$. We construct G_2-frame field from the vector field u as follows: Let $e_1 \in T_pS^2$, $e_4 = u$, then $e_5 = e_1e_4 = y_0e_2 - (ye_2)\varepsilon$. We set

$$\begin{aligned}
\tilde{e_2} &= \frac{(ye_1)}{|y|}\varepsilon,\\
e_3 &= e_1\tilde{e_2} = -\frac{y}{|y|}\varepsilon,\\
e_6 &= \tilde{e_2}e_4 = -|y|e_2 - \frac{y_0(ye_2)}{|y|}\varepsilon,\\
e_7 &= e_3e_4 = -|y|p - \frac{y_0(yp)}{|y|}\varepsilon.
\end{aligned}$$

Then $\{e_1, e_2, \cdots, e_7\}$ is the G_2-adapted frame at $p + y_0 \cdot 1 + y\,\varepsilon \in S^2 \times S^4$. By straightforward calculations, we get the local *Spin*(7) frame field along φ_{α_0} as follows

$$\begin{aligned}
n &= \frac{1}{2}\big((y_0p + (yp)\varepsilon) - \sqrt{-1}\,p\big),\\
f_1 = E_1^* &= \frac{1}{2}\big(e_1 - \sqrt{-1}\,(y_0e_2 - (ye_2)\varepsilon)\big),\\
f_2 = E_2^* &= \frac{1}{2}\Big(\frac{ye_1}{|y|}\varepsilon + \sqrt{-1}\,\Big(|y|e_2 + \frac{y_0}{|y|}(ye_2)\varepsilon\Big)\Big),\\
f_3 &= \frac{1}{2}\Big(-|y|1 + \frac{y_0y}{|y|}\varepsilon + \sqrt{-1}\,\Big(\frac{1}{|y|}(yp)\varepsilon\Big)\Big).
\end{aligned}$$

To calculate the *Spin*(7) invariants, we need the representation of co-frame as follows;

$$d\varphi_{\alpha_0} = \cos(\alpha_0)dp + \sin(\alpha_0)\big(dy_0 + (dy)\varepsilon\big) = \sum_{i=1}^{3} f_i\omega^i + \overline{f_i\omega^i}.$$

Therefore, we have

$$\begin{aligned}
\omega^1 &= 2\langle d\varphi_{\alpha_0}, \overline{f_1}\rangle\\
&= \cos(\alpha_0)\big(\langle dp, e_1\rangle + \sqrt{-1}\,\langle dp, e_2\rangle\big) - \sqrt{-1}\,\sin(\alpha_0)\langle\overline{y}dy, e_2\rangle,\\
\omega^2 &= 2\langle d\varphi_{\alpha_0}, \overline{f_2}\rangle\\
&= -\sqrt{-1}\,\cos(\alpha_0)|y|(\langle dp, e_2\rangle) + \frac{\sin(\alpha_0)}{|y|}\big(\langle\overline{y}dy, e_1\rangle - \sqrt{-1}\,\langle\overline{y}dy, e_2\rangle\big),\\
\omega^3 &= 2\langle d\varphi_{\alpha_0}, \overline{f_3}\rangle\\
&= \sin(\alpha_0)\Big(-|y|dy_0 + \frac{y_0}{|y|}\langle\overline{y}dy, 1\rangle - \frac{\sqrt{-1}}{|y|}\langle\overline{y}dy, p\rangle\Big).
\end{aligned}$$

By (1), we have

$$A = \overline{C} = \frac{1}{4}\left(\frac{1}{\sin(\alpha_0)} - \frac{\sqrt{-1}}{\cos(\alpha_0)}\right)\begin{pmatrix} y_0^2 - 1 & -y_0|y| & 0 \\ -y_0|y| & 1 - y_0^2 & 0 \\ 0 & 0 & 0 \end{pmatrix},$$

$$B = \frac{1}{4}\begin{pmatrix} \frac{1-y_0^2}{\sin(\alpha_0)} + \frac{1+y_0^2}{\cos(\alpha_0)} & y_0|y|\left(\frac{1}{\sin(\alpha_0)} - \frac{\sqrt{-1}}{\cos(\alpha_0)}\right) & 0 \\ y_0|y|\left(\frac{1}{\sin(\alpha_0)} - \frac{\sqrt{-1}}{\cos(\alpha_0)}\right) & \frac{1+y_0^2}{\sin(\alpha_0)} + \frac{1-y_0^2}{\cos(\alpha_0)} & 0 \\ 0 & 0 & \frac{2}{\sin(\alpha_0)} \end{pmatrix}.$$

Hence the $*$-scalar curvature τ^* of φ_{α_0} is given by

$$\tau^* = \frac{2(\cos^2(\alpha_0) + y_0^2)}{\sin^2(\alpha_0)\cos^2(\alpha_0)}.$$

Therefore the induced almost complex structure of φ_{α_0} is not homogeneous. Since $Spin(7)/SO(4) = G_3^+(\mathbf{R}^8)$, we see that the induced almost complex structure is unique (depend only on the radius of spheres). The equivalence classes (of the fixed radius) of the induced almost complex structures on $S^2 \times S^4$ is one point.

6.2. *On 1-parameter family of embeddings from $S^2 \times \mathbf{R}^4$ to $\mathfrak{C}$*

Since $V_3(\mathbf{R}^8) \simeq Spin(7)/SU(2)$, the imbedding from $S^2 \times \mathbf{R}^4$ to the octonions are obtained as a 1-parameter family φ_t, up to the action of $Spin(7)$, as follows ($\varphi_t : S^3 \times \mathbf{R}^4 \to \mathfrak{C}$)

$$\varphi_t(q, \tilde{y}) = qi\overline{q} + (-\sin(t) + \cos(t)\varepsilon)y_1 + (y_2 i + y_3 j + y_4 k)\varepsilon,$$

where $0 \leq t \leq \pi/2$, $q \in S^3 \subset \mathbf{H}$, $\tilde{y} = (y_1, y_2, y_3, y_4) \in \mathbf{R}^4$. The image $\varphi_0(S^3 \times \mathbf{R}^4)$ coincide with $S^2 \times \mathbf{R}^4$ as a Riemannian manifold. In particular the image of φ_0 satisfies

$$\varphi_0(S^3 \times \mathbf{R}^4) = S^2 \times \mathbf{R}^4 \subset \operatorname{Im}\mathbf{H} \oplus \mathbf{H}\varepsilon.$$

The induced almost Hermitian structure is homogeneous and satisfy Quasi-Kähler condition. Next we give the $Spin(7)$ frame field on the φ_t. To do this, we define the action of $Sp(1)$ as follows

$$\rho_{II}(q)(a + b\varepsilon) = qa\overline{q} + (qb\overline{q})\varepsilon$$

where $q \in S^3 \simeq Sp(1)$, $a + b\varepsilon \in \mathfrak{C}$. Then the $Spin(7)$ frame field along φ_t is given by

$$
\begin{aligned}
n &= \frac{1}{2}\Big\{\rho_{II}(q)(\cos(t)\cdot 1 + \sin(t)\varepsilon - \sqrt{-1}\,i)\Big\},\\
f_1 &= \frac{1}{2}\Big\{\rho_{II}(q)\big(j\varepsilon + \sqrt{-1}\,(\cos(t)\cdot k - \sin(t)k\varepsilon)\big)\Big\},\\
f_2 &= \frac{1}{2}\Big\{\rho_{II}(q)\big(j\varepsilon + \sqrt{-1}\,(\cos(t)\cdot k - \sin(t)k\varepsilon)\big)\Big\},\\
f_3 &= -\frac{1}{2}\Big\{\rho_{II}(q)(-\sin(t)\cdot 1 + \cos(t)\varepsilon + \sqrt{-1}\,\varepsilon)\Big\}.
\end{aligned}
$$

In order to represent to the coframe field $(\omega^1, \omega^2, \omega^3)$, We set the local one form μ_1, μ_2 on S^2

$$\mu_1 = \langle d(qi\overline{q}), qj\overline{q}\rangle, \quad \mu_2 = \langle d(qi\overline{q}), qk\overline{q}\rangle,$$

and $\operatorname{Im}\mathbf{H}$-valued 1-form $d\beta$ such that

$$d\beta = idy_1 + jdy_2 + kdy_3.$$

Then we have

$$
\begin{aligned}
\omega^1 &= \mu_1 - \sqrt{-1}\,\big(\cos(t)\mu_2 - \sin(t)\langle qk\overline{q}, d\beta\rangle\big),\\
\omega^2 &= \langle qj\overline{q}, d\beta\rangle + \sqrt{-1}\,\big(\sin(t)\mu_2 + \cos(t)\langle qk\overline{q}, d\beta\rangle\big),\\
\omega^3 &= dy_1 - \sqrt{-1}\,\langle qk\overline{q}, d\beta\rangle.
\end{aligned}
$$

Next, we calculate the 2nd fundamental forms. We see that

$$dn = -\frac{\sqrt{-1}}{2}dqi\overline{q} = -\frac{\sqrt{-1}}{2}(qj\overline{q}\otimes\mu_1 + qk\overline{q}\otimes\mu_2).$$

Therefore, we get

$$A = -C = \frac{\sqrt{-1}}{4}\begin{pmatrix} \sin^2(t) & \sin(t)\cos(t) & 0\\ \sin(t)\cos(t) & -\sin^2(t) & 0\\ 0 & 0 & 0\end{pmatrix},$$

$$B = \frac{\sqrt{-1}}{4}\begin{pmatrix} 1+\cos^2(t) & -\sin(t)\cos(t) & 0\\ -\sin(t)\cos(t) & \sin^2(t) & 0\\ 0 & 0 & 0\end{pmatrix}.$$

Hence the $*$-scalar curvature $\tau^* = -4\{\operatorname{tr} A\overline{A} - 2\operatorname{tr}{}^t\overline{B}B + \operatorname{tr} C\overline{C}\}$ is given by

$$\tau^* = 1 + \cos 2t \geq 0$$

We can show that the induced almost complex structures of φ_t are homogeneous for any t.

Proposition 6.1. *The induced homogeneous almost complex structures of φ_t depend on the parameter t.*

6.3. On 1-parameter family of embeddings from $S^1 \times S^2 \times \mathbf{R}^3$ to $\mathfrak{C}$

In this section, we consider the imbedding from $S^1 \times S^2 \times \mathbf{R}^3$ to the octonions, which are defined as a 1-parameter family φ_t, as follows

$$\begin{aligned}
&\varphi_t(e^{\imath\theta}, q, x_1, x_2, x_3)\\
&= \cos(\theta)\big(\cos(t)\cdot 1 - \sin(t)\cdot i\varepsilon\big) + x_1\big(\sin(t)\cdot 1 + \cos(t)\cdot i\varepsilon\big)\\
&\quad + qi\overline{q} + \sin(\theta)\varepsilon + x_2 j\varepsilon + x_3 k\varepsilon,
\end{aligned}$$

where $(e^{\imath\theta}, q, x_1, x_2, x_3) \in S^1 \times S^3 \times \mathbf{R}^3$ and $q \to qi\overline{q}$ is the Hopf map from S^3 to S^2 and $0 \le t \le \dfrac{\pi}{2}$. Then the *Spin*(7)-frame field along φ_t is given by

$$\begin{aligned}
n &= \frac{1}{2}\{\cos(\theta)\cos(t)\cdot 1 + \alpha_0\varepsilon - \sqrt{-1}\, qi\overline{q}\},\\
f_1 &= \frac{1}{2}\{qj\overline{q} + \sqrt{-1}\,(\cos(\theta)\cos(t)qk\overline{q} - (\alpha_0 qk\overline{q})\varepsilon\},\\
f_2 &= \frac{1}{2}\{-|\alpha_0|\cdot 1 + (\cos(\theta)\cos(t)/|\alpha_0|)(\alpha_0)\varepsilon + \sqrt{-1}\,\big((1/|\alpha_0|)(\alpha_0 qi\overline{q})\varepsilon\big)\},\\
f_3 &= -\frac{1}{2}\{(1/|\alpha_0|)(\alpha_0 qj\overline{q})\varepsilon\\
&\qquad - \sqrt{-1}\,(|\alpha_0|qi\overline{q} + (\cos(\theta)\cos(t)/|\alpha_0|)(\alpha_0 qk\overline{q})\varepsilon\},
\end{aligned}$$

where $\alpha_0 = \sin(\theta) - \cos(\theta)\sin(t)\cdot i$. Since

$$d\varphi_t = \sum_{i=1}^{3} f_i\omega^i + \overline{f_i}\,\overline{\omega^i},$$

the coframe ω_i of the *Spin*(7)-frame field are given by

$$\begin{aligned}
\omega^1 &= \mu_1 - \sqrt{-1}\,\Big\{\cos(\theta)\cos(t)\mu_2 - \sin(t)\langle iq, qk\rangle d\theta\\
&\quad - \cos(t)\sin(\theta)\langle iq, qk\rangle dx_1\\
&\quad - (\sin(\theta)\langle jq, qk\rangle + \cos(\theta)\sin(t)\langle kq, qk\rangle)dx_2\\
&\quad - (\sin(\theta)\langle kq, qk\rangle - \cos(\theta)\sin(t)\langle jq, qk\rangle)dx_3\Big\},\\
\omega^2 &= (1/|\alpha_0|)\Big\{(\sin(\theta)\cos(t) - \sqrt{-1}\,\sin(t)\langle iq, qi\rangle)d\theta\\
&\quad - (\sin(t) + \sqrt{-1}\,\sin(\theta)\cos(t)\langle iq, qi\rangle)dx_1\\
&\quad - \sqrt{-1}\,(\sin(\theta)\langle jq, qi\rangle + \cos(\theta)\sin(t)\langle kq, qi\rangle)dx_2
\end{aligned}$$

$$- \sqrt{-1}\,(\sin(\theta)\langle kq, qi\rangle - \cos(\theta)\sin(t)\langle jq, qi\rangle)dx_3\Big\},$$

$$\begin{aligned}\omega^3 = -(1/|\alpha_0|)\Big\{&\sqrt{-1}\,|\alpha_0|^2\mu_2\\ &+ \big(\sin(\theta)\langle iq, qj\rangle + \sqrt{-1}\,\cos(\theta)\sin(t)\langle iq, qk\rangle\big)d\theta\\ &+ \cos(t)\big(\sin(\theta)\langle iq, qj\rangle + \sqrt{-1}\,\cos(\theta)\cos(t)\langle iq, qk\rangle\big)dx_1\\ &+ \big(\sin(\theta)\langle jq, qj\rangle + \cos(\theta)\sin(t)\langle kq, qj\rangle\\ &\quad + \sqrt{-1}\,\cos(\theta)\cos(t)(\sin(\theta)\langle jq, qk\rangle + \cos(\theta)\sin(t)\langle kq, qk\rangle)\big)dx_2\\ &+ \sqrt{-1}\,\cos(\theta)\cos(t)\big(\sin(\theta)\langle kq, qk\rangle - \cos(\theta)\sin(t)\langle jq, qk\rangle\big)dx_3\Big\}.\end{aligned}$$

By a tedious calculation we verified that the $*$-scalar curvatures are given by

Theorem 6.1. $\tau^* = 2\cos^2(\theta)\cos^2(t)$.

We see that, for fixed $t \neq \pi/2 \bmod \pi$, the $*$-scalar curvatures are not constant on $S^1 \times S^2 \times \mathbf{R}^3$. Therefore the induced almost complex structures are not homogeneous. We can show that the induced almost complex structure for $t = \pi/2$ is also not homogeous. As a Riemannian manifold, this is homogeneous, however the automorphism groups of the induced almost Hermitian structures do not act transitively on $S^1 \times S^2 \times \mathbf{R}^3$.

References

1. R.L. Bryant, *Submanifolds and special structures on the octonions*, J. Diff. Geom. **17** (1982) 185–232.
2. R. Harvey and H.B. Lawson, *Calibrated geometries*, Acta Math. **148** (1982) 47–157.
3. H. Hashimoto. *Characteristic classes of oriented 6-dimensional submanifolds in the octonians*, Kodai Math. J. **16** (1993) 65–73.
4. ———, *Oriented 6-dimensional submanifolds in the octonions III*, Internat. J. Math and Math. Sci. **18** (1995) 111–120.

ON THE REAL HYPERSURFACES OF LOCALLY CONFORMALLY-KÄHLER MANIFOLDS*

M.J. HRISTOV

Department of Algebra and geometry, University of Veliko Tarnovo, Veliko Tarnovo, 5000, Bulgaria
E-mail: milenjh@yahoo.com

There are sixteen classes of real hypersurfaces of Kähler manifold, generated by the four basic classes W_1, W_2, W_4 and W_6 in the classification scheme for almost contact metric manifolds [1]. In [5] these classes are described in terms of the hypersurface second fundamental form. In the partial case when the ambient manifold is complex space form the identities for the Riemannian and for the Hermitian-like curvatures are introduced [6,9]. The paper deals with the classification problem for the real hypersurfaces of locally conformally Kähler manifolds.

Keywords: Hermitian manifolds; Locally conformally Kähler manifold; Almost contact metric manifolds; Contact conformal changes; Real hypersurfaces.

1. Preliminaries

1.1. *Locally conformally Kähler manifolds*

Let AH be the class of almost Hermitian manifolds $M^{2n}(J,g)$ of even real dimension $\dim_{\mathbb{R}} M = 2n$, $n \geq 2$, almost complex structure $J : J^2 = -\mathrm{id}$ and compatible Riemannian metric $g : g(Jx, Jy) = g(x,y)$, $\forall\, x, y \in \mathfrak{X}M$, ($\mathfrak{X}M$ is the set of C^∞ vector fields over M). Let ∇ be the Levi-Civita connection of g. In general $\nabla J \neq 0$. From viewpoint of the well known classification scheme [7] the class K of Kähler manifolds is the null-class $W_0 = K$: $\nabla J = 0$. In the table below, the four basic classes of almost Hermitian manifolds are described in terms of the essential complex components and equations in local Hermitian frame $\{Z_\alpha = e_\alpha - iJe_\alpha,\, Z_{\overline{\alpha}} = \overline{Z_\alpha},\, \alpha \in I = \{1, \dots n\},\, \overline{\alpha} \in \overline{I} = \{\overline{1}, \dots \overline{n}\}\}$ for:

*Partially supported by Scientific researches fund of "St. Cyril and St. Methodius" University of V. Tarnovo under contract RD620-02/2006.

- $\Omega(x, y) = g(x, Jy)$ — the fundamental (Kähler) 2-form,
- $F(x, y, z) = -(\nabla_x \Omega)(y, z) = -F(x, z, y) = -F(x, Jy, Jz)$ — the fundamental (0,3)-tensor field,
- $d\Omega = \Omega_{ABC}\, dz^A \wedge dz^B \wedge dz^C, \quad 3\Omega_{ABC} = -\underset{(A,B,C)}{\sigma} F(Z_A, Z_B, Z_C)$,
- $N = [J, J]$ — the Nijenhuis tensor,
- $\theta = \frac{1}{2(n-1)}\delta\Omega \circ J = \frac{1}{2(n-1)}\mathrm{div}\Omega \circ J = \frac{1}{2(n-1)} g^{AB} F(Z_A, Z_B, J*)$ — the Lee form.

<table>
<tr><td colspan="4">$W = AH = W_4 \oplus W_4^{\perp}$ — almost Hermitian manifolds
$M^{2n}(J, g) : J^2 = -\mathrm{id},\ g \circ J = g,\ n \geq 2$</td></tr>
<tr><td colspan="3">↙
$W_4^{\perp} = SK = W_3 \oplus W_3^{\perp}$
semi-Kähler
$\theta = 0,$
$F_{\overline{\alpha}\,\alpha\beta} = -3\Omega_{\overline{\alpha}\,\alpha\beta} = 0$</td><td>↘
$W_4 = CK$ — locally conformally Kähler
$\theta \neq 0,\ N = 0,$
$F_{\overline{\alpha}\,\beta\gamma} = -3\Omega_{\overline{\alpha}\,\beta\gamma}$
$= \frac{1}{n-1}(g_{\overline{\alpha}\,*} \wedge \theta)_{\beta\gamma}$</td></tr>
<tr><td colspan="2">↙
$W_3^{\perp} = QK = W_2 \oplus W_2^{\perp}$
quasi-Kähler
$\theta = 0,\ N \neq 0,$
$N_{\alpha\beta\gamma} = -6i\Omega_{\alpha\beta\gamma} + 2iF_{\gamma\beta\alpha}$</td><td>↘
$W_3 = SH$
special Hermitian
$\theta = 0,\ N = 0,$
$F_{\overline{\alpha}\,\beta\gamma} = -3\Omega_{\overline{\alpha}\,\beta\gamma}$</td><td></td></tr>
<tr><td>↙
$W_1 = W_2^{\perp} = NK$
nearly-Kähler
$\theta = 0,\ N \neq 0,$
$d\Omega = \nabla\Omega,$
$N_{\alpha\beta\gamma} = -4i\Omega_{\alpha\beta\gamma}$
$= 4iF_{\gamma\beta\alpha}$</td><td>↘
$W_2 = AK$
almost Kähler
$\theta = 0,\ N \neq 0,$
$d\Omega = 0,$
$N_{\alpha\beta\gamma} = 2iF_{\gamma\beta\alpha}$
$= 2i(F_{\alpha\beta\gamma} + F_{\beta\gamma\alpha})$</td><td></td><td></td></tr>
</table>

The class of Hermitian manifolds is $H = SH \oplus CK\,(= W_3 \oplus W_4)$ and is characterized by the conditions:

$$F(x,y,z) = F(Jx,Jy,z) \quad \text{i.e.} \quad F = -3d\Omega,\ \theta \neq 0 \iff N = 0.$$

Modulo a constant, the vector field

$$E = J\delta J, \quad \text{where} \quad \delta J = \operatorname{div} J = g^{AB}(\nabla_{Z_A} J)Z_B,$$

is the metric-dual to the Lee form $\theta : \theta(X) = g(E,X),\ \forall X \in \mathfrak{X}M$. It is called the *Lee vector field* and JE is called *anti-Lee vector field*. The section $\mathscr{L}_p = \operatorname{span}\{E_p, J_pE_p\}$, $p \in M$ will be called *Lee section* and the distribution $\mathscr{L} = \bigcup\limits_{p\in M} \mathscr{L}_p$ — *Lee distribution.*

A Hermitian manifold, in the class $CK = W_4 : 3d\Omega = \theta \wedge \Omega$ ($\Longrightarrow d\theta = 0 \iff \theta$ is locally exact, in $\dim_{\mathbb{R}} M \geq 6$), is said to be locally conformally Kählerian. If the Lee form θ is exact a W_4-manifold is said to be globally conformally Kählerian.

The geometric description of these manifolds is well known. Let $g' = \mathrm{e}^{2\sigma} g$ be conformal change of the Kähler metric g, where $\sigma \in \mathcal{F}M$ (the set of infinitely smooth $\mathbb{R}$-valued functions on M). Thus it's obtained the Hermitian manifold $M(J,g')$, which is locally conformally Kähler (CK-) manifold:

$$K = W_0 \ni M(J,g) \xrightarrow{g'=\mathrm{e}^{2\sigma}g} M(J,g') \in W_4 = CK.$$

The Levi-Civita connection ∇' of g' and ∇ of g are related by

$$\nabla'_x y = \nabla_x y + d\sigma(x).y + d\sigma(y).x - g(x,y).E,$$

$$E = (\theta \overset{loc}{=} d\sigma)^{\star}, \quad \forall x,y \in \mathfrak{X}M.$$

The corresponding Riemannian (0,4)-curvature tensors R' and R are related by

$$R' = \mathrm{e}^{2\sigma}[R - \psi(Q)], \quad \text{where} \quad Q = \frac{1}{2}\|d\sigma\|^2 g + \nabla d\sigma - d\sigma \otimes d\sigma$$

and ψ is the operator, transforming a symmetric $(0,2)$ tensor field Σ to curvature tensor $\psi(\Sigma)$ of type $(0,4)$:

$$\psi(\Sigma)(x,y,z,u) = \big[\Sigma_u \wedge g_z - \Sigma_z \wedge g_u\big](x,y),$$

where Σ_u, $g_z, \ldots$ denote the 1-forms $\Sigma_u(z) = \Sigma(u,z)$, $g_z(u) = g(z,u), \ldots$. For example $\psi\big(\frac{1}{2}g\big) = \pi_1$ is one of the basic $U(n)$-invariant curvature tensors, introduced in [11] and $g_E = \theta$ is the Lee-form. Thus $\psi(Q)$ is

$$\psi(Q)(x,y,z,u) = \|d\sigma\|^2.\pi_1(x,y,z,u)$$

$$+ \underbrace{[(\nabla_u d\sigma - d\sigma(u)d\sigma) \wedge g_z - (\nabla_z d\sigma - d\sigma(z)d\sigma) \wedge g_u](x,y)}_{=\Theta(x,y,z,u)}.$$

So, it is valid:

$$R' = \mathrm{e}^{2\sigma}\big[R - \|d\sigma\|^2.\pi_1 - \Theta\big].$$

The basic examples of CK-manifolds are well known (see [10,12]). The real hypersurfaces of Generalized Hopf manifold (CK-manifolds with parallel Lee form: $\nabla\theta = 0$) have been studied by many authors (see [4]).

1.2. *Contact conformally almost contact metric manifolds*

Let $M^{2n+1}(\varphi,\xi,\eta,g)$ be an almost contact metric manifold with fundamental 2-form Φ : $\Phi(x,y) = g(x,\varphi y)$ and fundamental (0,3)-tensor field $F = -\nabla\Phi$. The transformations $c(u,v)$, forming by functions $u, v \in \mathcal{F}M$:

$$M(\varphi,\xi,\eta,g) \xrightarrow{c(u,v)} M(\overline{\varphi} = \varphi,\, \overline{\xi} = \mathrm{e}^{-v}\xi,\, \overline{\eta} = \mathrm{e}^{v}\eta,\, \overline{g} = \mathrm{e}^{2u}hg + \mathrm{e}^{2v}\eta\otimes\eta),$$

are said to be *contact conformal transformations of the structure* (φ,ξ,η,g) and form the *contact conformal group* (see [2]).

The Levi-Civita connections ∇ and $\overline{\nabla}$ for g and $\overline{g}$ are related by [2]

$$\begin{aligned}2\overline{g}(\overline{\nabla}_x y, z) &= 2\mathrm{e}^{2u}g(\nabla_x y,z) + [(d\mathrm{e}^{2u})\otimes g](x,y,z) + [(d\mathrm{e}^{2u})\wedge g_x)](y,z)\\ &+ \big\{d(\mathrm{e}^{2v}-\mathrm{e}^{2u})\otimes\eta\otimes\eta + \eta\otimes[d(\mathrm{e}^{2v}-\mathrm{e}^{2u})\wedge\eta]\big\}(x,y,z)\\ &+ (\mathrm{e}^{2v}-\mathrm{e}^{2u})\big\{2\eta(\nabla_x y)\eta(z) + [(\mathcal{L}_\xi g)\otimes\eta](x,y,z)\\ &\qquad - 2[d\eta_z\otimes\eta](x,y) - 2[d\eta_z\otimes\eta](y,x)\big\}.\end{aligned}$$

The fundamental $(0,3)$-tensors $\overline{F} = -\overline{\nabla}\,\overline{\Phi}$ and $F = -\nabla\Phi$ are related by [2]

$$\begin{aligned}\overline{F}(x,y,z) &= \mathrm{e}^{2u}F(x,y,z) - \mathrm{e}^{2u}[\Phi_x\wedge du + g_x\wedge du\circ\varphi](y,z)\\ &- \mathrm{e}^{2v}[\eta\otimes(\eta\wedge dv\circ\varphi)](x,y,z)\\ &+ (\mathrm{e}^{2v}-\mathrm{e}^{2u})\big\{[\eta\wedge d\eta_x\circ\varphi](y,z)\\ &\qquad - [\eta\otimes(d\eta_{\varphi z} + d\eta_z\circ\varphi)](x,y)\big\}.\end{aligned} \tag{1.1}$$

In the sense of the classification scheme of Alexiev-Ganchev [1] (there are 12 basic classes W_i of almost contact metric manifolds) the contact conformal class (invariant under the action of contact conformal group) is the class $W_1\oplus W_2\oplus W_3\oplus W_9$ and the basic classes are described with respect to local complex frame in the following table.

class	essential $\mathbb{C}$-components and $\mathbb{C}$-equations	$\delta\eta$	$\delta\Phi\circ\varphi$	$\delta\Phi(\xi)$
W_1	$F_{00\alpha}=\omega_\alpha=2i\eta_{0\alpha}=iN_{0\alpha 0}$ $=i(\mathcal{L}_\xi g)_{\alpha 0}$	0	$\omega\circ\varphi$	0
W_2 α-Sasakian	$F_{\overline{\alpha}\beta 0}=\frac{f(\xi)}{2n}g_{\overline{\alpha}\beta}=-i\eta_{\overline{\alpha}\beta}$ $=F_{\beta\overline{\alpha}0}$	0	0	$f(\xi)$
W_3 α-Kenmotsu	$F_{\overline{\alpha}\beta 0}=-i\frac{f^*(\xi)}{2n}g_{\overline{\alpha}\beta}$ $=-\frac{i}{2}(\mathcal{L}_\xi g)_{\overline{\alpha}\beta}$ $=-\frac{3}{2}\Phi_{\overline{\alpha}\beta 0}=-F_{\beta\overline{\alpha}0}$	$-f^*(\xi)$	0	0
W_9	$F_{\overline{\alpha}\beta\gamma}=-3\Phi_{\overline{\alpha}\beta\gamma}$ $=2(\Phi_{\overline{\alpha}*}\wedge\theta)_{\beta\gamma}$	0	$f\circ\varphi$	0

Here:

- $\theta=-\frac{\delta\eta}{2n}\cdot\eta+\frac{1}{2(n-1)}\cdot\delta\Phi\circ\varphi$ is the Lee form;
- $\omega(z)=F(\xi,\xi,z)$, $f(z)=g^{ij}F(e_i,e_j,z)$, $f^*(z)=g^{ij}F(e_i,\varphi e_j,z)$ are the 1-forms, associated to F, where $T_pM=\mathrm{span}\{e_i,\varphi e_i,\xi\}$;
- $N=[\varphi,\varphi]+2d\eta\otimes\xi$ is the generalised Nijenhuis tensor;
- $\mathcal{L}$ denotes the Lie-differentiation;
- $\eta_{AB}=d\eta(Z_A,Z_B)$, $\Phi_{ABC}=d\Phi(Z_A,Z_B,Z_C)$.

For any $M^{2n+1}(\varphi,\xi,\eta,g)\in W_1\oplus W_2\oplus W_3\oplus W_9$ and $n\geq 2$ it follows:
i) $d\Phi=\theta\wedge\Phi$, ii) $d\eta=(\omega\circ\varphi)\wedge\eta+\frac{\delta\Phi\circ\varphi}{n}.\Phi$, iii) $d\theta=0$, $d(\omega\circ\varphi)=0$.

The maximal subgroups G_i of the contact conformal group G, preserving corresponding subclasses W_i and the local conformal structure of W_i are well known [2,3]:

- $G_0=\{c(u,v):\ du=0,\ dv=\xi(v).\eta\}$ preserves the class $W_0:\nabla\varphi\equiv 0$ of cosymplectic manifolds: $W_0=G_0(W_0)$;
- $G_1=\{c(u,v):\ du=0\}$ preserves the class W_1 (i.e. the class of manifolds, locally G_1-contact conformally equivalent to cosymplectic manifold):

$$W_1=G_1(W_1),\quad W_1\overset{loc}{=}G_1(W_0);$$

- $G_{Sas} = \{c(u,v) : \ du = 0, \ v = 2u\}$ preserves the class W_2^0 of Sasakian manifolds: $d\Phi = \eta \wedge \Phi$:

$$W_2^0 = G_{Sas}(W_2^0);$$

- $G_2 = \{c(u,v) : \ du = 0, \ dv = 0\}$ (the group of D-homotheties, i.e. the homotheties on the contact distribution $D = \operatorname{Ker}\eta$) preserves the class W_2 of α-Sasakian manifolds (i.e. the class of manifolds, locally D-homothetic equivalent to Sasakian manifod.):

$$W_2 = G_2(W_2), \quad W_2 \overset{loc}{=} G_2(W_2^0),$$
$$W_1 \oplus W_2 \overset{loc}{=} G_1(W_2) \overset{loc}{=} G_1 \circ G_(W_2^0) = G_1(W_2^0);$$

- $G_3 = \{c(u,v) : \ du = \xi(u).\eta, \ dv = \xi(v).\eta\}$ preserves the class W_3 of α-Kenmotsu manifolds (i.e. the class of manifolds, locally G_3-contact conformally equivalent to cosymplectic manifold):

$$W_3 = G_3(W_3), \quad W_3 \overset{loc}{=} G_3(W_0), \quad W_2 \oplus W_3 \overset{loc}{=} G_3(W_2) \overset{loc}{=} G_3 \circ G_2(W_2^0);$$

- $G_9 = \{c(u,v) : \ du(\xi) = 0, \ dv = \xi(v).\eta\}$ preserves the class W_9 (i.e. the class of manifolds, locally G_9-contact conformally equivalent to cosymplectic manifold):

$$W_9 = G_9(W_9), \quad W_9 \overset{loc}{=} G_9(W_0);$$

- $G_u = \{c(u,v) : \ u = v\}$ — the subgroup of usual conformal changes locally generates the direct sums: $W_1 \oplus W_3 \oplus W_9$ and $W_1 \oplus W_2 \oplus W_3 \oplus W_9$

$$W_1 \oplus W_3 \oplus W_9 \overset{loc}{=} G_u(W_1) \overset{loc}{=} G_u \circ G_1(W_0),$$
$$W_1 \oplus W_2 \oplus W_3 \oplus W_9 \overset{loc}{=} G_u(W_1 \oplus W_2) \overset{loc}{=} G_u \circ G_1(W_2).$$

2. Real hypersurfaces of locally conformally-Kähler manifolds

Over a real ($\mathbb{R}$-)hypersurface M^{2n+1} of an almost Hermitian manifold $\overline{M}^{2n+2}(J,G)$ in a natural way arises an almost contact metric structure (φ,ξ,η,g): if N is the unit vector field, normal to M^{2n+1}, then

$$\xi = -JN, \quad g = G_{|M}, \quad \varphi = J - \eta \otimes N, \quad \eta = g_\xi : \eta(X) = g(\xi, X)$$

for all $X \in \mathfrak{X}M^{2n+1}$. Let ∇ and ∇' be the Levi-Chivita connections on M^{2n+1} and $\overline{M}^{2n+2}$ respectively. We denote by Φ the fundamental 2-form of the structure (φ,ξ,η,g)

$$\Phi(X,Y) = g(X,\varphi Y), \quad X,Y \in \mathfrak{X}M^{2n+1} \quad \text{and} \quad F' = -\nabla'\Phi, \ F = -\nabla\Phi.$$

To any point $p \in M$ the following spaces and vectors are attached:

$$T_p\overline{M} = T_pM \oplus N_p, \quad N_p \perp T_pM \quad \text{and} \quad T_pM = D_p \oplus \{\xi_p\}, \quad \xi_p \perp D_p.$$

The distribution $D = \{D_p = \operatorname{Ker}\eta_p,\ p \in M\}$ is the *contact distribution*, and $\{\xi\} = \{\operatorname{Im}\eta_p.\xi_p,\ p \in M\}$ is the *vertical distribution* of M. There are two shape operators A and $\mathcal{A}$, acting in M as follows:

$$\begin{aligned} A: T_p\overline{M} &\longrightarrow T_pM \qquad\qquad & \mathcal{A}: T_pM &\longrightarrow D_p \\ X &\longmapsto AX = -\nabla'_X N, & X &\longmapsto \mathcal{A}X = \nabla_X \xi. \end{aligned}$$

If $h(X,Y) = g(AX,Y)$, $X,Y \in \mathfrak{X}M^{2n+1}$ is the second fundamental tensor of M^{2n+1}, then the Gauss and Weingarten formulas

$$\nabla'_X Y = \nabla_X Y + h(X,Y)N, \qquad \nabla'_X N = -AX$$

imply

$$\begin{aligned} &(\nabla'_X \eta)Y = (\nabla_X \eta)Y, \\ &(\nabla'_X J)Y = (\nabla_X \varphi)Y + \{h(X,\varphi Y) + (\nabla_X \eta)Y\}N \\ &\qquad\qquad\qquad\qquad - \eta(Y)AX + h(X,Y)\xi, \\ &F'(X,Y,Z) = F(X,Y,Z) + \eta(Z)h(X,Y) - \eta(Y)h(X,Z). \end{aligned} \tag{2.1}$$

If $\overline{M}^{2n+2}(J,G)$ is Kählerian ($\nabla' J = 0$), then $M^{2n+1}(\varphi,\xi,\eta,g)$ belongs to $W_1 \oplus W_2 \oplus W_4 \oplus W_6$ and there are 16 classes of $\mathbb{R}$-hypersurfaces [5], generated by these four basic classes. The $\mathbb{R}$-hypersurfaces in each of the 16 classes are described by the second fundamental tensor. When the ambient space is complex space form, the equalities for the Riemannian and for the Hermitian-like curvatures over the basic classes $W_{1,2,4,6}$ are obtained [6,9].

Further let $M^{2n+1}(\varphi,\xi,\eta,g)$ be $\mathbb{R}$-hypersurface of a Kähler manifold $\overline{M}^{2n+2}(J,G)$, oriented by a unit normal vector field N. Let $G' = e^{2\sigma}G$, $\sigma \in \mathcal{F}\overline{M}$, be conformal change of the metric and let $\overline{M}^{2n+2}(J,G')$ be the corresponding locally conformally Kähler manifold. Over the $\mathbb{R}$-hypersurface $M^{2n+1}(\varphi,\xi,\eta,g)$ acts the usual local contact conformal transformation $c_\sigma \in G_\sigma$. The image

$$c_\sigma(M^{2n+1}(\varphi,\xi,\eta,g)) = M^{2n+1}(\varphi,\ \xi' = e^{-\sigma}\xi,\ \eta' = e^{\sigma}\eta,\ g' = e^{2\sigma}(hg+\eta\otimes\eta))$$

is $\mathbb{R}$-hypersurface in $\overline{M}^{2n+2}(J,G')$, oriented by the image $N' = e^{-\sigma}N$ and belongs to the class

$$\begin{aligned} G_\sigma(W_1 \oplus W_2 \oplus W_4 \oplus W_6) &= G_\sigma(W_1 \oplus W_2) \oplus G_\sigma(W_4 \oplus W_6) \\ &= W_1 \oplus W_2 \oplus W_3 \oplus W_9 \oplus G_\sigma(W_4 \oplus W_6). \end{aligned}$$

To obtain the subclass $G_\sigma(W_4 \oplus W_6)$ we calculate the essential $\mathbb{C}$-components F'_{ABC} of the image of the essential $\mathbb{C}$-components $F_{a\,\beta 0}$, $a \in I \cup \overline{I}$. From (1.1) in the considered case: $u = v = \sigma$ we get the relation

$$F'(x,y,z) = \mathrm{e}^{2\sigma}\left\{ F(x,y,z) - \begin{vmatrix} \Phi(x,y) & \Phi(x,z) \\ d\sigma(y) & d\sigma(z) \end{vmatrix} \right.$$
$$\left. - \begin{vmatrix} (g+\eta\otimes\eta)(x,y) & (g+\eta\otimes\eta)(x,z) \\ d\sigma(\varphi y) & d\sigma(\varphi z) \end{vmatrix} \right\}.$$

The direct computations give the following essential $\mathbb{C}$-components F'_{ABC}:

$$F'_{\overline{\alpha}\,\beta 0} = -i\mathrm{e}^{2\sigma}[d\eta + d\sigma(\xi).g]_{\overline{\alpha}\,\beta} = F'_{\beta\overline{\alpha}\,0},$$
$$F'_{\overline{\alpha}\,\beta\gamma} = 2\mathrm{e}^{2\sigma}(\Phi_{\overline{\alpha}\,*} \wedge d\sigma)_{\beta\gamma},$$
$$F'_{\alpha\,\beta 0} = \mathrm{e}^{2\sigma}F_{\alpha\,\beta 0}$$

and so, locally, $G_\sigma(W_4 \oplus W_6) = W_4 \oplus W_6 \oplus W_9$.

Thus we get the following main result of this subsection

Theorem (Local statement). *There are* $2^6 = 64$ *classes of* $\mathbb{R}$*-hypersurfaces of a locally conformally Kähler manifold of real dimension* ≥ 6*, generated by the basic classes* W_i*,* $i = 1,2,3,4,6,9$ *of almost contact metric manifolds (each of the classes* W_i *contains the null-class* W_0 *of the cosimplectic manifolds:* $\nabla\varphi = 0$*). When the ambient locally conformally Kähler manifold is of real dimension* 4*, then the class* W_4 *is empty and there are 32 classes of* $\mathbb{R}$*-hypersurfaces, generated by the basic classes* W_i*,* $i = 1,2,3,6,9$.

By using (2.1) we get the second fundamental tensor for $\mathbb{R}$-hypersurfaces of CK-manifold.

Remark. As special cases of $\mathbb{R}$-hypersurfaces of CK-manifold one can to consider the "position" of the Lee vector field and/or Lee distribution with respect to the unit normal, oriented $\mathbb{R}$-hypersurface. This leads one to considerations as in [8] and can be done for each of the basic classes.

References

1. V. Alexiev and G. Ganchev, *On the classification of almost contact metric manifolds*, Math. and Educ. in Math., Proc. of the XV Spring Conf. of UBM. Sunny Beach, (1986) 155–161.
2. ———, *Contact conformal transformations of the almost contact metric structures*, Comptes rendus de l'Académie Bulgare de Sciences **39** (1986), 27–30.

3. ———, *Canonical conformal group and its invariants on an almost contact metric manifold*, Comp. Rend. Acad. Bulg. Sci. **42** (1989), 15–18.
4. S. Dragomir, *Generalized Hopf manifolds, locally conformal Kaehler structures and real hypersurfaces*, Kodai Math. J. **14** (1991), 366–391.
5. G. Ganchev and M. Hristov, *Real hypersurfaces of a Kähler manifold (the sixteen classes)*, Perspectives of Complex Analysis, Differential Geometry and Mathematical Physics, Proc. 5th Int. Workshop on Complex Structures and Vector Fields, Varna, 2000, Eds S. Dimiev and K. Sekigawa, World Scientific, Singapore, (2001) 147–158.
6. ———, *Riemannian curvatures of the four basic classes of real hypersurfaces of a complex space form*, Third Int. Conf. on Geometry, Integrability and Quantization, June 14–23, 2001, Varna, Bulgaria, Ivailo M. Mladenov and Gregory L. Naber, Editors, Coral Press, Sofia 2001, pp. 238–248.
7. A. Gray and L. Hervella, *The sixteen classes of almost hermitian manifolds and their linear invariants*, Ann. di Mat. pura ed appl. (IV), CXXIII, 1980, 35–58.
8. G. Ganchev and V. Mihova, *Kähler manifolds of quasi-constant holomorphic sectional curvatures*, arXiv:math.DG/0505671 v1, 31 May 2005.
9. M.J. Hristov, *Hermitian-like curvatures of the four basic classes of real hypersurfaces of a complex space form*, Trends in complex analysis, differential geometry and mathematical physics (St. Konstantin, 2003), 63–75, World Sci. Publishing, River Edge, NJ, 2003.
10. L. Ornea and L. Vanheche, *Harmonicity and minimality of vector fields and distributions on locally conformally Kähler and hyperkähler manifolds*, Bull. Belg. Math. Soc. Simon Stevin **12** (2005), 543–555.
11. F. Tricerri and L. Vanhecke, *Curvature tensors on an almost Hermitian manifolds*, TAMS **267** (1981), 365–398.
12. I. Vaisman, *Generalized Hopf manifolds*, Geom. Dedicata **13** (1982), 231–255.

NORMAL FRAMES AND LINEAR TRANSPORTS ALONG PATHS IN LINE BUNDLES. APPLICATIONS TO CLASSICAL ELECTRODYNAMICS

B.Z. Iliev

Laboratory of Mathematical Modeling in Physics
Institute for Nuclear Research and Nuclear Energy
Boul. Tzarigradsko chaussée 72, 1784 Sofia, Bulgaria
E-mail: bozho@inrne.bas.bg
http://theo.inrne.bas.bg/~bozho/

The definitions and some basic properties of the linear transports along paths in vector bundles and the normal frames for them are recalled. The formalism is specified on line bundles and applied to a geometrical description of the classical electrodynamics. The inertial frames for this theory are discussed.

Keywords: Normal frames; Frame fields; Linear transports along paths; Line bundles; Inertial frames; Classical electrodynamics.

1. Introduction

The transports along paths in vector bundles [1] are one of the possible generalizations of the parallel transports in these bundles. They are a useful tool for a geometric formulation of quantum mechanics [2]. The frames normal for them are defined as ones in which the transports' matrices are the identity matrix. The significance of the normal frames (and coordinates) for the physics is a result of the assertion that they are the mathematical concept representing the physical notion of an 'inertial frame of reference' [3,4]. From here it follows that the (strong) equivalence principle in gravity physics is a provable theorem [3] and that the scope of its validity can be enlarged to include the gauge theories [4]; in particular, the classical electrodynamics [5].

The paper contains a partial review of the theory of linear transports along paths in vector bundles and the frames normal for them (sections 2

2000 MSC Numbers: 53B99, 53C99, 53Z05, 55R25, 83D05
2003 PACS Numbers: 02.40.Vh, 11.15-q, 04.50.+h, 04.90.+e

and 3). It is exemplified on 1-dimensional vector bundles (section 4). The formalism is applied to a geometric description of the classical electrodynamics and the inertial frames for it (section 5). Section 6 is devoted to a brief discussion of the inertial frames for the classical electromagnetic field. Section 7 closes the paper.

2. Linear transports along paths (brief review)

Let (E, π, B) be a complex [a] vector bundle [6,7] with bundle (total) space E, base B, which is a C^1 differentiable manifold, projection $\pi : E \to B$, and homeomorphic fibres $\pi^{-1}(x)$, $x \in B$.[b] By J and $\gamma : J \to B$ are denoted real interval and path in B, respectively. The paths are generally *not* supposed to be continuous or differentiable unless their differentiability class is stated explicitly. If γ is a C^1 path, the vector field tangent to it is denoted by $\dot{\gamma}$.

Definition 2.1. A *linear transport along paths* in the bundle (E, π, B) is a map L assigning to every path γ a map L^γ, *transport along* γ, such that $L^\gamma \colon (s,t) \mapsto L^\gamma_{s\to t}$ where the map $L^\gamma_{s\to t} \colon \pi^{-1}(\gamma(s)) \to \pi^{-1}(\gamma(t))$, $s, t \in J$, called *transport along* γ *from* s *to* t, has the properties:

$$L^\gamma_{s\to t} \circ L^\gamma_{r\to s} = L^\gamma_{r\to t}, \qquad r, s, t \in J, \tag{1}$$

$$L^\gamma_{s\to s} = \mathsf{id}_{\pi^{-1}(\gamma(s))}, \qquad s \in J, \tag{2}$$

$$L^\gamma_{s\to t}(\lambda u + \mu v) = \lambda L^\gamma_{s\to t} u + \mu L^\gamma_{s\to t} v, \qquad \lambda, \mu \in \mathbb{C},\ u, v \in \pi^{-1}(\gamma(s)), \tag{3}$$

where $\circ$ (id_X) denotes composition of maps (the identity map of a set X).

Let $\{e_i(s;\gamma)\}$ be a C^1 basis in $\pi^{-1}(\gamma(s))$, $s \in J$.[c] So, along $\gamma\colon J \to B$ we have a set $\{e_i\}$ of bases on $\pi^{-1}(\gamma(J))$ such that the liftings $\gamma \mapsto e_i(\cdot,\gamma)$ of paths are of class C^1.

Define the *matrix* $\boldsymbol{L}(t, s; \gamma) := \left[L^i{}_j(t, s; \gamma)\right]$ of L by $L^\gamma_{s\to t}(e_i(s;\gamma)) =: L^j{}_i(t, s; \gamma) e_j(t; \gamma)$, $s, t \in J$. A change $\{e_i(s;\gamma)\} \mapsto \{e'_i(s;\gamma) := A^j_i(s;\gamma) \times e_j(s;\gamma)\}$ via of a non-degenerate matrix $A(s;\gamma) := \left[A^j_i(s;\gamma)\right]$ implies

$$\boldsymbol{L}(t, s; \gamma) \mapsto \boldsymbol{L}'(t, s; \gamma) = A^{-1}(t;\gamma) \boldsymbol{L}(t, s; \gamma) A(s;\gamma). \tag{4}$$

Proposition 2.1. *A non-degenerate matrix-valued function* $\boldsymbol{L} : (t, s; \gamma) \mapsto \boldsymbol{L}(t, s; \gamma)$ *is a matrix of some linear transport along paths* L *(in a given field*

[a] All of our definitions and results hold also for real vector bundles. Most of them are valid for vector bundles over more general fields too but this is inessential for the following.
[b] When writing $x \in X$, X being a set, we mean "for all x in X" if the point x is not specified (fixed, given) and is considered as an argument or a variable.
[c] Here and henceforth the Latin indices run from 1 to $\dim \pi^{-1}(x)$, $x \in B$. We also assume the Einstein summation rule on indices repeated on different levels.

$\{e_i\}$ of bases along γ) iff

$$\boldsymbol{L}(t,s;\gamma) = \boldsymbol{F}^{-1}(t;\gamma)\boldsymbol{F}(s;\gamma) \tag{5}$$

where $\boldsymbol{F} : (t;\gamma) \mapsto \boldsymbol{F}(t;\gamma)$ is a non-degenerate matrix-valued function.

Proposition 2.2. *If the matrix $\boldsymbol{L}$ of a linear transport L along paths has a representation $\boldsymbol{L}(t,s;\gamma) = {}^*\boldsymbol{F}^{-1}(t;\gamma)\,{}^*\boldsymbol{F}(s;\gamma)$ for some matrix-valued function ${}^*\boldsymbol{F}(s;\gamma)$, then all matrix-valued functions $\boldsymbol{F}$ representing $\boldsymbol{L}$ via Eq. (5) are given by $\boldsymbol{F}(s;\gamma) = \boldsymbol{D}^{-1}(\gamma)\,{}^*\boldsymbol{F}(s;\gamma)$ where $\boldsymbol{D}(\gamma)$ is a non-degenerate matrix depending only on γ.*

Let $\{e_i(s;\gamma)\}$ be C^1 field of bases along $\gamma : J \to B$. The derivation $D : \gamma \mapsto D^\gamma : s \mapsto D^\gamma_s$, associated to L, acts on a C^1 lifting of paths λ as

$$D^\gamma_s \lambda = \left[\frac{\mathrm{d}\lambda^i_\gamma(s)}{\mathrm{d}s} + {\Gamma^i}_j(s;\gamma)\lambda^j_\gamma(s)\right] e_i(s;\gamma). \tag{6}$$

Here the *(2-index) coefficients* ${\Gamma^i}_j$ of the linear transport L are defined by

$${\Gamma^i}_j(s;\gamma) := \left.\frac{\partial {L^i}_j(s,t;\gamma)}{\partial t}\right|_{t=s} = -\left.\frac{\partial {L^i}_j(s,t;\gamma)}{\partial s}\right|_{t=s}. \tag{7}$$

If a matrix $\boldsymbol{F}$ determines a transport L according to proposition 2.1, then

$$\boldsymbol{\Gamma}(s;\gamma) := \left[{\Gamma^i}_j(s;\gamma)\right] = \left.\frac{\partial \boldsymbol{L}(s,t;\gamma)}{\partial t}\right|_{t=s} = \boldsymbol{F}^{-1}(s;\gamma)\frac{\mathrm{d}\boldsymbol{F}(s;\gamma)}{\mathrm{d}s}. \tag{8}$$

A change $\{e_i\} \to \{e'_i = A^j_i e_i\}$ of the bases along a path γ with a non-degenerate C^1 matrix-valued function $A(s;\gamma) := \left[A^j_i(s;\gamma)\right]$ implies $\boldsymbol{\Gamma}(s;\gamma) = \left[{\Gamma^i}_j(s;\gamma)\right] \mapsto \boldsymbol{\Gamma}'(s;\gamma) = \left[{\Gamma'^i}_j(s;\gamma)\right]$ with

$$\boldsymbol{\Gamma}'(s;\gamma) = A^{-1}(s;\gamma)\boldsymbol{\Gamma}(s;\gamma)A(s;\gamma) + A^{-1}(s;\gamma)\frac{\mathrm{d}A(s;\gamma)}{\mathrm{d}s}. \tag{9}$$

3. Normal frames for linear transports

Let a linear transport L along paths be given in a vector bundle (E,π,B), $U \subseteq B$ be an arbitrary subset in B, and $\gamma : J \to U$ be a path in U.

Definition 3.1. A *frame field (of bases)* in $\pi^{-1}(\gamma(J))$ is called *normal along γ for L* if the matrix of L in it is the identity matrix along the given path γ. A *frame field (of bases)* defined on U is called *normal on U for L* if it is normal along every path $\gamma\colon J \to U$ in U. A *linear transport along paths* (or *along a path γ*) is called *Euclidean along some (or the given) path γ* if it admits a frame normal along γ. A *linear transport along paths* is called *Euclidean* on U if it admits frame(s) normal on U.

Proposition 3.1. *The following statements are equivalent in a given frame $\{e_i\}$ over $U \subseteq B$:*

(i) *The matrix of L is the identity matrix on U, i.e. $\boldsymbol{L}(t,s;\gamma) = \mathbb{1}$ along every path γ in U.*

(ii) *The matrix of L along every $\gamma : J \to U$ depends only on γ, i.e. it is independent of the points at which it is calculated: $\boldsymbol{L}(t,s;\gamma) = C(\gamma)$ where C is a matrix-valued function of γ.*

(iii) *If E is a C^1 manifold, the coefficients ${\Gamma^i}_j(s;\gamma)$ of L vanish on U, i.e. $\boldsymbol{\Gamma}(s;\gamma) = 0$ along every path γ in U.*

Corollary 3.1. *Every linear transport along paths is Euclidean along every fixed path without self-intersections.*

Theorem 3.1. *A linear transport along paths admits frames normal on some set (along a given path) if and only if its action along every path in this set (along the given path) depends only on the initial and final point of the transportation but not on the particular path connecting these points. So, a transport is Euclidean on $U \subseteq B$ iff it is path-independent on U.*

Proposition 3.2. *Let L be a linear transport along paths in (E, π, M), E and M being C^1 manifolds, and L be Euclidean on $U \subseteq M$ (resp. along a C^1 path $\gamma : J \to M$). Then the matrix $\boldsymbol{\Gamma}$ of its coefficients has the representation*

$$\boldsymbol{\Gamma}(s;\gamma) = \sum_{\mu=1}^{\dim M} \Gamma_\mu(\gamma(s))\dot\gamma^\mu(s) \equiv \Gamma_\mu(\gamma(s))\dot\gamma^\mu(s) \tag{10}$$

in any frame $\{e_i\}$ along every (resp. the given) C^1 path $\gamma\colon J \to U$, where $\Gamma_\mu = \left[{\Gamma^i}_{j\mu}\right]_{i,j=1}^{\dim \pi^{-1}(x)}$ are some matrix-valued functions, defined on an open set V containing U (resp. $\gamma(J)$) or equal to it, and $\dot\gamma^\mu$ are the components of $\dot\gamma$ in some frame $\{E_\mu\}$ along γ in the bundle space tangent to M, $\dot\gamma = \dot\gamma^\mu E_\mu$. The functions ${\Gamma^i}_{j\mu}$ are termed 3-index coefficients *of L.*

Let U be an open set. The changes $\{E_\mu\} \mapsto \{E'_\mu = B^\nu_\mu E_\nu\}$, $B = \left[B^\nu_\mu\right]$ being non-degenerate matrix-valued function, and $\{e_i|_x\} \mapsto \{e'_i|_x = A^j_i(x) e_j|_x\}$, with $A := \left[A^j_i\right]_{i,j=1}^{\dim \pi^{-1}(x)}$ being non-degenerate and of class C^1, imply (see Eq. (9) and Eq. (10)) that Γ_μ transforms into Γ'_μ such that

$$\Gamma'_\mu = B^\nu_\mu A^{-1}\Gamma_\nu A + A^{-1}E'_\mu(A) = B^\nu_\mu A^{-1}\big(\Gamma_\nu A + E_\nu(A)\big). \tag{11}$$

Theorem 3.2. *A linear transport L along paths is Euclidean on a submanifold N of M if and only if in every frame $\{e_i\}$, in the bundle space over N,* *the matrix of its coefficients has a representation Eq. (10) along every* C^1 *path in* N *and, for every* $p_0 \in N$ *and a chart* (V, x) *of* M *such that* $V \ni p_0$ *and* $x(p) = (x^1(p), \dots, x^{\dim N}(p), t_0^{\dim N+1}, \dots, t_0^{\dim M})$ *for every* $p \in N \cap V$ *and constant numbers* $t_0^{\dim N+1}, \dots, t_0^{\dim M}$*, the equalities*

$$\big(R^N_{\alpha\beta}(-\Gamma_1, \dots, -\Gamma_{\dim N})\big)(p) = 0, \qquad \alpha, \beta = 1, \dots, \dim N, \tag{12}$$

where $R^N_{\alpha\beta}(-\Gamma_1, \dots, -\Gamma_{\dim N}) := R_{\alpha\beta}(-\Gamma_1, \dots, -\Gamma_{\dim M}) = -\frac{\partial \Gamma_\alpha}{\partial x^\beta} - \frac{\partial \Gamma_\beta}{\partial x^\alpha} + \Gamma_\alpha \Gamma_\beta - \Gamma_\beta \Gamma_\alpha$ *hold for all* $p \in N \cap V$*. Here* $\Gamma_1, \dots, \Gamma_{\dim N}$ *are the first* $\dim N$ *of the matrices of the 3-index coefficients of* L *in the frame* $\{\frac{\partial}{\partial x^\mu}\}$ *in the tangent bundle space over* $N \cap V$*.*

4. Linear transports and normal frames in line bundles

Let (E, π, M) be one-dimensional vector bundle over a C^1 manifold M; such bundles are called line bundles. The fibre of (E, π, M) can be identified with $\mathbb{C}$ ($\mathbb{R}$ in the real case) and the fibre $\pi^{-1}(x)$ over $x \in M$ is an isomorphic image of $\mathbb{C}$ ($\mathbb{R}$). Let $\gamma : J \to M$ be of class C^1 and L be a transport in (E, π, M). A frame $\{e\}$ along γ consists of a single non-zero vector field $e : (s;\gamma) \to e(s;\gamma) \in \pi^{-1}(\gamma(s)) \backslash \{0\}$, $s \in J$, and in it the matrix of L^γ at $(t, s) \in J \times J$ is a number $\boldsymbol{L}(t, s; \gamma) \in \mathbb{C}$, $L^\gamma_{s \to t}(ue(s;\gamma)) = u\boldsymbol{L}(t, s; \gamma) e(t; \gamma)$ for $u \in \mathbb{C}$ and $s, t \in J$. By proposition 2.1, the general form of $\boldsymbol{L}$ is

$$\boldsymbol{L}(t, s; \gamma) = \frac{f(s;\gamma)}{f(t;\gamma)} \tag{13}$$

where $f : (s;\gamma) \mapsto f(s;\gamma) \in \mathbb{C} \backslash \{0\}$ is defined up to multiplication with a function of γ (proposition 2.2). Due to Eq. (8), the matrix of the coefficient(s) of L is $\boldsymbol{\Gamma}(s;\gamma) = \frac{\partial \boldsymbol{L}(t,s;\gamma)}{\partial s}\Big|_{t=s} = \frac{1}{f(s;\gamma)} \frac{\mathrm{d} f(s;\gamma)}{\mathrm{d}s} = \frac{\mathrm{d}}{\mathrm{d}s}\big[\ln(f(s;\gamma)\big]$ and [1, Eq. (2.31)] takes the form $\boldsymbol{L}(t, s; \gamma) = \exp\Big(-\int_s^t \boldsymbol{\Gamma}(\sigma;\gamma) \mathrm{d}\sigma\Big)$.

A change $e(s;\gamma) \mapsto e'(s;\gamma) = a(s;\gamma) e(s;\gamma)$, with $a(s;\gamma) \in \mathbb{C} \backslash \{0\}$, of the frame $\{e\}$ implies (see Eq. (4) and Eq. (9)) $\boldsymbol{L}(t, s; \gamma) \mapsto \boldsymbol{L}'(t, s; \gamma) = \frac{a(s;\gamma)}{a(t;\gamma)} \boldsymbol{L}(t, s; \gamma)$ and $\boldsymbol{\Gamma}(s;\gamma) \mapsto \boldsymbol{\Gamma}'(s;\gamma) = \boldsymbol{\Gamma}(s;\gamma) + \frac{\mathrm{d}}{\mathrm{d}s}\big[\ln(a(s;\gamma))\big]$.

The action of the derivation D generated by L is $D^\gamma_s \lambda = \Big(\frac{\mathrm{d}\lambda_\gamma(s)}{\mathrm{d}s} + \boldsymbol{\Gamma}(s;\gamma)\lambda_\gamma(s)\Big) e(s;\gamma)$ where $\lambda \in \mathrm{PLift}^1(E, \pi, M)$ and Eq. (6) was used.

A frame $\{e\}$ is normal for L along γ (resp. on U) iff in that frame equation Eq. (13) holds with $f(s;\gamma) = f_0(\gamma)$ where $\gamma : J \to M$ (resp. $\gamma : J \to U$) and $f_0 : \gamma \mapsto f_0(\gamma) \in \mathbb{C} \backslash \{0\}$ (see proposition 3.1). Since, in a frame normal along γ (resp. on U), it is fulfilled $\boldsymbol{L}(t, s; \gamma) = \mathbf{1}$, $\boldsymbol{\Gamma}(s;\gamma) = 0$

for a given path γ (every path in U), in every frame $\{e' = ae\}$, we have

$$\boldsymbol{L}'(t,s;\gamma) = \frac{a(s;\gamma)}{a(t;\gamma)}, \quad \boldsymbol{\Gamma}'(s;\gamma) = \frac{\mathrm{d}}{\mathrm{d}s}\big[\ln(a(s;\gamma))\big]. \tag{14}$$

In addition, for Euclidean on $U \subseteq M$ transport L, the representation

$$\boldsymbol{\Gamma}'(s;\gamma) = \Gamma'_\mu(\gamma(s))\dot{\gamma}'^{\mu}(s) \tag{15}$$

holds for every C^1 path $\gamma\colon J \to U$ and some $\Gamma'_\mu : V \to \mathbb{C}$ with V being an open set such that $V \supseteq U$ (proposition 3.2). Therefore (see theorem 3.1 and [1, theorem 4.2]) Eq. (14) holds for $a(s;\gamma) = a_0(\gamma(s))$ for some $a_0 : U \to \mathbb{C}\backslash\{0\}$. So Eq. (15) can be satisfied if we choose

$$\Gamma'_\mu = E_\mu(a) \qquad \text{for some } a : V \to \mathbb{C},\ a|_U = a_0 \tag{16}$$

with $\{E_\mu\}$ being a frame in the bundle space tangent to M. If U is not an open set, this choice of Γ'_μ is not necessary as Eq. (15) will be preserved, if to the r.h.s. of Eq. (16) is added a function G'_μ such that $G'_\mu\dot{\gamma}'^{\mu} = 0$.

Due to Eq. (11), the functions Γ_μ and Γ'_μ in two pairs of frames $(\{e\},\{E_\mu\})$ and $(\{e' = ae\},\{E'_\mu = B^\nu_\mu E_\nu\})$, respectively, are linked via

$$\Gamma'_\mu = B^\nu_\mu\Gamma_\nu + \frac{1}{a}E'_\mu(a) = B^\nu_\mu\big(\Gamma_\nu + E_\nu(\ln a)\big). \tag{17}$$

So, with respect to changes of the frames in the tangent bundle space over M, when $a = 1$, they behave like the components of a one-form.

Frames normal along injective paths always exist (corollary 3.1), but on an arbitrary submanifold $N \subseteq M$ they exist iff the functions Γ_μ satisfy the conditions Eq. (12) with $x \in N$ in the coordinates described in theorem 3.2.

5. Bundle description of the classical electromagnetic field

Recall [8,9], the classical electromagnetic field is described via a real 1-form A over a real 4-manifold M (endowed with a (pseudo-)Riemannian metric g and) representing the space-time model.[d] The electromagnetic field itself is represented by the two-form $F = \mathrm{d}A$, where "d" denotes the exterior derivative operator, with local components (in some local coordinates $\{x^\mu\}$)

$$F_{\mu\nu} = -\frac{\partial A_\mu}{\partial x^\nu} + \frac{\partial A_\nu}{\partial x^\mu}. \tag{18}$$

The Maxwell equations describing it, and its (minimal) interactions with other objects are invariant under a gauge transformation

$$A_\mu \mapsto A'_\mu = A_\mu + \frac{\partial\lambda}{\partial x^\mu} \tag{19}$$

[d] The particular choice of M is insignificant for the following.

or $A \mapsto A' = A + \mathrm{d}\lambda$, where λ is a C^2 function. The electromagnetic field is invariant under simultaneous changes of the local coordinate frame, $E_\mu = \frac{\partial}{\partial x^\mu} \mapsto E'_\mu = B^\nu_\mu E_\nu$ with $B^\nu_\mu := \frac{\partial x^\nu}{\partial x'^\mu}$, and a gauge transformation Eq. (19):

$$A_\mu \mapsto A'_\mu = B^\nu_\mu A_\nu + E'_\mu(\lambda) = B^\nu_\mu \Big(A_\nu + \frac{\partial \lambda}{\partial x^\nu} \Big). \qquad (20)$$

The similarity between Eq. (20) and Eq. (17) implies the idea of identifying the electromagnetic potentials A_μ with the matrices Γ_μ of the 3-index coefficients of some transport in a line bundle (E, π, M).

Let M be a real 4-manifold, representing the space-time model, and (E, π, M) be a line bundle over it.[e] We *identify the potentials A_μ of an electromagnetic field with the coefficients of a linear transport L along paths in (E, π, M) whose matrix has the representation Eq. (15) (along every path and in every pair of frames).* Hence, the 3-index coefficients of L are uniquely defined and supposed to be (arbitrarily) fixed in some pair of frames.

Since the 3-index coefficients of linear transport are defined in a pair of frames $(\{e\}, \{E_\mu\})$, $\{e\}$ in the bundle space E and $\{E_\mu\}$ in the tangent bundle space $T(M)$, the change Eq. (20) expresses simply the transformation of A_μ under the pair of changes $e \mapsto e' = ae$ and $E_\mu \mapsto E'_\mu = B^\nu_\mu E_\nu$ and follows from Eq. (17) for $a = \mathrm{e}^\lambda$. It should be emphasized, now the (pure) gauge transformation Eq. (19) appears as a special case of Eq. (20), corresponding to a change of the frame in E and a fixed frame in $T(M)$.[f] So Eq. (19) is directly incorporated in the definition of the field potential A. This conclusion is in contrast to classical electrodynamics where the change Eq. (19) is a simple observation of 'additional' invariance of the field.

Defining the electromagnetic field (strength) by $F = \mathrm{d}A$, the Eq. (18) remains valid in a coordinate frame $\{E_\mu = \partial/\partial x^\mu\}$. Since A and F possess all of the properties they must have in classical electrodynamics, they represent an equivalent description of electromagnetic field. The only difference with respect to the classical description is the clear geometrical meaning of these quantities, as a consequence of which an electromagnetic field can be identified with a linear transport in a line bundle over the space-time. The proposed treatment of electromagnetic field is equivalent to the mod-

---

[e] The consideration of the real case does not change the above results with the exception that $\mathbb{C}$ should be replaced by $\mathbb{R}$.

[f] Cf. a similar conclusion in [10, p. 178], in which a gauge transformation, in a general gauge theory, is interpreted as a change in fibre coordinates of a principle bundle.

ern one of gauge theories (see, e.g., [11] or [12]), where the electromagnetic potentials are regarded as coefficients of a suitable linear connection.

In the approach proposed, the different gauge conditions, which are frequently used, find a natural interpretation as a partial fix of the class of frames in the bundle space employed.

6. Normal and inertial frames

Comparing Eq. (18) with Eq. (12), we get [g]

$$F_{\mu\nu} = R_{\mu\nu}(-A_0, -A_1, -A_2, -A_3). \tag{21}$$

Thus, the electromagnetic tensor F is completely responsible for the existence of frames normal for L (theorem 3.2). If N is a submanifold of M, frames normal on N for L exist iff in the special coordinates $\{x^\mu\}$, described in theorem 3.2, is valid $F_{\alpha\beta}|_U = 0$ for $\alpha, \beta = 1, \dots, \dim N$. According to theorem 3.1, an electromagnetic field admits frames normal on $U \subseteq M$ iff the transport L corresponding to it is path-independent on U. Thus, if L is path-dependent on U, the field does not admit frames normal on U. This result is the classical analogue of a quantum Aharonov-Bohm effect [13,14], whose essence is that the electromagnetic potentials directly, not only through the field tensor F, can give rise to observable physical results.

Suppose L is Euclidean on a neighborhood $U \subseteq M$. Eq. (21) and theorem 3.2, imply $F|_U = \mathrm{d}A|_U = 0$, i.e. on U the electromagnetic field strength vanishes and hence the field is a pure gauge on U,

$$A_\mu|_U = \frac{\partial f_0}{\partial x^\mu}\bigg|_U \tag{22}$$

for some C^1 function f_0 defined on an open set containing U or equal to it. By proposition 3.1, in a frame $\{e'\}$ normal on U for L vanish the 2-index coefficients of L along any path γ in U:

$$\mathbf{\Gamma}'(s;\gamma) = A'_\mu(\gamma(s))\dot{\gamma}^\mu(s) = 0 \tag{23}$$

for every $\gamma : J \to U$ and $s \in J$. Using Eq. (22), it is trivial to see that any transformation Eq. (20) with $\lambda = -f_0$ transforms A_μ into A'_μ such that

$$A'_\mu|_U = 0. \tag{24}$$

Hence, by Eq. (23), the frame $\{e' = \mathrm{e}^{-f_0}e\}$ in E is normal for L on U. Therefore, in the frame $\{e'\}$, vanish not only the 2-index coefficients of L

[g] In this section, we assume the Greek indices to run over the range $0, 1, 2, 3$.

but also its *3-index* ones, i.e. $\{e'\}$ is a frame *strong normal* on U for L. Applying Eq. (20) one can verify, *all frames strong normal on a neighborhood U for L are obtainable from $\{e'\}$ by multiplying its vector e' by a function f such that $\frac{\partial f}{\partial x^\mu}\big|_U = 0$*, i.e. they are $\{b\,e^{-f_0}e\}$ with $b \in \mathbb{R}\backslash\{0\}$ as U is a neighborhood. Thus, every frame normal on a neighborhood U for L is strong normal on U for L and vice versa.

A frame (of reference) in the bundle space, in which Eq. (24) holds on $U \subseteq M$, is called *inertial on U for the electromagnetic field.* The frames inertial on U for a given electromagnetic field are the ones in which its potentials vanish on U.

In a frame inertial on $U \subseteq M$ for an electromagnetic field it is not only a pure gauge, but in such a frame its potentials vanish on U. Therfore (see also [15–17]) there existent frames inertial at a single point and/or along paths without self-intersections for every electromagnetic field; on m-submanifolds with $m \geq 2$ such frames exist only as an exception.

Let there be given a physical system consisting of pure or, possibly, interacting gravitational and electromagnetic fields which are described via, respectively, a linear connection ∇ in the tangent bundle $(T(M), \pi_T, M)$ over the space-time M and a linear transport along paths in a line bundle (E, π_E, M) over M. On one hand, the frames inertial for an electromagnetic field, if any, in the bundle space E are completely independent of any frame in the bundle space $T(M)$ tangent to M. On the other hand, the frames inertial for the gravity field, i.e. the ones normal for ∇, if any, are frames in $T(M)$ and have nothing in common with the frames in E. Consequently, if there is a frame $\{E_\mu\}$ in $T(M)$ inertial on $U \subseteq M$ for the gravity field and a frame $\{e\}$ in E inertial on the same set U for the electromagnetic field, the frame $\{e \times E_\mu\} = \{(e, E_\mu)\}$ in the bundle space of the bundle $(E \times T(M), \pi_E \times \pi_T, M \times M)$ over $M \times M$ can be called simply *inertial* on U (for the system of gravity and electromagnetic fields). Thus, in an inertial frame, if any, the potentials of both, gravity and electromagnetic, fields vanish. Relying on the results obtained in this work, as well as on the ones in [3,15–17], we can assert the existence of inertial frames at every single space-time point and/or along every path without self-intersections in it. On submanifolds of dimension higher than one, inertial frames exist only for some exceptional configurations of the fields which can be described on the base of the results in the cited works.

7. Conclusion

In the present paper was exemplified the theory of linear transports along paths and the frames normal for them on line bundles. The application of the results to the classical electrodynamics gives rise to a geometric interpretation of the electromagnetic field as a linear transport in a line bundle and to an introduction of inertial frames for this field. As pointed in [5], the linear transport, describing the electromagnetic field, is in fact the parallel transport generated by the linear connection describing it in the known its geometrical interpretation [18].

The coincidence of the normal and inertial frames for the electromagnetic field expresses the equivalence principle for that field [5]. This principle is a provable theorem and it is always valid at any single point or along given path (without selfintersections) as these are the only cases when normal frames for a linear connection or transport always exist.

The considerations in this work confirm our opinion that the frames (and possibly coordinates) in bundle spaces, in which some physical fields 'live', should be regarded as parts of the frames of references with respect to which a physical system is investigated.[h] For an electromagnetic field, these are the one-vector field frames $\{e\}$ in the bundle space E of the line bundle (E, π, M) in which the field is describe via a transport L. With respect to $\{e\}$ is defined the coefficient of L and with respect to a pair $(\{e\}, \{E_\mu\})$, with $\{E_\mu\}$ being a frame in the bundle tangent to the spacetime, are defined the (3-index) coefficients of L which coincide with the components A_μ of the 4-vector potential of the electromagnetic field. Since A_μ are observable (if one beliefs in the Aharonov-Bohm effect) and $\{E_\mu\}$ is an essential path of the frames of reference, one can conclude that $\{e\}$ should be a part of the frame of reference.

The work on this paper was partially supported by the National Science Fund of Bulgaria under Grant No. F 1515/2005.

References

1. B.Z. Iliev, JP Journal of Geometry and Topology **5**, 187 (2005), http://arXiv.org e-Print archive, E-print No. gr-qc/9809084, 1998.
2. ———, *Fibre bundle formulation of nonrelativistic quantum mechanics* (full version), http://arXiv.org e-Print archive, E-print No. quant-ph/0004041 (2000).

[h] Cf. [19, sect. 1] where similar ideas relative to conventions concerning physical laboratories can be found.

3. ———, Journal of Geometry and Physics **24**, 209 (1998), DOI No. 10.1016/S0393-0440(97)00011-9, http://arXiv.org e-Print archive, E-print No. gr-qc/9806062, June 1998.
4. ———, Journal of Geometry and Physics **45**, 24 (February 2003), DOI No. 10.1016/S0393-0440(02)00119-5, http://arXiv.org e-Print archive, E-print No. hep-th/0110194, October 2001.
5. ———, Equivalence principle in classical electrodynamics, in *Has the last word been said on classical electrodynamics? — New Horizons*, eds. A. Chubykalo, V. Onoochin, A. Espinoza and R. Smirnov-Rueda (Rinton Press, Princeton, NJ, 2004) pp. 385–403. http://arXiv.org e-Print archive, E-print No. gr-qc/0303002, March 2003.
6. W.A. Poor, *Differential geometric structures*, McGraw-Hill Book Company Inc., New York, 1981.
7. W. Greub, S. Halperin and R. Vanstone, *De Rham cohomology of manifolds and vector bundles*, Connections, Curvature, and Cohomology, Vol. 1, Academic Press, New York and London, 1972.
8. L.D. Landau and E.M. Lifshitz, *Classical theory of fields*, Course of theoretical physics, Vol. II, 5 edn., Pergamon Press, Oxford, 1967. Translation from Russian, Nauka, Moscow, 1973.
9. W. Drechsler and M.E. Mayer, *Fibre bundle techniques in gauge theories*, Lecture notes in physics, Vol. 67, Springer-Verlag, Berlin-Heidelberg-New York, 1977.
10. C. Nash and S. Sen, *Topology and Geometry for physicists*, Academic Press, London-New York, 1983.
11. N.P. Konopleva and V.N. Popov, *Gauge fields*, second edn., Hardwood Academic Publishers, Chur-London-New York, 1981. Translation from Russian: Atomizdat, Moscow, 1972 (1 ed.), 1980 (2 ed.).
12. J. Baez and J.P. Muniain, *Gauge fields, knots and gravity*, Series in knots and Everything, Vol. 4, World Scientific, Singapore-New Jersey-London-Hong Kong, 1994.
13. Y. Aharonov and D. Bohm, Phys. Rev. **115**, p. 485 (1959).
14. H.J. Bernstein and A.V. Phillips, Scientific America **245**, 94 (July 1981),
15. B.Z. Iliev, Journal of Physics A: Mathematical and General **29**, 6895 (1996), DOI No. 10.1088/0305-4470/29/21/020, http://arXiv.org e-Print archive, E-print No. gr-qc/9608019, 1998.
16. ———, Journal of Physics A: Mathematical and General **30**, 4327 (1997), DOI No.10.1088/0305-4470/30/12/019, http://arXiv.org e-Print archive, E-print No. gr-qc/9709053, 1997.
17. ———, Journal of Physics A: Mathematical and General **31**, 1287 (January 1998), DOI No. 10.1088/0305-4470/31/4/016, http://arXiv.org e-Print archive, E-print No. gr-qc/9805088, May 1998.
18. M. Göckeler and T. Schücker, *Differential geometry, gauge theories, and gravity*, Cambridge Univ. Press, Cambridge, 1987.
19. I. Fatibene, M. Rerraris and M. Francaviglia, International journal of geometric methods in modern physics **1**, 443 (2004).

EXISTENCE AND UNIQUENESS RESULTS FOR THE SCHRÖDINGER — POISSON SYSTEM BELOW THE ENERGY NORM

A.M. IVANOV

Department of Applied Mathematics, Technical University of Sofia,
8 "Kliment Ohridski" Str., 1756 Sofia, Bulgaria
E-mail: aivanov@tu-sofia.bg

G.P. VENKOV

Department of Applied Mathematics, Technical University of Sofia,
8 "Kliment Ohridski" Str., 1756 Sofia, Bulgaria
E-mail: gvenkov@tu-sofia.bg

We consider the problems of local and global in time existence of the solution to the Schrödinger-Poisson system (or the Hartree equation) in $\mathbb{R}^3$. We prove that if the solution at $t = 0$ has a bounded $H^{1/2}$-norm, then it remains bounded in some time interval $[0, T]$, which together with the charge conservation law immediately imply global existence. Thus, we extend the previous results for this type of nonlinear Schrödinger equation.

Keywords: Schrödinger-Poisson system; Hartree equation; local and global existence; Sobolev space.

1. Introduction

The Schrödinger-Poisson system can be written in terms of the wave function ψ and the potential V as follows

$$i\partial_t\psi + \Delta\psi = V\psi, \quad (t, x) \in \mathbb{R}_+ \times \mathbb{R}^3, \tag{1}$$

$$\Delta V = -4\pi|\psi|^2, \tag{2}$$

where the $(-)$ sign in the Poisson equation (2) corresponds to the repulsive type of the single-quantum state interaction. The above system appears in Quantum Mechanics, the analysis of the electron transport under particular contexts and the semiconductor theory.

Solving the second equation with respect to $V = V(|\psi|^2)$ and substituting into the first one we obtain the following single equation

$$i\partial_t\psi + \Delta\psi = (|x|^{-1} * |\psi|^2)\psi, \tag{3}$$

where $*$ denotes the usual convolution operator in $\mathbb{R}^3$. Equation (3) is known as the Schrödinger equation with nonlocal nonlinearity of Coulomb type or the Hartree equation and arises, for instance, as an effective description of boson stars (see for more details Born [2] and Lieb and Simon [10]).

In general, the Hartree nonlinearities $(|x|^{-\gamma} * |\psi|^2)\psi$, $0 < \gamma < 3$ are by far most difficult to handle, due mainly to the presence of the convolution; the power nonlinearities $|\psi|^{p-1}\psi$, $p > 1$ are quite manageable and can be handled by the standard theory of semilinear Schrödinger equation. The analysis divides into the following three cases: the short-range case $\gamma > 1$, the long-range case $0 < \gamma < 1$, and the borderline (or the critical) case $\gamma = 1$.

Although there is a large literature on the Cauchy problem for the Schrödinger-Poisson system (and Hartree equation), there still remains the question for the optimal energy space, which ensures the local and global existence of solutions with arbitrary initial data.

Let $H^s(\mathbb{R}^3)$ denotes the inhomogeneous Sobolev space of positive index s and $\dot{H}^s(\mathbb{R}^3)$ denotes the corresponding homogeneous space with norms

$$\|f\|_{H^s} = \|\langle\nabla\rangle^s f\|_{L^2(\mathbb{R}^3)}; \qquad \|f\|_{\dot{H}^s} = \||\nabla|^s f\|_{L^2(\mathbb{R}^3)},$$

respectively. We shall consider the Cauchy problem for the system (1)–(2) with initial data

$$\psi(0, x) = \psi_0. \tag{4}$$

The local existence definition that we shall use reads as follow: for any choice of initial data $\psi_0 \in H^s$, there exists a positive time $T = T(\|\psi_0\|_{H^s})$ depending only on the norm of the initial data, such that a solution to the Cauchy problem exists on the time interval $[0, T]$ and is unique in a certain Banach space of functions $X_T \subset C^0([0, T], H^s)$. If $T = \infty$ we say that a solution to the Cauchy problem exists globally.

For our purposes we restrict ourselves to initial data in H^s with $0 < s < 1$, and make some comments for the limiting cases where $s = 0, 1$. It is known (see Ginibre and Velo [5]), that the initial value problem (1)–(2) is locally well-posed in H^s when $s > \frac{1}{2}$, and local in time solutions enjoy charge conservation law

$$\|\psi(t, \cdot)\|_{L^2} = \|\psi_0(\cdot)\|_{L^2}. \tag{5}$$

Moreover, H^1 solutions enjoy conservation of the energy

$$E(\psi)(t) = \frac{1}{2}\|\psi(t,\cdot)\|^2_{\dot{H}^1} + \frac{1}{4}\int_{\mathbb{R}^3} V(t,x)|\psi(t,x)|^2 dx = E(\psi)(0), \qquad (6)$$

which together with (5) and the local theory immediately yields global in time well-posedness with initial data in H^1. One of the first results concerning similar problem is the paper of Chadam and Glasey [4], where they prove global existence for the Hartree equation with an external Coulomb potential in the energy space H^1. A small data global existence and scattering is proved in the work of Hayashi, Naumkin and Ozawa [6] in the space $H^{\gamma,0} \cap H^{0,\gamma}$ for $\frac{1}{2} < \gamma < \frac{n}{2}$, where $H^{m,s}$ denotes the weighted Sobolev space.

Another approach for proving local/global existence for the Schrödinger-Poisson system (and Hartree equation) is the use of Strichartz-type inequalities. In [3] Castella establishes global existence result for solutions in the class $C^0(\mathbb{R}, L^2(\mathbb{R}^3)) \cap L^q(\mathbb{R}, L^p(\mathbb{R}^3))$ for range of the admissible indices $p > 3$, $\frac{1}{q} = \frac{3}{4} - \frac{3}{2p}$ and initial data in L^2. Note that, due to the Sobolev embedding $H^s \subset L^2 \cap L^p$ in $\mathbb{R}^3$ with $p > 3$, implies $s = \frac{3}{2} - \frac{3}{p} > \frac{1}{2}$.

The main purpose of the present work is to construct a local in time theory for the homogeneous Sobolev space $\dot{H}^{\frac{1}{2}}$ without using conservation laws and the Strichartz (i.e., linear space-time) estimates for the free propagator $e^{it\Delta}$. We intend to prove our results by a standard fixed-point method applied to an integral version of (1)–(2). The proof employs sharp estimates (e. g., Kato [7] inequality (15) below) for obtaining local Lipschitz continuity of the nonlinearities of Hartree type. Then, replacing $\dot{H}^{\frac{1}{2}}$ with $H^{\frac{1}{2}}$, we succeed to establish global existence for the Schrödinger-Poisson system with arbitrary initial data in $H^{\frac{1}{2}}$, which improves the H^s- theory with $s > \frac{1}{2}$.

2. Local existence result in $\dot{H}^{\frac{1}{2}}$

For $R > 0$ and $T > 0$, let $\dot{X}_T$ be a ball of radius R in the function space $C^0([0,T], \dot{H}^{\frac{1}{2}}(\mathbb{R}^3))$, defined by

$$\dot{X}_T = \left\{ \psi \in C^0([0,T], \dot{H}^{\frac{1}{2}}(\mathbb{R}^3));\ \|\psi\|_{L^\infty([0,T],\dot{H}^{\frac{1}{2}})} \le R \right\}. \qquad (7)$$

Our first result is the following theorem.

Theorem 2.1. *There exists a constant $T > 0$, such that for any $\psi_0 \in \dot{H}^{\frac{1}{2}}(\mathbb{R}^3)$ with $\|\psi_0\|_{\dot{H}^{\frac{1}{2}}} \le R$, there exists unique solution $\psi \in \dot{X}_T$ to* (1)–(2), *which satisfies the initial condition* (4).

Proof. Assuming the initial data $\psi_0 \in \dot{H}^{\frac{1}{2}}$ and $\psi \in \dot{X}_T$ and using the Duhamel's integral representation we define the map $\Phi : \psi \mapsto \Phi(\psi)$ by

$$\Phi(\psi)(t) = U(t)\psi_0 - i\int_0^t U(t-s)V\psi(s)ds : \ 0 < t \leq T, \tag{8}$$

where $U(t) = e^{it\Delta}$ is the free Schrödinger group. Then, the following estimate holds

$$\|\Phi(\psi)\|_{L^\infty([0,T],\dot{H}^{\frac{1}{2}})} \leq C\Big(\|\psi_0\|_{\dot{H}^{\frac{1}{2}}} + \|V\psi\|_{L^1([0,T],\dot{H}^{\frac{1}{2}})}\Big). \tag{9}$$

Using the Leibnitz rule for fractional derivative [9], together with Kato-Ponce type estimates [8]

$$\begin{gathered}\|fg\|_{\dot{H}^s} \leq C\Big(\|f\|_{L^{p_1}}\||\nabla|^s g\|_{L^{p_2}} + \||\nabla|^s f\|_{L^{p_3}}\|g\|_{L^{p_4}}\Big),\\ s \geq 0, \quad \frac{1}{p_1} + \frac{1}{p_2} = \frac{1}{p_3} + \frac{1}{p_4} = \frac{1}{2}, \quad p_2, p_3 \in (1,\infty),\end{gathered} \tag{10}$$

the space estimate of the Hartree nonlinearity becomes

$$\|V\psi\|_{\dot{H}^{\frac{1}{2}}} \leq C\Big(\|\psi\|_{L^3}\||\nabla|^{\frac{1}{2}}V\|_{L^6} + \|\psi\|_{\dot{H}^{\frac{1}{2}}}\|V\|_{L^\infty}\Big). \tag{11}$$

Thanks to Sobolev inequality

$$\|\psi\|_{L^p} \leq C\|\psi\|_{\dot{H}^s}, \quad p > 1, \quad \frac{1}{p} = \frac{1}{2} - \frac{s}{3}, \tag{12}$$

we get

$$\|\psi\|_{L^3} \leq C\|\psi\|_{\dot{H}^{\frac{1}{2}}}. \tag{13}$$

For the L^∞ norm of the potential V, we obtain the first crucial estimate

$$\|V\|_{L^\infty} = \sup_{x\in\mathbb{R}^3}\int_{\mathbb{R}^3}\frac{|\psi(y)|^2}{|x-y|}dy \leq C\|(-\Delta_{x-y})^{\frac{1}{4}}\psi\|_{L^2}^2 \leq C\|\psi\|_{\dot{H}^{\frac{1}{2}}}^2, \tag{14}$$

where we use the operator inequality (see Kato [7])

$$|x-y|^{-1} \leq \frac{\pi}{2}(-\Delta_{x-y})^{\frac{1}{2}} \tag{15}$$

and the translation invariance $\Delta_{x-y} = \Delta_y$ for all $x \in \mathbb{R}^3$.

The Gagliardo-Nirenberg inequality (see for instance Stein [12])

$$\||\nabla|^j g\|_{L^p} \leq C\||\nabla|^m g\|_{L^r}^a\|g\|_{L^q}^{1-a}, \tag{16}$$

holds for any $1 \leq q, r \leq \infty$ and any $0 \leq j < m$. Here $p \geq 1$ is such that $\frac{1}{p} = \frac{j}{3} + \big(\frac{1}{r} - \frac{m}{3}\big)a + \frac{1-a}{q}$ and the parameter a is any from the interval

$\frac{j}{m} \le a \le 1$. Rewriting the above inequality with $j = \frac{1}{2}$, $m = 2$, $p = 6$, $r = \frac{3}{2}$ and $q = \infty$ we obtain

$$\||\nabla|^{\frac{1}{2}} V\|_{L^6} \le C\||\nabla|^2 V\|^a_{L^{\frac{3}{2}}} \|V\|^{1-a}_{L^\infty} = C\| - \Delta V\|^a_{L^{\frac{3}{2}}} \|V\|^{1-a}_{L^\infty}, \tag{17}$$

which holds for any parameter $a \in [1/4, 1]$. By using (13), (14), and the fact $\Delta V = -4\pi|\psi|^2$, we get the estimate

$$\||\nabla|^{\frac{1}{2}} V\|_{L^6} \le C\|\psi\|^{2a}_{L^3} \|V\|^{1-a}_{L^\infty} \le C\|\psi\|^{2a}_{\dot{H}^{\frac{1}{2}}} \|\psi\|^{2-2a}_{\dot{H}^{\frac{1}{2}}} = C\|\psi\|^2_{\dot{H}^{\frac{1}{2}}}. \tag{18}$$

Collecting the above estimates yields

$$\|V\psi\|_{\dot{H}^{\frac{1}{2}}} \le C\|\psi\|^3_{\dot{H}^{\frac{1}{2}}}, \tag{19}$$

which after integration over the finite slab $[0, T]$, leads to the following local estimate of the nonlinear term

$$\|V\psi\|_{L^1([0,T],\dot{H}^{\frac{1}{2}})} \le CTR^3. \tag{20}$$

Therefore we can rewrite (9) as

$$\|\Phi(\psi)\|_{L^\infty([0,T],\dot{H}^{\frac{1}{2}})} \le C\big(\|\psi_0\|_{\dot{H}^{\frac{1}{2}}} + TR^3\big). \tag{21}$$

In order to prove the contraction property for the map Φ, we shall proceed in the same way as for (20). Let $\psi, \phi \in \dot{X}_T$ and define the potential $V_{\psi\phi}$ by

$$\Delta V_{\psi\phi} = -4\pi\psi\bar{\phi}, \tag{22}$$

which is equivalent in $\mathbb{R}^3$ to

$$V_{\psi\phi} = \frac{1}{|x|} * \psi\bar{\phi}. \tag{23}$$

We can estimate

$$\begin{aligned} \|V_{\psi\psi}\psi - V_{\phi\phi}\phi\|_{\dot{H}^{\frac{1}{2}}} &\le \|(V_{\psi\psi} - V_{\phi\psi})\psi\|_{\dot{H}^{\frac{1}{2}}} \\ &\quad + \|(V_{\phi\psi} - V_{\phi\phi})\psi\|_{\dot{H}^{\frac{1}{2}}} + \|V_{\phi\phi}(\psi - \phi)\|_{\dot{H}^{\frac{1}{2}}} \\ &= A + B + D. \end{aligned} \tag{24}$$

For the first quantity, as in (11) we have

$$\begin{aligned} A &= \|(V_{\psi\psi} - V_{\phi\psi})\psi\|_{\dot{H}^{\frac{1}{2}}} \\ &\le C\big(\|\psi\|_{L^3}\||\nabla|^{\frac{1}{2}}(V_{\psi\psi} - V_{\phi\psi})\|_{L^6} + \|\psi\|_{\dot{H}^{\frac{1}{2}}}\|V_{\psi\psi} - V_{\phi\psi}\|_{L^\infty}\big) \end{aligned} \tag{25}$$

Definition (22) and Kato estimate (14) imply

$$\|V_{\psi\psi} - V_{\phi\psi}\|_{L^\infty} \le C\|\psi - \phi\|_{\dot{H}^{\frac{1}{2}}} \|\psi\|^2_{\dot{H}^{\frac{1}{2}}}, \tag{26}$$

while from Gagliardo-Nirenberg inequality (18) we write

$$\||\nabla|^{\frac{1}{2}}(V_{\psi\psi} - V_{\phi\psi})\|_{L^6} \leq C\|\psi - \phi\|_{\dot{H}^{\frac{1}{2}}} \|\psi\|^2_{\dot{H}^{\frac{1}{2}}} \tag{27}$$

and finally, using (13) we obtain

$$A \leq C\|\psi - \phi\|_{\dot{H}^{\frac{1}{2}}} \|\psi\|^2_{\dot{H}^{\frac{1}{2}}}. \tag{28}$$

Repeating the same procedure, we can bound the remaining quantities in (24) as follows

$$B \leq C\|\psi - \phi\|_{\dot{H}^{\frac{1}{2}}} \|\psi\|_{\dot{H}^{\frac{1}{2}}} \|\phi\|_{\dot{H}^{\frac{1}{2}}} \tag{29}$$

and

$$D \leq C\|\psi - \phi\|_{\dot{H}^{\frac{1}{2}}} \|\phi\|^2_{\dot{H}^{\frac{1}{2}}}. \tag{30}$$

Thus, we finally obtain

$$\|V_{\psi\psi}\psi - V_{\phi\phi}\phi\|_{L^1([0,T],\dot{H}^{\frac{1}{2}})} \leq CTR^2\|\psi - \phi\|_{L^\infty([0,T],\dot{H}^{\frac{1}{2}})}. \tag{31}$$

Choosing $C\|\psi_0\|_{\dot{H}^{\frac{1}{2}}} \leq \frac{R}{2}$ and $T \leq \frac{1}{2CR^2}$, we see that the map $\Phi : \psi \mapsto \Phi(\psi)$ is a contraction on $\dot{X}_T$ and hence $\Phi(t)$ has a unique fixed point ψ for any time $t \leq T(\|\psi_0\|_{\dot{H}^{\frac{1}{2}}})$. □

Of significant importance is the observation that the homogeneous Sobolev norms are not invariant under time translation and thus, we can not extend our local existence result globally in time. In the next section we shall establish global existence for the Schrödinger-Poisson system (1)–(2) by assuming at least charge conservation.

3. Global existence result in $H^{\frac{1}{2}}$

From the arguments mentioned above, the proof of global existence requires the use of at least one of the conservation laws (5) and (6). Therefore, we assume conservation of charge and consider the inhomogeneous Sobolev space $H^{\frac{1}{2}}$. Let X_T be a ball of radius $R > 0$ in the function space $C^0([0,T], H^{\frac{1}{2}}(\mathbb{R}^3))$, defined by

$$X_T = \left\{ f \in C^0([0,T], H^{\frac{1}{2}}(\mathbb{R}^3));\ \|f\|_{L^\infty([0,T],H^{\frac{1}{2}})} \leq R \right\}. \tag{32}$$

Our second main result concerns the global existence below the energy space.

Theorem 3.1. *For any $\psi_0 \in H^{\frac{1}{2}}(\mathbb{R}^3)$ with $\|\psi_0\|_{H^{\frac{1}{2}}} \leq R$, there exists unique solution $\psi \in X_T$ to (1)–(2), which satisfies the initial condition (4).*

Proof. We start by showing that the nonlinearity $V\psi$ is locally Lipschitz continuous from $L^1([0,T],H^{\frac{1}{2}})$ into $C^0([0,T],H^{\frac{1}{2}})$. In other words, we shall prove that for all $\psi,\phi \in X_T$ holds the following estimate

$$\|V_{\psi\psi}\psi - V_{\phi\phi}\phi\|_{L^1([0,T],H^{\frac{1}{2}})} \leq C(T,R)\|\psi-\phi\|_{L^\infty([0,T],H^{\frac{1}{2}})}. \tag{33}$$

From the equivalence

$$\|f\|_{L^2} + \||\nabla|^{\frac{1}{2}}f\|_{L^2} \lesssim \|f\|_{H^{\frac{1}{2}}} \lesssim \|f\|_{L^2} + \||\nabla|^{\frac{1}{2}}f\|_{L^2}, \tag{34}$$

it is sufficient to estimate the quantities

$$\|V_{\psi\psi}\psi - V_{\phi\phi}\phi\|_{L^2}, \qquad \||\nabla|^{\frac{1}{2}}(V_{\psi\psi}\psi - V_{\phi\phi}\phi)\|_{L^2}. \tag{35}$$

The second one is estimated in the previous section by (24)–(30) and it is obvious that

$$\||\nabla|^{\frac{1}{2}}(V_{\psi\psi}\psi - V_{\phi\phi}\phi)\|_{L^2} \leq C\big(\|\psi\|^2_{H^{\frac{1}{2}}} + \|\phi\|^2_{H^{\frac{1}{2}}}\big)\|\psi-\phi\|_{H^{\frac{1}{2}}}. \tag{36}$$

For the first quantity in (35), applying Hölder inequality we have

$$\begin{aligned}
&\|V_{\psi\psi}\psi - V_{\phi\phi}\phi\|_{L^2}\\
&\leq \|(V_{\psi\psi}-V_{\phi\psi})\phi\|_{L^2} + \|(V_{\phi\psi}-V_{\phi\phi})\psi\|_{L^2} + \|V_{\phi\phi}(\psi-\phi)\|_{L^2}\\
&\leq C\big(\|(V_{\psi\psi}-V_{\phi\psi})\|_{L^6}\|\phi\|_{L^3} + \|(V_{\phi\psi}-V_{\phi\phi})\|_{L^6}\|\phi\|_{L^3}\\
&\quad + \|V_{\psi\psi}\|_{L^\infty}\|\phi-\psi\|_{L^2}\big).
\end{aligned} \tag{37}$$

Observing that $\frac{1}{|x|} \in L^{(3,\infty)}$, where $L^{(p,q)}$ are the Lorentz spaces and $L^{(p,\infty)}$ denotes the weak L^p space (see Berg and Löfström [1] and O'Neil [11] for more details) and using the weak Young inequality, we get

$$\|(V_{\psi\psi}-V_{\phi\psi})\|_{L^6} \leq C\left\|\frac{1}{|x|}\right\|_{L^{(3,\infty)}}\||\psi|^2-\phi\bar{\psi}\|_{L^{\frac{6}{5}}} \leq C\|\psi\|_{L^3}\|\phi-\psi\|_{L^2}, \tag{38}$$

which is the second crucial inequality in the present work. The remaining terms in (37) deserve no further comment, since they can be estimated in a similar fashion to all estimates derived so far. Thus, by integration in time, we derive (33) with a Lipschitz constant $C(T,R) = CTR^2 \leq \frac{1}{2}$.

Let $\psi_0 \in H^{\frac{1}{2}}$ be a fixed initial data of (1)–(2) and let us suppose that the solution ψ belongs to the Banach space X_T. Local existence and uniqueness of a solution $\psi \in X_T$ now follows by standard methods for evolution equations with locally Lipschitz nonlinearities, that is, the map $\Phi(t)$, defined by (8) has a fixed point for any time $0 < t \leq T(\|\psi_0\|_{H^{\frac{1}{2}}})$. Moreover, continuous dependence on initial data, i.e. the map $\psi_0 \mapsto \psi \in X_T$ is continuous for every compact interval $I \subset [0,T]$, can also be deduced by standard arguments for general theory on semilinear evolution equations.

The global existence result follows from the following remark, that is, if we choose the radius R of the ball $X_T \subset C^0([0,T], H^{\frac{1}{2}})$ to be constant during the time evolution and such that $R \geq 2C\|\psi_0\|_{H^{\frac{1}{2}}}$, then one can reiterate in time with initial data in $T, 2T, \ldots$ and cover the whole $[0, \infty)$. □

Acknowledgments

The authors are grateful to Prof. Vladimir Georgiev, University of Pisa, Italy, for helpful advices and the long and inspiring discussions during the preparation of the work. The responsibility for any errors is of course entirely ours.

References

1. J. Berg and J. Löfström, *Interpolation spaces*, Springer-Verlag, Berlin-Heidelberg-New York, 1976.
2. M. Born, *Atomic Physics*, Blackie and Son Limited, London, 1969.
3. F. Castella, M3AS **7** (1997), 1051–1083.
4. J. Chadam and R. Glassey, J. Math. Phys. **16** (1975), 1122–1130.
5. J. Ginibre and G. Velo, Math. Zeitschr. **170** (1980), 109–136.
6. H. Hayashi, P. Naumkin and T. Ozawa, SIAM J. Math. Anal. **29** (1998), 1256–1267.
7. T. Kato, *Perturbation Theory for Linear Operators*, 2nd edition, Springer-Verlag, Berlin, 1980.
8. T. Kato and G. Ponce, Comm. Pure Appl. Math. **41** (1988), 891–907.
9. C. Kenig, G. Ponce and L. Vega, Comm. Pure Appl. Math. **46** (1993), 527–620.
10. E. Lieb and B. Simon, Comm. Math. Phys. **53** (1977), 185–194.
11. R. O'Neil, Duke Math. J. **30** (1963), 129–142.
12. E. Stein, *Singular Integrals and Differentiability Properties of Functions*, Princeton Math. Ser. **30**, Princeton University Press, Princeton, N.J., 1970.

DOUBLE-COMPLEX QUADRATIC FORMS

B. KIRADJIEV

Foresty Thechnical University,
10, bul. Kliment Ohridsky
1756 Sofia, Bulgaria
E-mail: Kiradjiev2003@yahoo.com

In this paper a multidimensional version of the double-complex algebra $C(1,j)$ is developed. The problem about the double-complex quadratic forms, stated in [1], is considered. We find the natural holomorphic Q-geometry and complex Hermitian one following the classical theorem of Lagrange for complex quadratic forms Q (see for instance [2]). A remark about fourth-complex quadratic forms is included [3]. Let us note that in fact the double-complex algebra is an alternative version of the known bi-complex algebra [1].

Keywords: Double-complex n-vectors; Double-complex quadratic forms; Complex Hermitian quadratic forms; Double-complex Q-orthogonal geometry; Fourth-complex quadratic forms.

1. Double-complex numbers and double-complex vectors

We shall consider the algebra of double-complex numbers $C(1,j)$ constituted by the elements $a = z + jw$, where z and w are complex numbers and $j^2 = i$, $i^2 = -1$. The underlying vector space coincides with $C \times C = C^2$, the vector space of couples (z, w) of complex numbers. The conjugate a^* of a is defined as $a^* = z - jw$. It follows that in the algebra $C(1,j)$ we have $aa^* = z^2 - iw^2$. In $C \times C$ we have $a + a^* = 2z$, and $a - a^* = 2jw$.

By C^{2n} is denoted the Cartesian product of n times C^2, and respectively by $C^n(1,j)$ the Cartesian product of n times $C(1,j)$. Clearly $(C^2)^n$ is the underlying vector space of the algebra $C^n(1,j)$.

Let $\alpha := (\alpha^1, \alpha^2, \ldots, \alpha^n)$ be a sequence of n double-complex numbers. It defines a double-complex n-vector, denoted by α. The double-complex numbers α^k are called coordinates of α. We say that α is vector in $C^n(1,j)$. We shall consider a system $\alpha_1, \ldots, \alpha_m$ of m double-complex n-vectors. By

2000 MS Classification: 30G35, 31C99, 32W99

(α_k^l), $k = 1, \dots, m$, $l = 1, \dots, m$, is denoted the matrix of all coordinates of the given system. If this matrix is of maximal rank the mentioned system is linear independent. In the case when the $n \times n$ matrix (α_k^l), $k = 1, \dots, n$, $l = 1, \dots, n$, is of maximal rank, i.e. $\det(\alpha_k^l) \neq 0$, the sequence of double-complex vectors $(\alpha^1, \alpha^2, \dots, \alpha^n)$ is a system of linear independent double-complex n-vectors. Each maximal system of linear independent double-complex vectors in $C^n(1, j)$ is called a base of $C^n(1, j)$.

Now we will equipped the set of all double-complex n-vectors $C^n(1, j)$ with operations of addition and multiplication of its elements. Namely, we have

Proposition 1.1. *The set of all double-complex n-vectors is an algebra with respect to the following operations: if α and β are double-complex n-vectors*

$$\alpha + \beta := (\alpha^1 + \beta^1, \dots, \alpha^n + \beta^n),$$
$$\alpha\beta := (\alpha^1\beta^1, \dots, \alpha^n\beta^n).$$

Proof. It is easy to verify that with respect to the first operation (the addition) the set $C^n(1, j)$ is a commutative group with $O(0, \dots, 0)$ as a neutral element. The second operation is a commutative multiplication. Both of them define a structure of an algebra over the algebra $C(1, j)$ of double-complex numbers. □

Definition 1.1. We say that the double-complex n-vector α is a zero-divisor in the algebra $C^n(1, j)$ if there exist double-complex n-vector γ such that $\gamma\alpha = 0$.

Proposition 1.2. *The double-complex n-vector α is a zero-divisor in $C^n(1, j)$ iff for each coordinate $\alpha^k = z^k + jw^k$, we have $(z^k)^2 - i(w^k)^2 = 0$, $k = 1, \dots, n$.*

Proof. Let $\gamma = \{(u^k + jv^k)\}$ be a double-complex n-vector such that $\gamma\alpha = 0$. In coordinates the product $\gamma\alpha$ seems as follows $\gamma\alpha = \{(u^k + jv^k)(z^k + jw^k)\}$. Calculating we obtain the following homogenous linear system with respect to the given non-zero couple u^k, v^k

$$z^k u^k + i w^k v^k = 0, \quad w^k u^k + z^k v^k = 0.$$

This implies that the determinant of the coefficients of the considered system is equal to zero, namely $(z^k)^2 - i(w^k)^2 = 0$, $k = 1, \dots, n$. □

Definition 1.2. We say that α^* is a conjugate n-vector of the n-vector $\alpha = (\alpha^k)$ if $\alpha^* := ((\alpha^1)^*, (\alpha^2)^*, \dots, (\alpha^n)^*)$.

Proposition 1.3. *For every double-complex n-vector α we have*

$$\alpha\alpha^* = ((z^1)^2 - i(w^1)^2, \dots, (z^n)^2 - i(w^n)^2) \in C^n.$$

2. Bilinear and quadratic forms on $C^n(1, j)$

We shall consider C-bilinear symmetric forms on the algebra of double-complex n-vectors. Let $B(\alpha, \beta)$ be such form. This means that a mapping B is defined as follows

$$B : C^n(1, j) \times C^n(1, j) \to C, \quad (\alpha, \beta) \to B(\alpha, \beta),$$
$$\alpha, \beta \in C^n(1, j), \; B(\alpha, \beta) \in C,$$

and $B(\alpha, \beta) = B(\beta, \alpha)$ for every couple (α, β) of elements of $C^n(1, j)$.

Let $(\varepsilon_1, \varepsilon_2, \dots, \varepsilon_n)$ be a base in $C^n(1, j)$. Then we have

$$\alpha = \sum \alpha^l \varepsilon_l \quad \text{and} \quad \beta = \sum \beta^k \varepsilon_k,$$

and

$$B(\alpha, \beta) = \sum_l \sum_k \alpha^l \beta^k B(\varepsilon_l, \varepsilon_k).$$

Setting $B(\varepsilon_l, \varepsilon_k) = b_{lk}$ we can rewrite the above expression as follows

$$B(\alpha, \beta) = \sum_l \sum_k b_{lk} \alpha^l \beta^k \quad \text{with } b_{lk} = b_{kl} \in C.$$

Definition 2.1. We say that a C-quadratic form $Q(\alpha)$ over $C^n(1, j)$ is defined if there is a C-bilinear form $B(\alpha, \beta)$ such that $Q(\alpha) = B(\alpha, \alpha)$ for every double-complex vector α.

So, $Q(\alpha)$ is defined with the help of the symmetric bilinear form $B(\alpha, \beta)$. Inversely, we have

$$B(\alpha, \beta) = 1/2(Q(\alpha + \beta) - Q(\alpha) - Q(\beta)), \quad \forall \alpha, \beta \in C^n(1, j).$$

With $\alpha = \sum \alpha^l \varepsilon_l$ we obtain $Q(\alpha) = \sum \sum q_{lk} \alpha^l \alpha^k$, setting $q_{lk} = b_{lk}$.

A matrix representation for the quadratic form is obtained as ordinary:

$$Q(\alpha) = (\alpha^1, \dots, \alpha^n) Q(\alpha^1, \dots, \alpha^n)^\tau)$$

where Q denotes the matrix $\|q_{lk}\|$, $(\alpha^1, \dots, \alpha^n)^\tau$ is the column of the coordinates α^k and $(\alpha^1, \dots, \alpha^n)$ is its row.

3. Double-complex holomorphic Q-orthogonal geometry

Let $Q := Q(\alpha^1, \ldots \alpha^n)$ be a double-complex symmetric quadratic form. The double-complex coordinate representation of Q seems as follows

$$\begin{aligned} Q(\alpha^1, \ldots, \alpha^n) = q_{11}(\alpha^1)^2 &+ 2q_{12}\alpha^1\alpha^2 + \cdots + 2q_{1n}\alpha^1\alpha^n \\ &+ \ q_{22}(\alpha^2)^2 + \cdots + 2q_{2n}\alpha^2\alpha^n \\ &+ \cdots \\ &+ \ q_{nn}(a^n)^2, \end{aligned}$$

where $\alpha^k = z^k + jw^k$, $k = 1, \ldots, n$. Having in mind that

$$(\alpha^k)^2 = (z^k)^2 + i(w^k)^2 + j(2z^k w^k)$$

and

$$\alpha^k \alpha^l = (z^k + jw^k)(z^l + jw^l) = z^k z^l + iw^k w^l + j(z^k w^l + w^k z^l),$$

we obtain a complex-coordinate representation of Q,

$$\begin{aligned} Q(z^1, w^1, \ldots, z^n, w^n) &= \sum_k((z^k)^2 + i(w^k)^2) + \sum_{k,l}(z^k z^l + iw^k w^l) \\ &+ j\left[\sum_k(2z^k w^k) + \sum_{k,l}(z^k w^l + w^k z^l)\right]. \end{aligned}$$

Or $Q = Q_0 + jQ_1$, where Q_0 and Q_1 are complex-valued quadratic forms, respectively the even part and the odd part of Q:

$$Q_0 = \sum_k((z^k)^2 + i(w^k)^2) + \sum_{k,l}(z^k z^l + iw^k w^l),$$

$$Q_1 = \sum_k(2z^k w^k) + \sum_{k,l}(z^k w^l + w^k z^l).$$

3.1. *Double-complex holomorphicity*

A double-complex function $f(\alpha) = f_0(z, w) + jf_1(z, w)$, where $\alpha = z + jw$ is a double-complex variable, is a holomorphic double-complex function [1], if by definition its even part $f_0(z, w)$ and odd part $f_1(z, w)$ satisfy the following double-complex Cauchy-Riemann system of complex partial differential equations

$$\partial f_0/\partial z = \partial f_1/\partial w, \quad \partial f_0/\partial w = i\partial f_1/\partial z.$$

Theorem 3.1. *Each double-complex quadratic form*

$$Q(z^1, w^1, \ldots, z^k, w^k, \ldots, z^n, w^n)$$

is a separately double-complex holomorphic function with respect to every couple (z^k, w^k) of its coordinates.

Proof. We have $\alpha^k = z^k + jw^k$, and $(\alpha^k)^2 = (z^k)^2 + i(w^k)^2 + j(2z^k w^k)$. So, for the even part $(\alpha^k)^2_0$ we have $(\alpha^k)^2_0 = (z^k)^2 + i(w^k)^2$, and respectively for the odd part: $(\alpha^k)^2_1 = 2z^k w^l$. Calculating we obtain

$$\partial(\alpha^k)^2_0/\partial z^k = 2z^k, \qquad \partial(\alpha^k)^2_1/\partial w^k = 2z^k,$$
$$\partial(\alpha^k)^2_0/\partial z^k = 2iw^k, \qquad \partial(\alpha^k)^2_1/\partial z^k = 2w^k.$$

Clearly, the double-complex Cauchy-Riemann system is satisfied by $(\alpha^k)^2_0$ and $(\alpha^k)^2_1$. Analogousely, we verify that the even part and the odd part of each product $\alpha^k\alpha^l$ satisfy the double-complex Cauchy-Riemann system with respect to the couple (z^k, w^k). Following the same way we verify that

$$\partial Q_0/\partial z^k = \partial Q_1/\partial w_k, \quad \partial Q_0/\partial w^k = i\partial Q_1/\partial z^k. \qquad \square$$

Remark 3.1. In the above written equalities Q_0 and Q_1 are considered as C-valued functions of the complex variables (z^k, w^k), i.e.

$$Q(\ldots, z^k + jw^k, \ldots) = Q_0(\ldots, z^k, w^k, \ldots) + jQ_1(\ldots, z^k, w^k, \ldots).$$

Remark 3.2. Each double-complex quadratic form in $C^n(1, j)$ is represented as a couple of complex quadratic forms in C^{2n}, but the inverse is not true. It is to take in view the proved holomorphicity of the double-complex quadratic forms as a necessary condition.

Example 3.1. Let us take the couple $P_0 = z^2 + w^2$, $P_1 = 2zw$ of complex quadratic forms in C^2. It is easy to see that the couple (P_0, P_1) does not satisfies the double-complex Cauchy-Riemann system.

Remark 3.3. With the notion of analytic double-complex valued function of many double-complex variables in mind one can prove that if f is a continuous double-complex function in an open set $G \subset C^n(1, j)$, and is holomorphic in each couple of coordinates variables (z^k, w^k) separately, then f is an analytic double-complex function. For this purpose it is to use an Osgood type lemma for double-complex holomorphic functions (to be published). Applying this theorem for double-complex quadratic forms we obtain the each double-complex quadratic form is an analytic double-complex function with respect to all its coordinates.

3.2. *Q-orthogonal geometry*

With the help of the double-complex symmetric quadratic form Q we can develop a notion of Q-orthogonal base in $C^n(1,j)$. This is a base $\{\varepsilon_k\}$ for which $Q(\varepsilon_k, \varepsilon_l) = 0$ (or $B(\varepsilon_k, \varepsilon_l) = 0$, as $q_{kl} = b_{kl}$) for each couple (k,l) with $k \neq l$. A question arises whether every double-complex base in $C^n(1,j)$ admits a non-degenerate linear transformation in a Q-orthogonal base in $C^n(1,j)$? The answer is given by the following Lagrange type theorem.

Theorem 3.2. *For every double-complex quadratic form Q on $C^n(1,j)$ there exist a base $\{\varepsilon_1, \ldots, \varepsilon_n\}$ such that $Q(\varepsilon_k, \varepsilon_l) = 0$ when $k \neq l$.*

Proof. The proof follows the classical algorithm of Lagrange, adapted for double-complex variables. □

Corollary 3.1. *In a Q-orthogonal base the matrix of Q is diagonal.*

$$\begin{pmatrix} q_{11}(\alpha^1)^2 & 0 & \cdots & 0 \\ 0 & q_{22}(\alpha^2)^2 & \cdots & 0 \\ \vdots & \vdots & & \vdots \\ 0 & 0 & \cdots & q_{nn}(\alpha^n)^2 \end{pmatrix}.$$

For this form we have the so called canonical representation

$$Q(\alpha^1, \alpha^2, \ldots, \alpha^n) = q_{11}(\alpha^1)^2 + q_{22}(\alpha^2)^2 + \cdots + q_{nn}(\alpha^n)^2,$$

(in double-complex variables) and, respectively, in complex coordinates

$$Q = \sum_k q_{kk}\{(z^k)^2 + i(w^k)^2 + j(2z^k w^k)\}, \quad q_{kk} \in C.$$

In this Q-orthogonal representation the considered quadratic form is separately holomorphic. Indeed, $Q = Q_0 + jQ_1$ where

$$Q_0 = \sum q_{kk}\{(z^k)^2 + i(w^k)^2\} \quad \text{and} \quad Q_1 = 2\sum q_{kk} z^k w^k.$$

We see that $\partial Q_0/\partial z^k = \sum_k q_{kk} 2z^k$ and $\partial Q_1/\partial w^k = \sum_k q_{kk} 2z^k$, so $\partial Q_0/\partial z^k = \partial Q_1/\partial w^k$.

Analogousely, we see that $\partial Q_0/\partial w^k = i\partial Q_1/\partial z^k$.

Remark 3.4. It is to remark that all ternary, ..., n-ary forms F_n are holomorphic too, but it is not known if they possess a kind of F_n-geometric interpretation.

4. Complex-hermitian double-complex quadratic forms

A complex Hermitian double-complex quadratic form Q, or shortly double-complex Hermitian form, is defined as follows

$$Q(\alpha,\alpha)=\sum\sum q_{lk}\alpha^l(\alpha^k)^*,\quad q_{lk}=Q(\varepsilon_k,\varepsilon_l)\in C.$$

Theorem 4.1. *For every double-complex Hermitian quadratic form Q there exist a base in which Q has the following canonical form*

$$Q(\beta,\beta)=\sum\lambda_k\beta^k(\beta^k)^*,\quad \lambda_k\in C.$$

The proof can be obtained in the same way like in theorem 3.2.

In coordinates we receive

$$Q(\alpha,\alpha)=\sum_k\sum_l q_{lk}(z^lz^k-iw^lw^k)+j\sum_l\sum_k(z^kw^l-z^lw^k).$$

and

$$Q(\beta,\beta)=\sum_k\lambda_k((z^k)^2-i(w^k)^*).$$

5. Fourth-complex quadratic forms

The algebra of fourth-complex numbers $C(1,j,j2,j3)$ is introduced in [3]. This is a commutative and associative complex algebra, which elements are denoted as follows $\alpha=z_0+z_1j+z_2j^2+z_3j^3$,$j^4=i$. A treatment of the quadratic forms of fourth-complex variables in terms of quadruples of complex quadratic forms can be developed. Here we give the simplest example

$$\alpha^2=(z_0+z_1j+z_2j^2+z_3j^3)^2=Q_0+Q_1j+Q_2j^2+Q_3j^3,$$

where Q_k, $k=0,1,2,3$, are complex quadratic forms,

$$Q_0=z_0+iz_2^2+2iz_1z_3,\quad Q_1=2(z_0z_1+iz_2z_3),$$
$$Q_2=z_1^2+iz_3^2+2z_1z_2,\quad Q_3=2(z_0z_3+z_1z_2).$$

The algebra of multi fourth-complex vectors, say $C^n(1,j,j^2,j^3)$, appears naturally, and the corresponding to the above exposed properties of $C^n(1,j)$ too.

In view of the pages limitation here, we hope to present elsewhere a more detailed exposition with a classification of the double-complex and fourth-complex surfaces.

References

1. S. Dimiev, *Double complex analytic geometry*, to be published in the Proceedings of the Bedlevo Conference, 2006, Poland.
2. G.E. Shilov, *Mathematical Analysis (Finite Dimensional Linear Spaces)*, Moskow, 1969 (in Russian).
3. B. Kiradjiev and P. Stoev, *Holomorphic fourth-complex functions*, Tensor, N. S., **64** (2003), 99–104.

EFFECTS OF FINITE ARITHMETIC IN NUMERICAL COMPUTATIONS

M.M. KONSTANTINOV

University of Architecture and Civil Engineering, 1046 Sofia, Bulgaria
E-mail: mmk_fte@uacg.bg

P.H. PETKOV

Technical University of Sofia, 1756 Sofia, Bulgaria
E-mail: php@tu-sofia.bg

A number of myths is considered, which are popular among users of computational systems for numerical calculations. Several bad and good computational practices in finite arithmetic are also discussed.

Keywords: Numerical analysis; Finite arithmetic; Computational myths.

1. Introduction

The invention of the digital computer in the middle of XX century led to a change in the viewpoint on numerical algorithms. It became clear that many classical computational schemes are not suitable for direct implementation in finite arithmetic (FA) and this has been recognized by the great numerical analysts of the past such as J. Von Neumann and A. Turing.

A reliable computational procedure has to take into account the main factors determining the accuracy of the computed solution, namely 1) the properties of the FA and in particular its *rounding unit*; 2) the properties of the computational problem and in particular its *sensitivity*; 3) the properties of the numerical algorithm and in particular its *numerical stability*. But often these factors are not taken into account by users and even developers of mathematical software, who are not numerical analysts.

The aim of this tutorial paper is to highlight some of these problems. It is based on previous publications of the authors, some of them together with V. Mehrmann, N. Higham, N. Vulchanov and Z. Gancheva [1–3].

2. Myths in computational practice

There are sustainable myths [1,3,4] in computational practice which are popular even among experienced users. According to MYTH 1 *"Large errors in the computed solution are due to the effect of large number of small rounding errors done at each operation in FA"*. This is rarely true. If there is a large error then probably it is due to a small number of critical operations, when a numerical disaster has occurred. Such operation may be a catastrophic cancellation of true digits in subtraction of close numbers.

Let for example $a = 1.000000000\xi$ be an approximate number, where the first 10 digits are correct and the digit $\xi > 0$ is uncertain. Then the difference $a - 1 = 0.000000000\xi$ will contain no true significant digits. Another example of a simple computation with large error is the calculation of e by the formula $e \approx e_n := (1 + 1/n)^n$ for large n. This bad way to compute e is based on the fact that $e_n \to e$ for $n \to \infty$. In FA with rounding unit $\approx 10^{-16}$ we obtain the good approximation e_8 with relative error $|e - e_8|/e \approx 10^{-8}$ and the bad result $e_{17} = 1$ with $|e - e_{17}|/e \approx= 0.6321$. The reason is in the FA summation $1 + 10^{-17}$ which gives 1.

According to MYTH 2 *"Rounding errors are not important because they are small (a variant: because they are compensated)"*. The believers in this myth have at least heard about rounding errors. The next MYTH 3 *"A short computation free of cancellation, overflow and underflow, should be accurate"* seems like true statement but the two examples below [4] prove the opposite.

Consider the transformation of $x > 0$ into y by the relations $x_{k+1} = x_k^{1/2}$, $x_1 = x$, for $k = 1, 2, \ldots, n$ and $x_{k+1} = x_k^2$ for $k = n+1, n+2, \ldots, 2n$, $y := x_{2n+1}$. The theoretical result is $y = x$. But for moderate values of n of about 100, on most computer platforms and for most computing environments (as for December 2006), the computed result will be either 1 or something else with a very large relative error.

Consider next the solution of the linear vector equation $Ax = b$, where the matrix A is nonsingular, by LU decomposition $A = LU$ without pivoting. Here L is a lower triangular matrix with diagonal elements, equal to 1, while U is an upper triangular matrix with nonzero diagonal elements. Let $A = \begin{bmatrix} \omega & -1 \\ 1 & 1 \end{bmatrix} = \begin{bmatrix} 1 & 0 \\ \omega^{-1} & 1 \end{bmatrix} \begin{bmatrix} \omega & -1 \\ 0 & 1+\omega^{-1} \end{bmatrix} = LU$, $b = \begin{bmatrix} 1+\omega \\ 0 \end{bmatrix}$, where $\omega \notin \{-1, 0\}$. The solution is $x = [1, -1]^{\top}$. In FA, for $|\omega|$ sufficiently small, the quantity $1 + \omega^{-1}$ will be rounded to ω^{-1}. Thus the computed matrices will be $\widehat{L} = L$, $\widehat{U} = \begin{bmatrix} \omega & -1 \\ 0 & \omega^{-1} \end{bmatrix}$. The solution $\widehat{x}$ of the rounded system $\widehat{L}\widehat{U}\widehat{x} = b$ is $\widehat{x} = [0, -1 - \omega]^\top$ with relative error of 71 percent.

MYTH 4 *"Subtraction of close numbers in FA is dangerous because the relative error of subtraction is large"* deserves some attention. Indeed, the relative error in subtraction is not bounded. At the same time subtraction of close numbers is usually done exactly in FA. What happens is that if the close numbers are approximate (which is typical in FA) then left-most true digits are cancelled and possible inaccuracies in right-most digits become important. So useful information is lost even when the subtraction itself is done exactly. If the close numbers were exact, then the computed result would also be exact. Of course, cancellation may be harmless. For example, the FA operation $a+(b-c)$ will be accurate when $1 \gg |b-c|/|a|$, $a \neq 0$. Thus MYTH 5 *"Cancellation in subtraction of near numbers is always dangerous"* is indeed a myth.

Consider now MYTH 6 *"Increasing the precision of FA always increases the accuracy of the computed result"*. Sometimes this is not true, e.g. the sum $10^{79.5} + 1007 - 10^{99} + 999 - 10^{79.5} + 10^{99} = 2006$ will be computed as 0 on most computers in single, double and extended precision. But it is true that decreasing the rounding unit decreases the known bounds on the error of FA computations, since in many of them the rounding unit is a multiplier.

According to MYTH 7 *"Rounding errors are always harmful"*. Sometimes rounding errors may help. For example, the QR algorithm cannot start theoretically for certain matrices but due to rounding errors it actually starts.

Counterexamples to MYTH 8 *"The final result cannot be more accurate than the intermediate results"* are given in [4]. The direct computation of $y = (\exp(x) - 1)/x$ in FA for small $|x|$ will give erroneous results instead of the exact answer which is close to 1. At the same time setting $z = \exp(x)$ we may compute y as $(z-1)/\log(z)$. This seems a bad idea since here we evaluate two functions exp and log instead of exp only. However, this may produce acceptable results for y even when the intermediate results $z - 1$ and $\log(z)$ are very inaccurate. In MATLAB, Ver. 7 for $x = 1.115 \times 10^{-16}$ we obtain $(\exp(x) - 1)/x = 1.9914$ with relative error of 99 percent, and $(\exp(x) - 1)/\log(\exp(x)) = 1$. The latter result has relative error less than 10^{-16} although the intermediate results $\exp(x) - 1$ and $\log(\exp(x))$ have large errors.

There are many myths in solving linear and nonlinear equations $f(x) = 0$, where x and $f(x)$ are vectors of same size and the function f is continuous. Let $\widehat{x}$ be an approximate solution with residual $r(\widehat{x}) = \|f(\widehat{x})\|$.

The continuity of the function r and the fact that $r(x) = 0$ whenever $f(x) = 0$, are at the base of MYTH 9 *"The accuracy of the computed solution $\widehat{x}$ may be checked by the residual $r(\widehat{x})$ – the smaller the residual, the better the approximation"* and its close variant MYTH 10 *"For two approximate solutions the better one has smaller residual"*. That these myths fail for nonlinear equations is illustrated by the next example. The algebraic equation $f(x) = x^3 - 23.001x^2 + 143.022x - 121.021 = 0$ has a single real root $x = 1$. Let $\widehat{x}_1 = 0.99$ and $\widehat{x}_2 = 11.00$ be two approximations to this root, where only $\widehat{x}_1$ may in fact be considered as an approximation. But we have $r(\widehat{x}_1) = 1.0022$ and $r(\widehat{x}_2) = 0.1$. Thus the residual of the bad approximation $\widehat{x}_2$ (with relative error of 1000 percent) is 10 times less than the residual of the good approximation $\widehat{x}_1$ (with relative error of 1 percent)! The graph of the function f for $x \geq 0$ explains clearly what has happened.

But Myths 9 and 10 fail for linear equations $Ax = b$ as well although this fact is not very popular among experienced users. This may be observed in equations with ill–conditioned matrix A for which the condition number $\mathrm{cond}(A) = \|A\|\,\|A^{-1}\|$ of A is large. Consider the system $Ax = b$ with $A = \begin{bmatrix} \varepsilon^3 & 1 \\ 0 & 1 \end{bmatrix}$, $b = \begin{bmatrix} 1 \\ 1 \end{bmatrix}$, where $\varepsilon > 0$ is a small parameter. The solution is $x = [0, 1]^\top$. Let $\widehat{x}_1 = [0, 1 + \varepsilon]^\top$ be a good approximation to the solution and $\widehat{x}_2 = [1/\varepsilon, 1]^\top$ be a very bad one. The relative error of $\widehat{x}_1$ is $e_1 = \varepsilon \ll 1$, while this of $\widehat{x}_2$ is $e_2 = 1/\varepsilon \gg 1$. At the same time the residual for x_1 is $\varepsilon\sqrt{2}$, while the residual for x_2 is ε^2. For $\varepsilon \to 0$ the residual ε^2 of the bad solution $\widehat{x}_2$ (with relative error $1/\varepsilon$ tending to ∞) is arbitrarily smaller than the residual $\varepsilon\sqrt{2}$ of the good solution $\widehat{x}_1$ (with relative error ε tending to 0). Note that here $\mathrm{cond}(A)$ is of order $2\varepsilon^{-3}$. Thus we have the following rule.

"The accuracy check by residuals may be completely misleading for scalar and vector nonlinear equations as well as for linear vector equations".

For the equation $Ax = b$ with nonsingular matrix A and a vector $\widehat{x} \neq x$, we have (in the 2-norm) $1/\|A^{-1}\| \leq \|A\widehat{x} - b\|/\|\widehat{x} - x\| \leq \|A\|$ and these inequalities are reachable. Denote by $\widehat{x}_1$ and $\widehat{x}_2$ the vectors, for which $1/\|A^{-1}\| = \|A\widehat{x}_1 - b\|/\|\widehat{x}_1 - x\|$, $\|A\widehat{x}_2 - b\|/\|\widehat{x}_2 - x\| = \|A\|$. Then we have $e_1/e_2 = \mathrm{cond}(A)(r_1/r_2)$, where $e_k = \|\widehat{x}_k - x\|$, $r_k = \|A\widehat{x}_k - b\|$. This explains why the better approximation may have larger residual for ill-conditioned systems. Hence we have the following paradox.

"The accuracy check by the size of the residual may be successful when the equation is well — conditioned and hence the computed solution is good and there is nothing to check. But when the equation is ill — conditioned

and there is a danger of large errors then the check based on the residual may be misleading."

There are three sophisticated myths that may be useful. Consider MYTH 11 *"A reliable way to check the accuracy of the approximate solution is to repeat the computations with double, or some other extended precision"* and its procedural variant MYTH 12 *"If, after a repeated computation with extended precision, the first several digits in the approximate solutions computed in both ways coincide, then these digits are true".*

It is true, however, that if we use FA with sufficiently small rounding unit, we can achieve an arbitrary number of true digits in the computed result. And it is also true that for this purpose we shall need an arbitrary large computing time.

Experienced computer users know that if small changes in the data lead to large changes in the result (i.e. if the computational problem is very sensitive) then the computed solution may be contaminated with large errors. Sometimes this correct observation is reversed assuming that if, for a given set of small perturbations, the result is slightly changed, then the accuracy of the solution is satisfactory. Thus we come to the next MYTH 13 *"If the check of the result by repeated computations with slightly perturbed data gives slightly perturbed results then the computational procedure is reliable and the solution is computed with good accuracy".* The reason why this statement is a myth is that in sensitive problems there are insensitive manifolds along which large changes of the data cause small changes in the result. At the same time small perturbations of the data along other directions can change the result dramatically.

Consider for example the equation $Ax = b$ with $A = \begin{bmatrix} a+1 & a \\ a & a-1 \end{bmatrix}$, $b = [2a, 2a]$, where $a > 0$ is large. The matrix A is nonsingular with $\det(A) = -1$ and the solution is $x = [2a, -2a]^\top$. If we change the parameter a to $a + \delta$ then the relative change in both b and x will be $|\delta|/a$. The system looks very well conditioned. But if we take $b = [2a+1, 2a-1]^\top$ the solution becomes $x = [1, 1]^\top$. So a relative change of order $1/(2a)$ in the data causes a relative change of order $2a$ in the result – an amplification of order $4a^2$. This is due to the ill-conditioning of the problem with $\text{cond}(A) = 4a^2 + 2 + \text{O}(a^{-2})$.

The next statement is not only a myth but an useful heuristics.

"If a check of the result by several sets of randomly chosen small perturbations in the data and by several rounding units shows small perturbation in the computed results, then with a high degree of reliability we may expect that the computed solution is close to the exact one".

3. Bad and good computational practices

3.1. *Floating–point computations*

Consider the computational problem $Y = f(X)$, where the data X and the result Y are elements of normed spaces. We shall suppose that near X it is fulfilled that $\|f(X_1) - f(X_2)\| \leq L\|X_1 - X_2\|$, where $\|\cdot\|$ denotes the corresponding norm. The Lipschitz constant $L \geq 0$ is known as the *absolute condition number* of the computational problem. More precisely, $L = L(X, \delta)$ is the supremum of $\|f(X + \Delta) - f(X)\|/\|\Delta\|$ over all $\Delta \neq 0$ with $\|\Delta\| \leq \delta$.

Below we recall some facts about FA with a finite set $\mathbb{M} \subset \mathbb{R}$ of machine numbers. For definiteness we consider a floating–point FA with rounding unit $\mathbf{u}$ and standard range $R := [m, M]$, where $m \in \mathbb{M}$ is the minimum positive and $M \in \mathbb{M}$ is the maximum positive number that may be represented in FA. According to the IEEE Standard [5] we have $\mathbf{u} \simeq 1.1 \times 10^{-16}$, $m \simeq 4.9 \times 10^{-324}$, $M \simeq 1.8 \times 10^{308}$.

Real numbers $x \in (-m, m)$ are rounded to zero and this is called *underflow*. Numbers x with $|x| \in R$ are rounded to the nearest machine number $\text{fl}(x)$ with a rule to break ties if x is in the middle between two machine numbers. Finally, numbers with $|x| > M$ cannot be represented in FA. The latter phenomenon is called *overflow* and should be avoided. For $x \in [M^{-1}, M]$ we have $\text{fl}(x) = x(1 + \varepsilon)$, $|\varepsilon| \leq \mathbf{u}$.

According to the MAIN HYPOTHESIS OF FA *"If* $\circ$ *denotes an arithmetic operation and* $M^{-1} \leq |x|, |y|, |x \circ y| \leq M$ *then the computed value* $\text{fl}(x \circ y)$ *of* $x \circ y$ *satisfies* $\text{fl}(x \circ y) = (x \circ y)(1 + \varepsilon)$ *with* $|\varepsilon| \leq \mathbf{u}$*"*.

A dangerous operation in FA is the subtraction $x - y$ of close approximate numbers x and y. If $2y \geq x \geq y \geq 0$ then the computed result $\text{fl}(x - y)$ is obtained without rounding errors if the FA uses a guard digit. Nevertheless, even if the operands x and y are known with high accuracy, the computed result may be much less accurate due to the loss of significant left-most digits. This phenomenon is known as *cancellation*, or even *catastrophic cancellation*. For example, in FA the quantity $6(x - \sin x)/x^3$ will be computed wrongly as 0 for small x instead of the exact answer which is close to 1. The reason is cancellation in the subtraction of the close numbers $\text{fl}(x)$ and $\text{fl}(\sin(\text{fl}(x)))$.

There exist integer algorithms which allow computations which are exact to a large extent. For example, some computer systems for doing mathematics may work in a way which corresponds to a machine word with length of several thousand or even million decimal digits. However, this leads to a slow speed of computations.

Consider a computational algorithm for the computation of $Y = f(X)$ which produces the answer $\widehat{Y} = \widehat{f}(X)$ in FA. Here the expression $\widehat{f}(X)$ depends also on the parameters of FA. The computational algorithm is *numerically stable* if the computed result $\widehat{Y}$ is close to the exact result of a near problem in the sense that $\|\widehat{Y} - f(\widehat{X})\| \leq C_1\mathbf{u}\|Y\|$ and $\|\widehat{X} - X\| \leq C_2\mathbf{u}\|X\|$ within a tolerance of order $\mathbf{u}^2$, where $C_1, C_2 \geq 0$ depend on the algorithm. The above relations give the accuracy estimate $\|\widehat{Y} - Y\| \leq C_1\mathbf{u}\|Y\| + LC_2\mathbf{u}\|X\| + \mathrm{O}(\mathbf{u}^2)$. For $X \neq 0$ and $Y \neq 0$ we obtain an useful bound on the relative error in the computed solution

$$\frac{\|\widehat{Y} - Y\|}{\|Y\|} \leq \mathbf{u}(C_1 + C_2 C_{\mathrm{rel}}) + \mathrm{O}(\mathbf{u}^2),$$

where $C_{\mathrm{rel}} = L\|X\|/\|Y\|$ is the *relative condition number* of the problem $Y = f(X)$. This estimate reveals the three main factors determining the accuracy of the computed solution as follows.

(1) The properties of FA through the constant $\mathbf{u}$ and implicitly through the avoidance of over — and underflows.
(2) The properties of the computational problem through the constant C_{rel}.
(3) The properties of the computational algorithm through the constants C_1 and C_2.

It is often assumed that $C_1 = 0$ and $C_2 = 1$. In this case the accuracy estimate takes the simple form $\|\widehat{Y} - Y\|/\|Y\| \leq \mathbf{u}C_{\mathrm{rel}}$.

The above estimates correspond to a *good computational practice*, where the following heuristic rule of thumb may be useful.

"If $\mathbf{u}C_{\mathrm{rel}} < 1$ then one may expect about $-\lg(\mathbf{u}C_{\mathrm{rel}})$ true decimal digits in the computed solution".

When the actual relative error is much larger than $\mathbf{u}(C_1 + C_2C_{\mathrm{rel}})$ or $\mathbf{u}C_{\mathrm{rel}}$ we have an example of a bad computational practice. Bad computational practices may occur as an attempt to translate directly a certain computational scheme into a computational algorithm without taking into account the properties of FA.

In the next sections we consider bad and good computational practices in the main problems of numerical linear algebra, see e.g. [6].

3.2. *Computing determinants*

Let $A = [a_{ik}]$ be an $n \times n$ real or complex matrix with determinant $\det(A)$. Determinants provide an useful tool in matrix analysis but their computation may be difficult in FA, especially if one uses the expression

$\det(A) = \sum_{\mathbf{j}} s(\mathbf{j}) a_{1,j_1} a_{2,j_2} \cdots a_{n,j_n}$, where $\mathbf{j}$ is the permutation $k \mapsto j_k$ with sign $s(\mathbf{j}) = \pm 1$. The use of this formula may be a catastrophic way to find the determinant. Indeed, the sum contains in general $n!$ terms and the calculation of each term requires $n-1$ multiplications. Hence there are $(n-1)n!$ multiplications and the time for computing the determinant of a general matrix may be very large. Let we have a computer which performs 10^9 floating point operations (FLOPS) per second. Then the computation of the determinant of a general 25×25 matrix will require about 10^{10} years which is approximately the age of the Universe.

Of course, there are better ways to compute determinants. For instance one may use the LU decomposition $PA = LU$ of A with partial pivoting, where P is a permutation matrix, L is a lower triangular matrix with diagonal elements, equal to 1, and $U = [u_{ik}]$ is an upper triangular matrix. Here $\det(A) = s\, u_{11} u_{22} \cdots u_{nn}$, where $s = 1$ ($s = -1$) if the matrix P corresponds to an even (odd) permutation. Other decompositions, e.g. the QR decomposition $A = QRP$ with column pivoting, may also be used for computing $\det(A)$, where Q is an orthonormed matrix, $R = [r_{ij}]$ is an upper triangular matrix and P is a permutation matrix. In this case $\det(A) = \pm r_{11} r_{22} \cdots r_{nn}$. Finally, if the singular value decomposition of A is used then $\det(A) = \pm \sigma_1 \sigma_2 \cdots \sigma_n$, where σ_k are the singular values of A.

3.3. *Solving linear algebraic equations*

Consider the linear vector algebraic equation $Ax = b$, where A is $n \times n$ invertible matrix and b is n vector. Formally the solution is $x = A^{-1}b$ but the computation of x in this way is not a good idea. Another bad computational practice here is to use the Cramer formulae.

Instead, one may use the QR decomposition with column pivoting $A = QRP$, where Q is $n \times n$ matrix with orthonormed columns, R is upper triangular matrix with nonzero diagonal elements and P is a permutation matrix. The QR decomposition is obtained via a finite numerically stable algorithm. After that the elements of the vector $y = Px$ are computed by back substitution from the equation $Ry = Q^{\mathrm{H}}b$, where ^{H} denotes complex conjugate transposition, and the solution is given by $x = Py$. Another popular way to solve the equation is via the Gauss elimination method with pivoting. Both methods are not very expensive and require about $\mathrm{O}(n^3)$ FLOPS. Singular value decomposition may also be used but this is an expensive way to solve the equation requiring about $\mathrm{O}(n^4)$ FLOPS.

3.4. *Rank determination*

Consider a general matrix $A \in \mathbb{K}^{m\times n}$, $\mathbb{K} = \mathbb{R}$ or $\mathbb{C}$. There are several ways to define the rank of $A \neq 0$, e.g. $\rho = \mathrm{rank}(A)$ is the maximum size of the nonzero minors of A. This leads to a bad computational procedure based on computation of determinants in FA. In particular, the check for invertibility of the matrix $A \in \mathbb{K}^{n\times n}$ using the inequality $\det(A) \neq 0$ must be avoided.

Next, ρ may be defined as the number of linearly independent rows of A. This leads to a good way to compute the rank since the number of linearly independent rows may be detected using LU or QR decompositions of A. Let the QR decomposition of A be $A = QR = Q_1R_1$, where $R_1 \in \mathbb{K}^{\rho\times n}$ is upper triangular matrix with $\mathrm{rank}(R_1) = \rho$. Thus the rank of A is the number of the first nonzero rows of the factor R in the QR decomposition $A = QR$. For improving the accuracy in FA, a QR decomposition $A = QRP$ with column pivoting is used, where P is a permutation matrix.

The rank of the matrix A is also the number of its positive singular values. The singular values may be retrieved by the singular value decomposition (SVD) $A = USV^{\mathrm{H}}$, $S = \mathrm{diag}(\sigma_1, \sigma_2, \ldots, \sigma_\rho, 0) \in \mathbb{K}^{m\times n}$, of A, where $U \in \mathbb{K}^{m\times m}$ and $V \in \mathbb{K}^{n\times n}$ are orthonormed matrices and $\sigma_1 \geq \sigma_2 \geq \cdots \geq \sigma_\rho > 0$ are the positive singular values of A. They are also the square roots of the positive eigenvalues of the matrix $A^{\mathrm{H}}A$.

Using SVD, the Moore — Penrose pseudoinverse $A^\dagger \in \mathbb{K}^{n\times m}$ of $A \in \mathbb{K}^{m\times n}$ is written as $A^\dagger := 0_{n\times m}$ if $A = 0_{m\times n}$, and $A^\dagger = VS^\dagger U^{\mathrm{H}}$, $S^\dagger = \mathrm{diag}\left(\sigma_1^{-1}, \sigma_2^{-1}, \ldots, \sigma_\rho^{-1}, 0\right) \in \mathbb{K}^{n\times m}$, if $A \neq 0$. Note that $VS^\dagger U^{\mathrm{H}}$ may not be the SVD of the matrix $A^\dagger$.

The determination of $\mathrm{rank}(A)$ as the number of positive singular values of A is very reliable computational procedure. But there is a principal difficulty in the rank determination in FA: the rank may be a sensitive characteristic of a matrix in the following sense. Suppose that $\rho < p = \min\{m, n\}$. Then an arbitrary small perturbation E in A may increase the rank to p in the sense that $\mathrm{rank}(A + E) = p$ but perturbations with $\|E\| < \sigma_\rho$ cannot decrease the rank.

The computation of the rank of a general matrix $A \in \mathbb{K}^{m\times n}$ of moderate size cannot be done "by hand". But when the matrix is written in the computer memory it is rounded to the nearest machine matrix $\mathrm{fl}(A)$, where $\|\mathrm{fl}(A) - A\|$ is of order $\mathbf{u}\|A\|$. If the initial matrix A is not of full rank then the rounded matrix may well be of full rank p. Also, if the initial matrix arises in a mathematical model of a real system, then it may be subject to measurement and/or parametric uncertainties. In both cases the "theoretical" rank of the matrix shall remain unknown.

Let $\widehat{\sigma}_k$ be the positive singular values of $A \in \mathbb{K}^{m\times n}$ computed in FA with rounding unit $\mathbf{u}$. The rank determination is difficult when the singular values form a sequence with uniformly decreasing members, e.g. $\sigma_k \simeq 10^{1-k}\|A\|$. Here we must eventually delete the singular values which are not inherent for the problem and are due to the effects of round offs. For this purpose one may define the concept of numerical rank of a matrix. Let $\tau > 0$ be a (small) numerical threshold. Then the matrix A is of *numerical rank* $\mu = \mu(A,\tau)$ within a threshold τ when it has exactly μ singular values larger than τ, i.e. when $\sigma_\mu > \tau$ and $\sigma_{\mu+1} \leq \tau$ (if $\mu \leq p-1$). If $\sigma_1 \leq \tau$ the numerical rank is assumed as zero.

Usually the threshold is taken as $\tau = c\mathbf{u}\|A\|$ for some $c > 0$. Since we usually have only values $\widehat{\sigma}_k$, the numerical rank in FA is defined as the number of quantities $\widehat{\sigma}_k > c\mathbf{u}\widehat{\sigma}_1$.

In a similar way is defined the numerical pseudoinverse of a matrix $A \in \mathbb{K}^{m\times n}$ with SVD $A = USV^{\mathrm{H}}$. The *numerical pseudoinverse* $A_\tau^\dagger \in \mathbb{K}^{n\times m}$ of A within a threshold $\tau > 0$ is defined as $A_\tau^\dagger = 0_{n\times m}$ if $\sigma_1 \leq \tau$, and $A_\tau^\dagger = VS_\tau^\dagger U^{\mathrm{H}}$, $S_\tau^\dagger = \mathrm{diag}\left(\sigma_1^{-1}, \sigma_2^{-1}, \ldots, \sigma_\mu^{-1}, 0\right) \in \mathbb{K}^{n\times m}$, if $\sigma_1 > \tau$, where $\mu = \mu(A,\tau)$ is the numerical rank of A. In practice one may choose τ as $p^s\mathbf{u}\widehat{\sigma}_1$, where $s = 0, 1, 2$.

The rank determination and the computation of pseudoinverses based on the concept of numerical rank are examples of good computational practices. The rank determination using QR decomposition with column pivoting is also a good computational algorithm.

3.5. *Solving algebraic equations*

The algebraic equation $p(x) = x^n + a_{n-1}x^{n-1} + \cdots + a_1x + a_0 = 0$ over $\mathbb{R}$ or $\mathbb{C}$ may be solved in general by explicit formulae for $n \leq 4$. However, their direct use may lead to loss of accuracy in the computed result due to cancellation errors.

It is natural to set two basic requirements for a computational root finding scheme. First, it should compute certain approximations to the n roots of the equation whenever all coefficients a_k are in the standard range of FA. Second, the rounding errors in the computed roots should be of order of the root condition numbers multiplied by $\mathbf{u}$.

For $n = 2$ the solution is $x_{1,2} = \left(-a_1 \pm \sqrt{a_1^2 - 4a_0}\right)/2$. The direct use of this formula in FA is a bad algorithm since it may lead to overflow when $a_1^2 > M$, to underflow when $0 < a_1^2 < m$ and to catastrophic cancellation when $a_1^2 \gg 4|a_0|$. Unfortunately, this bad approach is sometimes given without warning as an exercise when the concept of algorithm is presented to the students.

There are several reliable approaches to solve the algebraic equation $p(x) = 0$, e.g. via the companion matrix C of the polynomial p. In this case the QR algorithm or some other sophisticated scheme (as the Kagström–Ruhe algorithm) is applied to compute the eigenvalues of C which are the roots of p.

3.6. *Computing eigenvalues of a matrix*

The eigenvalues of the $n \times n$ matrix A are the roots of its characteristic polynomial (c.p) $\det(\lambda I - A) = \lambda^n - c_1\lambda^{n-1} + \cdots + (-1)^n c_n = 0$, where c_k is the sum of principal k–th order minors of A. Due to this definition, sometimes the solution of the characteristic equation is presented as a way to find the collection of eigenvalues $\Lambda := \{\lambda_1, \lambda_2, \ldots, \lambda_n\}$ of A. This is in general a bad computational algorithm. Indeed, the computation of c_k for a general matrix A may be a difficult task and hence the computed values $\widehat{c}_k$ may be contaminated with large errors. Also, the roots of the c.p. may be very sensitive to changes in its coefficients even if the eigenvalues of the matrix A are not sensitive relative to changes in its entries. That is why we have the following rule.

"The numerical computation of the eigenvalues of a general matrix should not done via its c.p."

Modern numerical algorithms for computing Λ are based on transformation of A into Schur form $S = U^{\mathrm{H}}AU$, where U is unitary matrix and the matrix S is upper triangular with the eigenvalues of A on its diagonal. The reduction into Schur form is usually done by the QR algorithm of Francis–Kublanovskaya. The QR algorithm works well when the eigenvalues of A are not very sensitive to perturbations and in particular when this matrix has only linear elementary divisors. When the eigenvalues are sensitive they may be computed with large errors. Consider for example the matrix $A = \begin{bmatrix} -90001 & 769000 & -690000 \\ -810000 & 6909999 & -6200000 \\ -891000 & 7601000 & -6820001 \end{bmatrix}$ which has a single Jordan block with eigenvalue -1 and c.p. $\lambda^3 + 3\lambda^2 + 3\lambda + 1$. The command `eig(A)` from MATLAB, ver. 6.0 produces the wrong result $\widehat{\lambda}_{1,2} = -1.5994 \pm 1.0441\,\imath$, $\widehat{\lambda}_3 = 0.1989$. Here even a positive eigenvalue has been computed. The command `poly(A)` computes the wrong c.p. $\lambda^3 + 3.0000\lambda^2 + 3.0120\lambda - 0.7255$. The reason is the high sensitivity of the eigenvalues although the result -0.7255 instead of 1 for $-\det(A)$ is strange. In MATLAB, ver. 7.0 the computed results are slightly better, namely $\widehat{\lambda}_{1,2} = -1.4506 \pm 0.7837\,\imath$, $\widehat{\lambda}_3 = -0.0988$. The computed c.p. now is $\lambda^3 + 3.0000\lambda^2 + 3.0051\lambda + 0.2686$. Here again the free term is obtained with large error. Finally, the command `jordan(A)` from MATLAB, ver. 7.0 implementing symbolic computations obtains the correct Jordan form.

For this example the computation of the spectrum of A via its c.p. gives the correct answer in MATLAB environment because the multiplication of two elements of A is done in fact in integer arithmetic without rounding.

A good numerical way to find eigenvalues is to apply the QR algorithm or the algorithm of Kagström and Ruhe. A modification of this algorithm is implemented in the interactive system SYSLAB [7]. For the matrix A above SYSLAB computes the Jordan form with 9 true decimal digits.

References

1. N. Vulchanov and M. Konstantinov, *Modern Mathematical Methods for Computer Calculations. Part I: Foundations of Numerical Computations*, Studies in Mathematics, Vol. 1, Bulgarian Institute of Analyses and Research, Sofia, 1996, 2005.
2. N. Higham, M. Konstantinov, F. Mehrmann and P. Petkov, *IEEE Control Systems Magazine* **24**, 28 (2004).
3. M. Konstantinov, P. Petkov and Z. Gancheva, *Thirteen myths in numerical analysis*, in: Proc. 34 Spring Conference of the Union of Bulgarian Mathematicians, (Sofia, 2005).
4. N. Higham, *Accuracy and Stability of Numerical Algorithms*, SIAM, Philadelphia, 1996, 2002.
5. *IEEE Standard for Binary Floating-Point Arithmetic*, ANSI/IEEE Standard 754–1985, IEEE, New York, 1985.
6. G. Golub and C. Loan, *Matrix Computations*, The John Hopkins University Press, Baltimore, 1989.
7. N. Christov, P. Petkov and M. Konstantinov, *Journal of Automatics* **28**, 71 (1987).

EXACT SOLUTIONS OF THE MANAKOV SYSTEM

N.A. KOSTOV

*Institute of Electronics, Bulgarian Academy of Sciences,
72 Tsarigradsko chaussee, 1784 Sofia, Bulgaria*

We obtain periodic and quasi-periodic wave solutions of a set of coupled nonlinear Schrödinger equations (integrable Manakov model). One and two phase quasi-periodic solutions for the integrable Manakov system are given in terms of Weierstrass and Jacobian elliptic functions. The reduction of quasi-periodic solutions to elliptic functions and soliton solutions is analyzed.

Keywords: Soliton; Elliptic soliton; Manakov model; Periodic solitons.

1. Introduction

A comprehensive algebro-geometric integration of the Manakov system is developed in [1]. The associated spectral variety is a trigonal Riemann surface, which is described explicitly and the solutions of the equations are given in terms of θ-functions of the surface. Recently the Darboux-dressing transformations have been applied to the Lax pair associated with systems of coupled nonlinear wave equations (vector nonlinear Schrödinger-type equations and three resonant wave equations) [2,3]. Solutions with boomeronic and trapponic behaviour were investigated. The method of squared wavefunctions for the vector nonlinear Schrödinger equation is given in [4,5] for smooth enough potentials , which tend to zero fast enough for $x \to \pm\infty$ and in [6] for quasi-periodic case. The squared wavefunctions of the octet representation of $SU(3)$ are related to periodic solutions in terms of Weierstrass' elliptic functions. The aim of the present paper is to derive exact solutions of the Manakov system using inverse scattering transform method and appropriate ansatzs.

2. Basic equations and solutions

We consider coupled nonlinear Schrödinger equations of the form (Manakov model) [7]

$$\begin{aligned} \imath Q_{1t} + Q_{1xx} + \sigma(|Q_1|^2 + |Q_2|^2)Q_1 = 0, \\ \imath Q_{2t} + Q_{2xx} + \sigma(|Q_1|^2 + |Q_2|^2)Q_2 = 0, \end{aligned} \tag{1}$$

where $\sigma = 1$ for focussing case and $\sigma = -1$ for defocussing case. We seek solution of (1), $\sigma = 1$ in the form [8,9]

$$Q_j = q_j(z)\,\mathrm{e}^{i\Theta_j}, \quad j = 1, 2, \tag{2}$$

where $z = x - ct$, $\Theta_j = \Theta_j(z,t)$, with q_j, Θ_j real, where the functions Θ_j, $j = 1, 2$ behave as

$$\Theta_j = \frac{1}{2}cx + (a_j - \frac{1}{4}c^2)t - \mathcal{C}_j \int_0^z \frac{\mathrm{d}z'}{q_j^2(z')} + \Theta_{j0}.$$

Let

$$q_j(z) = \sqrt{A_j \wp(z + \omega') + B_j}, \tag{3}$$

and using the well known relations

$$\int_o^z \frac{dz'}{\wp(z) - \wp(\tilde{a}_j)} = \frac{1}{\wp'(\tilde{a}_j)} \left(2z\zeta(\tilde{a}_j) + \ln \frac{\sigma(z - \tilde{a}_j)}{\sigma(z + \tilde{a}_j)} \right), \tag{4}$$

and

$$\wp(z + \omega') - \wp(\tilde{a}_j) = -\frac{\sigma(z + \omega' + \tilde{a}_j)\sigma(z + \omega' - \tilde{a}_j)}{\sigma(z + \omega')^2 \sigma(\tilde{a}_j)^2}, \tag{5}$$

we obtain

$$\begin{aligned} Q_1 &= \sqrt{-A_1} \frac{\sigma(z + \omega' + \tilde{a}_1)}{\sigma(z + \omega')\sigma(\tilde{a}_1)} \exp\left(\frac{i}{2}cx + i(a_1 - \frac{1}{4}c^2)t - (z + \omega')\zeta(\tilde{a}_1) \right), \\ Q_2 &= \sqrt{-A_2} \frac{\sigma(z + \omega' + \tilde{a}_2)}{\sigma(z + \omega')\sigma(\tilde{a}_2)} \exp\left(\frac{i}{2}cx + i(a_2 - \frac{1}{4}c^2)t - (z + \omega')\zeta(\tilde{a}_2) \right), \end{aligned}$$

where

$$\begin{aligned} &\sum_{j=1}^{2} A_j = -2, \quad a_j = \sum_{k=1}^{2} B_k - \frac{B_j}{A_j}, \\ &\frac{\mathcal{C}_j^2}{A_j} = \frac{i}{2}\sqrt{4\lambda^3 - \lambda g_2 - g_3}_{|\lambda = -\frac{B_j}{A_j}}, \quad \wp(\tilde{a}_j) = -\frac{B_j}{A_j} = \hat{a}_j. \quad j = 1, 2. \end{aligned} \tag{6}$$

These solutions have the following interpretation in terms of spectral function

$$\psi(x,t,\lambda) = \sqrt{g(x,t)} \exp\left(\theta(x,t,\lambda) - \frac{i}{2}\sqrt{R(\lambda)} \int_0^z \frac{1}{g(z',\lambda)} dz' \right), \tag{7}$$

where $g(x,t) = \mu(x,t) - \lambda = \wp(z - \omega') - \lambda$, $\theta(x,t) = \frac{1}{2}cx + (\lambda + \sum_{k=1}^{2} B_k - \frac{1}{4}c^2)t$ and λ is the spectral parameter. Our solutions are obtained when λ is given at points $-B_j/A_j$ and $Q_j = \sqrt{A_j}\psi(x,t,-B_j/A_j)$. The dynamics is on the Riemann surface $\nu^2 = R(\lambda) = 4\lambda^3 - \lambda g_2 - g_3$ and the function $\mu(x,t)$ obey $\mu_x = \sqrt{R(\mu)}$.

One special solution of system (1), $\sigma = 1$ is written by

$$q_1 = C_1 \mathrm{cn}(\alpha z, k), \qquad q_2 = C_2 \mathrm{cn}(\alpha z, k), \tag{8}$$

where $\alpha^2 = a_1/(2k^2 - 1)$, $C_1^2 + C_2^2 = 2\alpha^2 k^2$, $a_1 = a_2 = a$ and in the limit $k \to 1$ we obtain well known *Manakov* soliton solution [7]

$$Q_1 = \frac{\sqrt{2a}\epsilon_1 \exp\left\{i\left(\frac{1}{2}c(x - x_0) + (a - \frac{1}{4}c^2)t\right)\right\}}{\mathrm{ch}(\sqrt{a}(x - x_0 - ct))},$$

$$Q_2 = \frac{\sqrt{2a}\epsilon_2 \exp\left\{i\left(\frac{1}{2}c(x - x_0) + (a - \frac{1}{4}c^2)t\right)\right\}}{\mathrm{ch}(\sqrt{a}(x - x_0 - ct))},$$

where we introduce the following notations

$$|\epsilon_1|^2 + |\epsilon_2|^2 = 1, \quad \zeta_1 = \frac{1}{2}c + i\sqrt{a} = \xi + i\eta, \tag{9}$$

where x_0 is the position of soliton, (ϵ_1, ϵ_2) are the components of polarization vector. One notes that the real part of ζ_1 i.e. $c/2$ gives us the soliton velocity while the imaginary part of ζ_1, i.e. $\sqrt{2a}$ gives the soliton amplitude and width. Next solution of the system (1), $\sigma = -1$ we obtain using the following ansatz [8,10,11]

$$q_i(\zeta) = \sqrt{A_i \wp^2(\zeta + \omega') + B_i \wp(\zeta + \omega') + C_i}, \quad i = 1,2 \tag{10}$$

then we have

$$\sum_{i=1}^{2} A_i = 0, \sum_{i=1}^{2} B_i = -6,\ a_i = \sum_{k=1}^{2} C_k - 3\frac{B_i}{A_i},\ C_i = \frac{B_i^2}{A_i} - \frac{1}{4}A_i g_2, \tag{11}$$

$$C_i^2 = \frac{A_i^2}{2 \cdot 3^3 \cdot 4}(4\lambda^5 + 27\lambda^2 g_3 + 27\lambda g_2^2 - 21\lambda^3 g_2 - 81 g_2 g_3), \tag{12}$$

where $\lambda = -3B_i/A_i$. The paremeters A_i, B_i, C_i, $i = 1,2$ are expressed in terms of a_i by

$$A_1 = \frac{18}{a_1 - a_2}, \quad A_2 = \frac{18}{a_2 - a_1},$$

$$B_1 = -\frac{6(a_1 - \Delta)}{a_1 - a_2}, \quad B_2 = -\frac{6(a_2 - \Delta)}{a_2 - a_1},$$

$$C_1 = \frac{2(a_1 - \Delta)^2}{a_1 - a_2} - \frac{9}{2}\frac{g_2}{a_1 - a_2}, \quad C_2 = \frac{2(a_2 - \Delta)^2}{a_2 - a_1} - \frac{9}{2}\frac{g_2}{a_2 - a_1},$$

and $\Delta = \frac{2}{5}a_1 + \frac{2}{5}a_2$. One can obtain the special solution [12]

$$q_1 = C\mathrm{sn}(\alpha z, k)\mathrm{dn}(\alpha z, k), \qquad q_2 = C\mathrm{cn}(\alpha z, k)\mathrm{dn}(\alpha z, k), \tag{13}$$

where

$$\alpha^2 = \frac{1}{15}(4a_2 - a_1), \quad C^2 = \frac{2}{5}(4a_2 - a_1), \quad k^2 = 5\frac{a_2 - a_1}{4a_2 - a_1}, \tag{14}$$

in the limit $k \to 1$ we have the following soliton solution

$$Q_1 = \sqrt{6a}\epsilon_1 \exp\left\{i\left(\frac{1}{2}c(x - x_0) + (a - \frac{1}{4}c^2)t\right)\right\}$$
$$\times \mathrm{th}(\sqrt{a}(x - x_0 - ct))\mathrm{sech}(\sqrt{a}(x - x_0 - ct)),$$
$$Q_2 = \frac{\sqrt{6a}\epsilon_2 \exp\left\{i\left(\frac{1}{2}c(x - x_0) + (4a - \frac{1}{4}c^2)t\right)\right\}}{\mathrm{ch}^2(\sqrt{a}(x - x_0 - ct))},$$

where we introduce the following notations

$$|\epsilon_1|^2 = |\epsilon_2|^2 = 1, \quad a_1 = a, \, a_2 = 4a_1. \tag{15}$$

Another special solution of (1) , $\sigma = -1$ is written by

$$q_1 = C\left(\frac{1}{3}C_1 - k^2\mathrm{sn}^2(\alpha z, k)\right), \tag{16}$$

$$q_2 = C\mathrm{sn}(\alpha z, k)\mathrm{dn}(\alpha z, k), \tag{17}$$

where

$$\alpha^2 = \frac{1}{15}(a_2^2 - a_1^2), \quad C^2 = -\frac{6}{5}(a_1 + a_2),$$
$$k^2 = \frac{\sqrt{\frac{5}{3}(a_1^2 - a_2^2)} + 2a_1 - 3a_2}{2\sqrt{\frac{5}{3}(a_1^2 - a_2^2)}}, \quad \frac{1}{3}C_1 = \frac{1}{2} - \frac{1}{2}\left(\frac{5}{3}\frac{a_1 - a_2}{a_1 + a_2}\right)^{1/2}.$$

in the limit $k \to 1$ we obtain the following soliton solution

$$Q_1 = \sqrt{\frac{9a}{4}}\epsilon_1 \exp\left\{i\left(\frac{1}{2}c(x - x_0) - (a + \frac{1}{4}c^2)t\right)\right\}$$
$$\times \left(-\frac{2}{3} + \mathrm{sech}^2(\sqrt{\frac{a}{8}}(x - x_0 - ct)\right),$$
$$Q_2 = \sqrt{\frac{9a}{4}}\epsilon_2 \exp\left\{i\left(\frac{1}{2}c(x - x_0) - (\frac{7}{8}a + \frac{1}{4}c^2)t\right)\right\}$$
$$\times \mathrm{th}(\sqrt{\frac{a}{8}}(x - x_0 - ct)\mathrm{sech}(\sqrt{\frac{a}{8}}(x - x_0 - ct)),$$

where $|\epsilon_1|^2 = |\epsilon_2{}^2 = 1$, $a_1 = -a$, $a_2 = -\frac{7}{8}a$.

3. Exact solutions of the Manakov system using Lax pairs method

The system (1) can be solved by the inverse scattering transform method [7] and is written as zero-curvature representation form by

$$U_t - V_x + [U, V] = 0, \tag{18}$$

where U, V are 3×3 matrices defined as

$$U = \begin{pmatrix} -i\lambda & \frac{Q_1}{\sqrt{2}} & \frac{Q_2}{\sqrt{2}} \\ -\frac{\sigma}{\sqrt{2}}Q_1^* & i\lambda & 0 \\ -\frac{\sigma}{\sqrt{2}}Q_2^* & 0 & i\lambda \end{pmatrix}, \qquad V = V_0\lambda^2 + V_1\lambda + V_2 \tag{19}$$

and

$$V_0 = \begin{pmatrix} -2i & 0 & 0 \\ 0 & 2i & 0 \\ 0 & 0 & 2i \end{pmatrix}, \qquad V_1 = \begin{pmatrix} 0 & Q_1\sqrt{2} & Q_2\sqrt{2} \\ -\sigma Q_1^*\sqrt{2} & 0 & 0 \\ -\sigma Q_2^*\sqrt{2} & 0 & 0 \end{pmatrix},$$

$$V_2 = \begin{pmatrix} \frac{i}{2}\sigma(|Q_1|^2 + |Q_2|^2) & i\frac{\sqrt{2}}{2}Q_{1x} & i\frac{\sqrt{2}}{2}Q_{2x} \\ i\sigma\frac{\sqrt{2}}{2}Q_{1x}^* & -\frac{1}{2}i\sigma|Q_1|^2 & -\frac{i}{2}\sigma Q_1^*Q_2 \\ i\sigma\frac{\sqrt{2}}{2}Q_{2x}^* & -\frac{i}{2}\sigma Q_2^*Q_1 & -\frac{1}{2}i\sigma|Q_2|^2 \end{pmatrix},$$

where λ is the spectral parameter. The necessary Lax pair is

$$\Psi_x = U\Psi, \qquad \Psi_t = V\Psi. \tag{20}$$

Next we define 3×3 matrix $M(x, t, \lambda)$, which is solution of $M_x = [U, M]$ and

$$L = \begin{pmatrix} A(x,t,\lambda) & B^1(x,t,\lambda) & B^2(x,t,\lambda) \\ C^1(x,t,\lambda) & D^{22}(x,t,\lambda) & D^{23}(x,t,\lambda) \\ C^2(x,t,\lambda) & D^{32}(x,t,\lambda) & D^{33}(x,t,\lambda) \end{pmatrix},$$

or in explicit form

$$A_x = \frac{1}{\sqrt{2}}\left(Q_1C^1 + Q_2C^2 + \sigma Q_1^*B^1 + \sigma Q_2^*B^2\right),$$

$$B_x^1 = -2i\lambda B^1 + (D^{22} - A)\frac{Q_1}{\sqrt{2}} + D^{32}\frac{Q_2}{\sqrt{2}},$$

$$B_x^2 = -2i\lambda B^2 + (D^{33} - A)\frac{Q_2}{\sqrt{2}} + D^{23}\frac{Q_1}{\sqrt{2}},$$

$$C_x^1 = 2i\lambda C^1 + (D^{22} - A)\sigma\frac{Q_1^*}{\sqrt{2}} + D^{23}\sigma\frac{Q_2^*}{\sqrt{2}},$$

$$C_x^2 = 2i\lambda C^2 + (D^{33} - A)\sigma\frac{Q_2^*}{\sqrt{2}} + D^{32}\sigma\frac{Q_1^*}{\sqrt{2}},$$

$$D_x^{22} = -\frac{1}{\sqrt{2}}(\sigma Q_1^* B^1 + Q_1 C^1), \quad D_x^{32} = -\frac{1}{\sqrt{2}}(\sigma Q_2^* B^1 + Q_1 C^2),$$
$$D_x^{23} = -\frac{1}{\sqrt{2}}(\sigma Q_1^* B^2 + Q_2 C^1), \quad D_x^{33} = -\frac{1}{\sqrt{2}}(\sigma Q_2^* B^2 + Q_1 C^2),$$

and define $\{A_l(x,t)\}$, $0 \le l \le n+1$, $\{B_l^k(x,t)\}$, $\{C_l^k(x,t)\}$, $0 \le l \le n$, $k = 1, 2$, $\{D_l^{22}(x,t)\}_{0 \le l \le n+1}$, $\{D_l^{33}(x,t)\}_{0 \le l \le n+1}$, $\{D_l^{23}(x,t)\}_{0 \le l \le n}$, $\{D_l^{32}(x,t)\}_{0 \le l \le n}$, recursively by

$$A_{l+1,x} = \frac{1}{\sqrt{2}}\left(Q_1 C_l^1 + Q_2 C_l^2 + \sigma Q_1^* B_l^1 + \sigma Q_2^* B_l^2\right),$$
$$B_{l,x}^1 = -2i\lambda B_{l+1}^1 + (D_{l+1}^{22} - A_{l+1})\frac{Q_1}{\sqrt{2}} + D_{l+1}^{32}\frac{Q_2}{\sqrt{2}},$$
$$B_{l,x}^2 = -2i\lambda B_{l+1}^2 + (D_{l+1}^{33} - A_{l+1})\frac{Q_2}{\sqrt{2}} + D_{l+1}^{23}\frac{Q_1}{\sqrt{2}},$$
$$C_{l,x}^1 = 2i\lambda C_{l+1}^1 + (D_{l+1}^{22} - A_{l+1})\sigma\frac{Q_1^*}{\sqrt{2}} + D_{l+1}^{23}\sigma\frac{Q_2^*}{\sqrt{2}},$$
$$C_{l,x}^2 = 2i\lambda C_{l+1}^2 + (D_{l+1}^{33} - A_{l+1})\sigma\frac{Q_2^*}{\sqrt{2}} + D_{l+1}^{32}\sigma\frac{Q_1^*}{\sqrt{2}},$$
$$D_{l+1,x}^{22} = -\frac{1}{\sqrt{2}}(\sigma Q_1^* B_l^1 + Q_1 C_l^1), \quad D_{l+1,x}^{32} = -\frac{1}{\sqrt{2}}(\sigma Q_2^* B_l^1 + Q_1 C_l^2),$$
$$D_{l+1,x}^{23} = -\frac{1}{\sqrt{2}}(\sigma Q_1^* B_l^2 + Q_2 C_l^1), \quad D_{l+1,x}^{33} = -\frac{1}{\sqrt{2}}(\sigma Q_2^* B_l^2 + Q_1 C_l^2),$$

with initial conditions

$$A_0 = -2i, \quad D_0^{22} = 2i, \quad D_0^{33} = 2i, \quad D_0^{23} = 0, \quad D_0^{32} = 0,$$
$$A_1 = 0, \quad D_1^{22} = 0, \quad D_1^{33} = 0, \quad D_1^{23} = 0, \quad D_1^{32} = 0.$$

Explicitly, one computes

$$B_0^1 = \sqrt{2}Q_1, \quad B_0^2 = \sqrt{2}Q_2, \quad C_0^1 = -\sqrt{2}\sigma Q_1^*, \quad C_0^2 = -\sqrt{2}\sigma Q_2^*$$
$$A_2 = \frac{i}{2}(|Q_1^*|^2 + |Q_2^*|^2), \quad B_0^1 = \frac{1}{\sqrt{2}}Q_{1,x}, \quad B_0^2 = \frac{1}{\sqrt{2}}Q_{2,x},$$
$$C_0^1 = \frac{1}{\sqrt{2}}Q_{1,x}^*, \; C_0^2 = \frac{1}{\sqrt{2}}Q_{2,x}^*, \; D_2^{22} = -\frac{i}{2}|Q_1|^2 + ia_1, \; D_2^{23} = -\frac{i}{2}Q_1^* Q_2,$$
$$D_2^{33} = -\frac{i}{2}|Q_2|^2 + ia_2, \quad D_2^{32} = -\frac{i}{2}Q_2^* Q_1.$$

The Lax representation yields a trigonal nonhyperelliptic curve $V = (\nu, \lambda)$ $\det(L(\lambda) - \nu \mathbf{I}_3) = 0$ where $\mathbf{I}_3$ is the 3×3 unit matrix. For the second flow the curve is given explicitly by

$$
\begin{aligned}
&(\nu+2\imath\lambda^2)(\nu-2\imath\lambda^2)^2+(A_0\lambda+B_0)(\nu-2\imath\lambda^2)\\
&-\imath C_0+4D_0\lambda^2-E_0\lambda-F_0\nu^2-\imath H_0\lambda^4=0,
\end{aligned}
\tag{21}
$$

where

$$
\begin{aligned}
A_0 &= \imath\sigma\left(\sum_{k=1}^{2} Q_k^* Q_{k,x}-Q_{k,x}^* Q_k\right), \quad B_0=\tilde{H}-a_1a_2\\
\tilde{H} &= \frac{1}{2}\sigma\sum_{k=1}^{2}|Q_{k,x}|^2+\frac{1}{4}\left(\sum_{k=1}^{2}|Q_k|^2\right)^2-\frac{1}{2}\sigma\sum_{k=1}^{2}a_k|Q_k|^2\\
C_0 &= \frac{1}{4}Q_{1,x}Q_{2,x}^*Q_1^*Q_2+\frac{1}{4}Q_{2,x}Q_{1,x}^*Q_2^*Q_1-\frac{1}{4}|Q_{1,x}^*|^2|Q_2^*|^2-\frac{1}{4}|Q_{2,x}^*|^2|Q_1^*|^2\\
&-\frac{1}{2}\sigma\left(|Q_1|^2+|Q_2|^2\right)\left(a_1a_2-\frac{1}{2}\sigma a_2|Q_1|^2-\frac{1}{2}\sigma a_1|Q_2|^2\right)\\
&+\frac{1}{2}\sigma a_2|Q_{1,x}|^2+\frac{1}{2}\sigma a_1|Q_{2,x}|^2,\\
E_0 &= \sigma(Q_{1,x}^*Q_1-Q_1^*Q_{1,x})a_2+\sigma(Q_{2,x}^*Q_2-Q_2^*Q_{2,x})a_1,\\
D_0 &= \imath a_1a_2, \quad F_0=\imath(a_1+a_2), \quad H_0=-4\imath F_0.
\end{aligned}
\tag{22}
$$

The series of birational transformations

$$
\nu=\nu_0+2i\lambda^2,\ \nu_0=\frac{\frac{1}{2}E_0-\frac{1}{2}\lambda_1+\frac{1}{4}\nu_1}{4i\lambda_1^2-4iF_0+D_0},\ \lambda=\nu_1,\ \nu_1=4w,\ z=i\lambda_1, \tag{23}
$$

yields a hyperelliptic curve of genus two defined by canonical form

$$
w^2=4z^5+\sigma_4z^4+\sigma_3z^3+\sigma_2z^2+\sigma_1z+\sigma_0, \tag{24}
$$

where the *moduli* of the curve σ_i are expressible in terms of physical parameters - level of energy $\tilde{H}$ and constants $a_1, a_2, \tilde{C}_1, \tilde{C}_2$ as follows

$$
\begin{aligned}
\sigma_4 &= -8(a_1+a_2),\\
\sigma_3 &= -4\tilde{H}+4(a_1+a_2)^2+8a_1a_2,\\
\sigma_2 &= 4\tilde{H}(a_1+a_2)-4\tilde{G}-C_1^2-C_2^2-8a_1a_2(a_1+a_2),\\
\sigma_1 &= 4\tilde{G}(a_1+a_2)-4a_1a_2\tilde{H}+2\tilde{C}_1^2a_2+2\tilde{C}_2^2a_1+4a_1^2a_2^2,\\
\sigma_0 &= -4a_1a_2\tilde{G}-\tilde{C}_1^2a_2^2-\tilde{C}_2^2a_1^2,
\end{aligned}
$$

where

$$
\begin{aligned}
\tilde{G} &= \frac{1}{8}\left(Q_1Q_{2,x}^*-Q_2^*Q_{1,x}\right)\left(Q_1^*Q_{2,x}-Q_2Q_{1,x}^*\right)\\
&+\frac{1}{8}\left(Q_2Q_{1,x}-Q_1Q_{2,x}\right)\left(Q_2^*Q_{1,x}^*-Q_1^*Q_{2,x}^*\right)
\end{aligned}
$$

$$+\frac{1}{2}\sigma\left(|Q_1|^2+|Q_2|^2\right)\left(a_1a_2-\frac{1}{2}\sigma a_2|Q_1|^2-\frac{1}{2}\sigma a_1|Q_2|^2\right)$$
$$-\frac{1}{2}\sigma a_2|Q_{1,x}|^2-\frac{1}{2}\sigma a_1|Q_{2,x}|^2,\ \tilde{C}_k^2=\frac{1}{4}|\left(Q_kQ_{k,x}^*-Q_k^*Q_{k,x}\right)|^2,\ k=1,2.$$

The first transformation in (23) transforms initial *sigular* spectral curve to nonsingular trigonal curve of genus two. The second two transformations in (23) transfom nonsingular curve to canonical Weierstrass form of hyperelliptic curve of genus two. This curve is not contained among the curves investigated in [1], so the solutions parametrized by this curve are new.

4. Alternative Theta-functional integration

Let us consider the nonstationary Schrödinger equation with a real smooth finite-gap potential $u(x,t)$

$$(i\partial_t+\partial_x^2-u(x,t))\psi(x,t,\lambda)=0. \tag{25}$$

For every finite smooth finite-gap potential the following expansion of $u(x,t)$ in terms of squared eigenfunctions $\psi(x,t,\lambda)$ hold on [13], $u(x,t)=-\sum_{i=1}^{g}|\alpha_i|^2|\psi(x,t,p_i)|^2+C$ where α_i, C are constants parametrized in general by nonhyperelliptic Riemann surface K of genus g and p_i are some fixed points on K. This approach to the Manakov model has been at first used in [14]. The quasi-periodic solutions in terms of theta functions are obtained in [13]. In particular case when K is hyperelliptic Riemann surface we can construct the following special Baker-AKhiezer function $\psi(x,t,\lambda)=\sqrt{\mathcal{F}(z,\lambda)}\exp(\Theta)$ where

$$\Theta=\frac{1}{2}\imath cx+\imath(\lambda-\frac{1}{4}c^2)t-\frac{1}{2}\left(\int_0^z\frac{\nu(\lambda)}{\mathcal{F}(z',\lambda)}dz'\right),$$

and $z=x-ct$, $\mathcal{F}(z,\lambda)$ is the generalised Hermite or the standard Hermite polynomial [15]. A brief computation reveals that the BA function solves Hill's equation

$$\left(\frac{d^2}{dz^2}-u(z)\right)\tilde{\psi}=\lambda\tilde{\psi},\quad \tilde{\psi}(z,\lambda)=\sqrt{\mathcal{F}(z,\lambda)}\exp\left(-\frac{1}{2}\int_0^z\frac{\nu(\lambda)}{\mathcal{F}(z',\lambda)}dz'\right).$$

For the special points $a_i-\Delta\ i=1,2$ in closed intervals $[\lambda_{2i-1},\lambda_{2i}]$, $i=1,2$ the functions $Q_i=\alpha_i\psi(x,t,\lambda=a_i-\Delta)$ are solutions of (1). We undertake the shift of the spectral parameter [15], $\lambda\longrightarrow\lambda+\Delta$, $\Delta=\frac{2}{5}a_1+\frac{2}{5}a_2$ to make initial curve compatible with the Lax representation. The final formula for the solutions of the system (1) then reads

$$Q_1(x,t) = \sqrt{2\frac{\mathcal{F}(x,a_1-\Delta)}{a_1-a_2}}\exp\left\{\imath a_1 t - \frac{1}{2}\nu(a_i-\Delta)\int^x \frac{\mathrm{d}x'}{\mathcal{F}(x',a_1-\Delta)}\right\},$$
$$Q_2(x,t) = \sqrt{2\frac{\mathcal{F}(x,a_2-\Delta)}{a_2-a_1}}\exp\left\{\imath a_2 t - \frac{1}{2}\nu(a_2-\Delta)\int^x \frac{\mathrm{d}x'}{\mathcal{F}(x',a_2-\Delta)}\right\}. \tag{26}$$

5. Conclusions

New exact solutions of the Manakov model were derived using different methods: Lax pair method and ansatz method. The solution was explicitly given in terms of Weierstrass and Jacobian elliptic functions. The relation to previously derived solutions are given.

Acknowledgements

The present work is supported by the National Science Foundation of Bulgaria, contract No F-1410. I also thank Prof. V.S. Gerdjikov for the careful reading of my paper and for useful remarks.

References

1. J.N. Elgin, V.Z. Enolskii and A.R. Its, *Effective integration of the nonlinear vector Schrodinger equation*, Physica D: Nonlinear Phenomena, **225**, (2007) 127–152.
2. V. Gerdjikov and T. Valchev, *Breather Solutions of N-wave Equations, Eighth International Conference on Geometry*, Integrability and Quantization June 914, 2006, Varna, Bulgaria , Editors: I.M. Mladenov and M. de Len, SOFTEX, Sofia (2006), 1–17
3. A. Degasperis and S. Lombardo, *Multicomponent integrable wave equations: I. Darboux-dressing transformation*, J. Phys. A: Math. Theor. **40** (2007) 961–977.
4. V.S. Gerdjikov, *Complete Integrability, Gauge Equivalence and Lax Representations of the Inhomogeneous Nonlinear Evolution Equations*, Theor. Math. Phys. **92** (1992) 374–386.
5. ———, *Basic aspects of soliton theory*, Sixth International Conference on Geometry, Integrability and Quantization June 310, 2004, Varna, Bulgaria, Editors: I.M. Mladenov and A. C. Hirshfeld, SOFTEX, Sofia (2004), 1–48.
6. H.J. Shin, *Squared wavefunctions approach to periodic solutions of vector nonlinear Schrödinger equation*, J. Phys. A: Math. Gen. **36** (2003) 4113–4126.
7. S.V. Manakov, *On the theory of two-dimensional stationary self-focusing of electromagnetic waves*, Sov. Phys. JETP **38** (1974) 248–253.

8. A.V. Porubov and D.F. Parker, *Some general periodic solutions to coupled nonlinear Schrödinger equation*, Wave Motion **29** (1999) 97–109.
9. N.A. Kostov, V.Z. Enolskii, V.S. Gerdjikov, V.V. Konotop and M. Salerno, *Two-component Bose-Einstein condensates in periodic potential*, Phys. Rev E **70**, (2004) 056617.
10. V. Enolskii and N. Kostov, *On the geometry of elliptic solitons*, Acta Applicandae Math. **36** (1994) 57–86.
11. N. Kostov and M. Uzunov, *New kinds of periodical waves in birefringent optical fibers*, Opt. Commun. **89** (1992) 389–392.
12. M. Florjanzyk and R. Tremblay, *Periodic and solitary waves in bimodal optical fibers*, Phys. Lett. A **25** (1989) 34–36.
13. I.M. Krichever, *Spectral theory of "finite-gap" nonstationary Schrödinger operators, Nonstationary Peierls model*, Funk. Analiz. Pril. **20**, 42–54, 1986 (in Russian).
14. I.V. Cherednik, *Differential equations of the Baker-Akhiezer functions of algebraic curves*, Funct. Anal. Appl. **12**, 195–203, 1978.
15. P. Christiansen, J. Eilbeck, V. Enolskii and N. Kostov, *Quasi-periodic and periodic solutions for Manakov type systems of coupled nonlinear Schrodinger equations*, Proc. Royal Soc. London A. **456** (2000) 2263–2281.

THE MANAKOV SYSTEM AS TWO MOVING INTERACTING CURVES

N.A. KOSTOV

Institute of Electronics, Bulgarian Academy of Sciences, 72 Tsarigradsko chaussee, 1784 Sofia, Bulgaria

R. DANDOLOFF[†], V.S. GERDJIKOV[‡] and G.G. GRAHOVSKI[†‡]

[†] *Université de Cergy-Pontoise, 2 avenue, A. Chauvin, F-95302, Cergy-Pontoise Cedex, France*

[†‡] *Institute for Nuclear Research and Nuclear Energy, Bulgarian Academy of Sciences, 72 Tsarigradsko chaussee, 1784 Sofia, Bulgaria*

The two time-dependent Schrödinger equations in a potential $V(s,u)$, u denoting time, can be interpreted geometrically as a moving interacting curves whose Fermi-Walker phase density is given by $-(\partial V/\partial s)$. The Manakov model appears as two moving interacting curves using extended da Rios system and two Hasimoto transformations.

Keywords: Soliton curves and surfaces; Manakov model; Periodic solitons.

1. Introduction

In recent years, there has been a large interest in the applications of the Frenet-Serret equations [1,2] for a space curve in various contexts, and interesting connections between geometry and integrable nonlinear evolution equations have been revealed. The subject of how space curves evolve in time is of great interest and has been investigated by many authors. Hasimoto [3] showed that the evolution of a thin vortex filament regarded as a moving space curve can be mapped to the nonlinear Schrödinger equation (NLSE):

$$i\Psi_u + \Psi_{ss} + \frac{1}{2}|\Psi|^2\Psi = 0. \tag{1}$$

Here, u and s are time and space variables, respectively, subscripts denote partial derivatives. Lamb [4] used Hasimoto transformation to connect other motions of curves to the mKdV and sine-Gordon equations. Langer and Perline [5] showed that the dynamics of non-stretching vortex filament in $\mathbb{R}^3$ leads to the NLS hierarchy. Motions of curves on S^2 and S^3 were considered by Doliwa and Santini [6]. Lakshmanan [7] interpreted the dynamics of a nonlinear string of fixed length in $\mathbb{R}^3$ through the consideration of the motion of an arbitrary rigid body along it, deriving the AKNS spectral problem without spectral parameter. Recently, Nakayama [8] showed that the defocusing nonlinear Schrödinger equation, the Regge-Lund equation, a coupled system of KdV equations and their hyperbolic type arise from motions of curves on hyperbola in the Minkowski space. Recently the connection between motion of space or plane curves and integrable equations has drawn wide interest and many results have been obtained [9,10,13–17].

2. Preliminaries

2.1. *The Manakov model*

Time-dependent Schrödinger equation in potential $V(s,u)$

$$i\Psi_u + \Psi_{ss} - V(s,u)\Psi = 0, \tag{2}$$

goes into the NLS eq. (1) if the potential $V(s,u) = -1/2|\psi(s,u)|^2$. Similarly, a set of two time-dependent Schrödinger equations:

$$i\Psi_{1,u} + \Psi_{1,ss} - V(s,u)\Psi_1 = 0, \quad i\Psi_{2,u} + \Psi_{2,ss} - V(s,u)\Psi_2 = 0. \tag{3}$$

where $V(s,u) = -|\Psi_1|^2 - |\Psi_2|^2$, can be viewed as the Manakov system:

$$i\Psi_{1,u} + \Psi_{1,ss} + (|\Psi_1|^2 + |\Psi_2|^2)\Psi_1 = 0, \tag{4}$$

$$i\Psi_{2,u} + \Psi_{2,ss} + (|\Psi_1|^2 + |\Psi_2|^2)\Psi_2 = 0. \tag{5}$$

It is convenient to use two Hasimoto transformations[3]

$$\Psi_i = \kappa_i(s,u)\exp\left[i\int^s \tau_i(s',u)ds'\right], \quad i = 1,2, \tag{6}$$

in Eqs. (4), (5). Equating imaginary and real parts, this leads to the coupled partial differential equations (extended daRios system [18])

$$\kappa_{i,u} = -(\kappa_i\tau_i)_s - \kappa_{i,s}\tau_i, \quad i = 1,2, \tag{7}$$

$$\tau_{i,u} = \left[\frac{\kappa_{i,ss}}{\kappa_i} - \tau_i^2\right]_s - V(s,u)_s, \tag{8}$$

where

$$V(s,u) = -|\Psi_1|^2 - |\Psi_2|^2 = -\kappa_1^2 - \kappa_2^2. \tag{9}$$

2.2. Soliton curves

A three dimensional space curve is described in parametric form by a position vectors $\mathbf{r}_i(s)$, $i = 1, 2$, where s is the arclength. Let $\mathbf{t}_i = (\partial \mathbf{r}_i/\partial s)$, $i = 1, 2$ be the two unit tangent vectors along the two curves. At a given instant of time the triads of unit orthonormal vectors $(\mathbf{t}_i, \mathbf{n}_i, \mathbf{b}_i)$, where $\mathbf{n}_i$ and $\mathbf{b}_i$ denote the normals and binormals, respectively, satisfy the Frenet-Serret equations for two curves

$$\mathbf{t}_{i,s} = \kappa_i \mathbf{n}_\mathbf{i}, \quad \mathbf{n}_{i,s} = -\kappa_i \mathbf{t}_i + \tau_i \mathbf{b}_i, \quad \mathbf{b}_{i,s} = -\tau_i \mathbf{n}_i, \quad i = 1, 2, \tag{10}$$

κ_i and τ_i denote, respectively the two curvatures and torsions of the curves. A moving curves are described by $r_i(s, u)$, where u denote time. The temporal evolution of two triads corresponding to a given value s can be written in the general form as

$$\mathbf{t}_{i,u} = g_i \mathbf{n}_i + h_i \mathbf{b}_i, \quad \mathbf{n}_{i,u} = -g_i \mathbf{t}_i + \tau_i^0 \mathbf{b}_i, \quad \mathbf{b}_{i,u} = -h_i \mathbf{t}_i - \tau_i^0 \mathbf{n}_i, \tag{11}$$

where the coefficients g_i, h_i and τ_i^0, as well as κ_i and τ_i, are functions of s and u.

$$\begin{pmatrix} \mathbf{t}_i \\ \mathbf{n}_i \\ \mathbf{b}_i \end{pmatrix}_s = \begin{pmatrix} 0 & \kappa_i & 0 \\ -\kappa_i & 0 & \tau_i \\ 0 & -\tau_i & 0 \end{pmatrix} \begin{pmatrix} \mathbf{t}_i \\ \mathbf{n}_i \\ \mathbf{b}_i \end{pmatrix}, \quad \begin{pmatrix} \mathbf{t}_i \\ \mathbf{n}_i \\ \mathbf{b}_i \end{pmatrix}_u = \begin{pmatrix} 0 & g_i & h_i \\ -g_i & 0 & \tau_i^0 \\ -h_i & -\tau_i^0 & 0 \end{pmatrix} \begin{pmatrix} \mathbf{t}_i \\ \mathbf{n}_i \\ \mathbf{b}_i \end{pmatrix}.$$

Introduce

$$L_i = \begin{pmatrix} 0 & \kappa_i & 0 \\ -\kappa_i & 0 & \tau_i \\ 0 & -\tau_i & 0 \end{pmatrix}, \quad M_i = \begin{pmatrix} 0 & g_i & h_i \\ -g_i & 0 & \tau_i^0 \\ -h_i & -\tau_i^0 & 0 \end{pmatrix} \tag{12}$$

and $\Delta \mathbf{t}_i \equiv (\mathbf{t}_{i,su} - \mathbf{t}_{i,us})$, $\Delta \mathbf{n}_i \equiv (\mathbf{n}_{i,su} - \mathbf{n}_{i,us})$, and $\Delta \mathbf{b}_i \equiv (\mathbf{b}_{i,su} - \mathbf{b}_{i,us})$. Then

$$\begin{aligned} \begin{pmatrix} \Delta \mathbf{t}_i \\ \Delta \mathbf{n}_i \\ \Delta \mathbf{b}_i \end{pmatrix} &= (\partial_s M_i - \partial_u L_i + [L_i, M_i]) \begin{pmatrix} \mathbf{t}_i \\ \mathbf{n}_i \\ \mathbf{b}_i \end{pmatrix} \\ &= \begin{pmatrix} 0 & \alpha_i^1 & \alpha_i^2 \\ -\alpha_i^1 & 0 & \alpha_i^3 \\ -\alpha_i^2 & -\alpha_i^3 & 0 \end{pmatrix} \begin{pmatrix} \mathbf{t}_i \\ \mathbf{n}_i \\ \mathbf{b}_i \end{pmatrix}, \end{aligned} \tag{13}$$

where

$$\alpha_i^1 = \kappa_{i,u} g_{i,s} + h_i \tau_i, \; \alpha_i^2 = -h_{i,s} + \kappa_i \tau_i^0 - g_i \tau_i, \; \alpha_i^3 = \tau_{i,u} - \tau_{i,s} - \kappa_i h_i, \tag{14}$$

$$\kappa_{i,u} = g_{i,s} - h_i \tau_i, \qquad \tau_i^0 = (h_{i,s} + g_i \tau_i)/\kappa_i. \tag{15}$$

A generic curve evolution must satisfy the geometric constraints

$$\Delta \mathbf{t}_i \cdot (\Delta \mathbf{n}_i \times \Delta \mathbf{b}_i) = 0, \tag{16}$$

i.e., $\Delta \mathbf{t}_i$, $\Delta \mathbf{n}_i$ and $\Delta \mathbf{b}_i$ must remain coplanar vectors under time involution. Further, since Eq. (16) is automatically satisfied for $\Delta \mathbf{t}_i = 0$, we see that $\Delta \mathbf{n}_i$ and $\Delta \mathbf{b}_i$ need not necessarily vanish. In addition, we see from (16) that $\Delta \mathbf{t}_i = 0$ implies $\alpha_i^1 = \alpha_i^2 = 0$, so that

$$\Delta \mathbf{n}_i = \alpha_i^3 \Delta \mathbf{b}_i, \quad \Delta \mathbf{b}_i = \alpha_i^3 \Delta \mathbf{n}_i \quad g_i = -\kappa_i \tau_i, \qquad h_i = \kappa_{i,s}. \tag{17}$$

Substituting these in the second equation in (15) gives

$$\tau_i^0 = \left[\frac{\kappa_{i,ss}}{\kappa_i} - \tau_i^2 \right], \tag{18}$$

hence Eq. (7) yields $(\tau_{i,u} - \tau_{i,s}^0) = -V(s,u)_s = (\kappa_1^2 + \kappa_2^2)_s$. Next there is an underlying angle anholonomy [19,20] or 'Fermi-Walker phase' $\delta \Phi^{FW} = (\tau_{i,u} - \tau_{i,s}^0) ds du$ with respect to its original orientation, when s and u change along an infinitesimal closed path of area $dsdu$.

3. Integration of the extended da Rios system

The coupled nonlinear equations (7), (8) constitute the extended Da Rios system as derived in [18]. The solutions of (7), (8) with $\kappa(\xi)$ and $\tau = \tau(\xi)$, where $\xi = s - cu$ are simple. On substitution, we obtain

$$c\,\kappa_{i,\xi} = 2\kappa_{i,\xi}\tau_i + \kappa_i \tau_{i,\xi}, \quad \xi = s - cu, i = 1,2, \tag{19}$$

$$-c\tau_{i,\xi} = \left[-\tau_i^2 + \frac{\kappa_{i,\xi\xi}}{\kappa_i} + \kappa_1^2 + \kappa_2^2 \right]_\xi, \qquad \tau_i = \frac{c}{2}, \tag{20}$$

where we use the boundary condition $\kappa_i \to 0$, $i = 1,2$ as $s \to \infty$. Hence κ_i obey the nonlinear oscillator equations

$$\kappa_{i,\xi\xi} + \left(\sum_{j=1}^{2} \kappa_j^2 \right) \kappa_i = a_i \kappa_i, \quad i = 1,2, \tag{21}$$

where a_i, $i = 1,2$ are arbitrary constants.

Example 1. One soliton solutions of the Manakov system are given by

$$\Psi_1 = \sqrt{2a}\,\epsilon_1 e^{i(\frac{1}{2}c(s-s_0)+(a-\frac{1}{4}c^2)u)} \operatorname{sech}(\sqrt{a}(s - s_0 - ct)) \tag{22}$$

$$\Psi_2 = \sqrt{2a}\,\epsilon_2 e^{i(\frac{1}{2}c(s-s_0)+(a-\frac{1}{4}c^2)u)} \operatorname{sech}(\sqrt{a}(s - s_0 - ct)), \tag{23}$$

and $|\epsilon_1|^2 + |\epsilon_2|^2 = 1$

We first note that for Manakov system, the expressions for the curvatures $\kappa_i, i = 1,2$ and the torsions $\tau_i, i = 1,2$ for the moving curves corresponding to a one soliton solutions of the Manakov system are given by

$$\kappa^2 = \kappa_1^2 + \kappa_2^2 = \sqrt{2a}\,\mathrm{sech}(\sqrt{a}(s - s_0 - ct)), \quad \tau_1 = \tau_2 = \frac{1}{2}c, \tag{24}$$

and

$$\kappa_1 = \sqrt{2a}\,\epsilon_1 \mathrm{sech}(\sqrt{a}(s - s_0 - ct)), \quad \kappa_2 = \sqrt{2a}\,\epsilon_2 \mathrm{sech}(\sqrt{a}(s - s_0 - ct)).$$

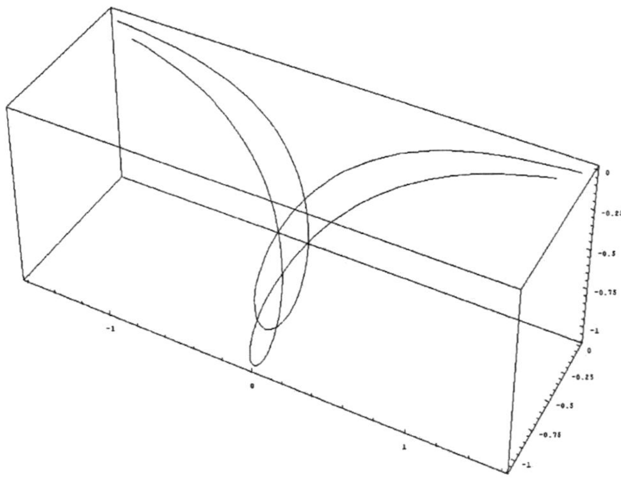

Fig. 1. Two curves (28) of one soliton solution of Manakov system, $\epsilon_1 = \sqrt{2}/\sqrt{3}$, $\epsilon_2 = 1/\sqrt{3}$

Example 2. One special solution of Manakov system is written by

$$\kappa_1 = C_1 \mathrm{cn}(\alpha\xi, k), \qquad \kappa_2 = C_2 \mathrm{cn}(\alpha\xi, k), \tag{25}$$

where

$$\alpha^2 = \frac{a_1}{2k^2 - 1}, \quad C_1^2 + C_2^2 = 2\alpha^2 k^2, \quad a_1 = a_2 = a\,. \tag{26}$$

In the limit $k \to 1$ we obtain the well known *Manakov* soliton solution

$$\Psi_1 = \frac{\sqrt{2a}\epsilon_1 \exp\left\{i\left(\frac{1}{2}c(s - s_0) + (a - \frac{1}{4}c^2)u\right)\right\}}{\mathrm{ch}(\sqrt{a}(s - s_0 - ct))},$$

$$\Psi_2 = \frac{\sqrt{2a}\epsilon_2 \exp\left\{i\left(\frac{1}{2}c(s - s_0) + (a - \frac{1}{4}c^2)u\right)\right\}}{\mathrm{ch}(\sqrt{a}(s - s_0 - ct))}.$$

Here we introduce the following notations

$$|\epsilon_1|^2 + |\epsilon_2|^2 = 1, \quad \zeta_1 = \frac{1}{2}c + i\sqrt{a} = \xi_1 + i\eta_1, \tag{27}$$

where s_0 is the position of soliton, (ϵ_1, ϵ_2) are the components of polarization vector. The real part of ζ_1 i.e. $c/2$ gives us the soliton velocity while the imaginary part of ζ_1, i.e. $\sqrt{2a}$, gives the soliton amplitude and width.

Example 3. Integrating (10) for two unit tangent vectors along the curves $\mathbf{t}_i = (\partial \mathbf{r}_i/\partial s), i = 1, 2$ for position vectors $\mathbf{r}_i(s), i = 1, 2$ we obtain

$$\mathbf{r}_j = \begin{pmatrix} \frac{s}{2} - \frac{\epsilon_j}{\epsilon_j^2 + \frac{1}{2}c} \tanh\left(\epsilon_j(s - cu)\right) \\ -\frac{\epsilon_j}{\epsilon_j^2 + \frac{1}{2}c} \operatorname{sech}(\epsilon_j(x - cu)) \cos\left(\frac{1}{2}cs + (\epsilon_j^2 - \frac{1}{4}c^2)u\right) \\ -\frac{\epsilon_j}{\epsilon_j^2 + \frac{1}{2}c} \operatorname{sech}(\epsilon_j(x - cu)) \sin\left(\frac{1}{2}cs + (\epsilon_j^2 - \frac{1}{4}c^2)u\right) \end{pmatrix}, \quad j = 1, 2, \tag{28}$$

and $\epsilon_1 = \cos\alpha$, $\epsilon_2 = \sin\alpha$, where α is arbitrary positive number.

Example 4. Let $u(x) = 6\wp(\xi + \omega')$ be the two-gap Lamé potential with simple periodic spectrum (see for example [21])

$$\lambda_0 = -\sqrt{3g_2},\ \lambda_1 = -3e_0,\ \lambda_2 = -3e_1,\ \lambda_3 = -3e_2,\ \lambda_4 = \sqrt{3g_2}, \tag{29}$$

and the corresponding Hermite polynomial have the form

$$F(\wp(\xi + \omega'), \lambda) = \lambda^2 - 3\wp(\xi + \omega')\lambda + 9\wp^2(\xi + \omega') - \frac{9}{4}g_2. \tag{30}$$

Consider the genus 2 nonlinear anisotropic oscillator (21) with Hamiltonian

$$H = \frac{1}{2}(p_1^2 + p_2^2) + \frac{1}{4}(\kappa_1^2 + \kappa_2^2)^2 - \frac{1}{2}(a_1\kappa_1^2 + a_2\kappa_2^2), \tag{31}$$

where (κ_i, p_i), $i = 1, 2$ are canonical variables with $p_i = \kappa_{i,x}$ and a_1, a_2 are arbitrary constants. The simple solutions of these system are given in terms of Hermite polynomial

$$\kappa_1^2 = 2\frac{F(x, \tilde{\lambda}_1)}{\tilde{\lambda}_2 - \tilde{\lambda}_1}, \quad \kappa_2^2 = 2\frac{F(x, \tilde{\lambda}_2)}{\tilde{\lambda}_1 - \tilde{\lambda}_2}. \tag{32}$$

Let us list the corresponding solutions

(A) Periodic solutions in terms of single Jacobian elliptic function

The nonlinear anisotropic oscillator admits the following solutions:

$$\kappa_1 = C_1 \operatorname{sn}(\alpha\xi, k), \qquad \kappa_2 = C_2 \operatorname{cn}(\alpha\xi, k). \tag{33}$$

Here the amplitudes C_1, C_2 and the temporal pulse-width $1/\alpha$ are defined by the parameters a_1 and a_2 as follows:

$$\alpha^2 k^2 = a_2 - a_1, \quad C_1^2 = a_2 + \alpha^2 - 2\alpha^2 k^2, \quad C_2^2 = a_1 + \alpha^2 + \alpha^2 k^2, \tag{34}$$

where $0 < k < 1$.

Following our spectral method it is clear, that the solutions (33) are associated with eigenvalues $\lambda_2 = -e_2$ and $\lambda_3 = -e_3$ of one — gap Lamé potential.

(B) Periodic solutions in terms of products of Jacobian elliptic functions Another solution is defined by [22]

$$\kappa_1 = C\mathrm{dn}(\alpha\xi, k)\mathrm{sn}(\alpha\xi, k), \qquad \kappa_2 = C\mathrm{dn}(\alpha\xi, k)\mathrm{cn}(\alpha\xi, k), \tag{35}$$

where sn, cn, dn are the standard Jacobian elliptic functions, k is the modulus of the elliptic functions $0 < k < 1$. The wave characteristic parameters: amplitude C, temporal pulse-width $1/\alpha$ and k are related to the physical parameters and, k through the following dispersion relations

$$C^2 = \frac{2}{5}(4a_2 - a_1), \quad \alpha^2 = \frac{1}{15}(4a_2 - a_1), \quad k^2 = \frac{5(a_2 - a_1)}{4a_2 - a_1}. \tag{36}$$

We have found the following solutions of the nonlinear oscillator [23]

$$\kappa_1 = C\alpha^2 k^2 \mathrm{cn}(\alpha\xi, k)\mathrm{sn}(\alpha\xi, k), \quad \kappa_2 = C\alpha^2 \mathrm{dn}^2(\alpha\xi, k) + C_1 \tag{37}$$

where C, C_1, α and k are expressed through parameters a_1 and a_2 by the following relations

$$C^2 = \frac{18}{a_2 - a_1}, \quad \alpha^2 = \frac{1}{10}\left(2a_2 - 3a_1 + \sqrt{\frac{5}{3}(a_2^2 - a_1^2)}\right)$$

$$k^2 = \frac{2\sqrt{\frac{5}{3}(a_2^2 - a_1^2)}}{\sqrt{\frac{5}{3}(a_2^2 - a_1^2)} + 2a_2 - 3a_1}, \quad C_1 = \frac{C}{30}(4a_1 - a_2). \tag{38}$$

(C) Periodic solutions associated with the two-gap Treibich-Verdier potentials. Below we construct the two periodic solutions associated with the Treibich-Verdier potential. Let us consider the potential

$$u(x) = 6\wp(\xi + \omega') + 2\frac{(e_1 - e_2)(e_1 - e_3)}{\wp(\xi + \omega') - e_1}, \tag{39}$$

and construct the solution in terms of Lamé polynomials associated with the eigenvalues $\tilde{\lambda}_1, \tilde{\lambda}_2$, $\tilde{\lambda}_1 > \tilde{\lambda}_2$

$$\tilde{\lambda}_1 = e_2 + 2e_1 + 2\sqrt{(e_1 - e_2)(7e_1 + 2e_2)}, \tag{40}$$
$$\tilde{\lambda}_2 = e_3 + 2e_1 + 2\sqrt{(e_1 - e_3)(7e_1 + 2e_3)}.$$

The finite and real solutions q_1, q_2 have the form

$$\kappa_1 = C_1\mathrm{sn}(\xi, k)\mathrm{dn}(\xi, k) + C_2\mathrm{sd}(\xi, k), \ \kappa_2 = C_3\mathrm{cn}(\xi, k)\mathrm{dn}(\xi, k) + C_4\mathrm{cd}(z, k),$$

where C_i, $i = 1, \dots 4$ are constants and have important geometrical interpretation [21] and sd, cd, are standard Jacobian elliptic functions. The concrete expressions in terms of $k, \tilde{\lambda}_1, \tilde{\lambda}_2$ are given in [24,25].

In a similar way we can find the elliptic solution associated with the eigenvalues

$$\tilde{\lambda}_1 = e_2 + 2e_1 + 2\sqrt{(e_1 - e_2)(7e_1 + 2e_2)}, \quad \tilde{\lambda}_2 = -6e_1. \tag{41}$$

We have

$$\kappa_1 = \tilde{C}_1 \mathrm{dn}^2(\xi, k), \qquad \kappa_2 = C_1 \mathrm{sn}(\xi, k)\mathrm{dn}(\xi, k) + C_2 \mathrm{sd}(\xi, k), \tag{42}$$

where $\tilde{C}_1, C_1, C_2$ are given in [24,25].

The general formula for elliptic solutions of genus 2 nonlinear anisotropic oscillator is given by [24]

$$\kappa_1^2 = \frac{1}{\tilde{\lambda}_2 - \tilde{\lambda}_1}\Bigg(2\tilde{\lambda}_1^2 + 2\tilde{\lambda}_1 \sum_{i=1}^{N} \wp(\xi - x_i) + 6 \sum_{1 \le i < j \le N} \wp(\xi - x_i)\wp(\xi - x_j) - \frac{N g_2}{4} + \sum_{1 \le i < j \le 5} \lambda_i \lambda_j\Bigg),$$

$$\kappa_2^2 = \frac{1}{\tilde{\lambda}_1 - \tilde{\lambda}_2}\Bigg(2\tilde{\lambda}_2^2 + 2\tilde{\lambda}_2 \sum_{i=1}^{N} \wp(\xi - x_i) + 6 \sum_{1 \le i < j \le N} \wp(\xi - x_i)\wp(\xi - x_j) - \frac{N g_2}{4} + \sum_{1 \le i < j \le 5} \lambda_i \lambda_j\Bigg),$$

where x_i are solutions of equations $\sum_{i \neq j} \wp'(x_i - x_j) = 0$, $j = 1, \dots, N$.

4. Extended da Rios-Betchov system

Following Betchov we can derive the system of equations, which may be reduced to those for a two fictitious gases with negative pressures accompanied with two complicated nonlinear dispersive stresses. Introducing four new variables $\rho_1 = \kappa_1^2$, $\rho_1 = \kappa_1^2$, $u_1 = 2\tau_1$, $u_2 = 2\tau_2$ using extended Da Rios system (7), (8) we obtain

$$\frac{\partial \rho_1}{\partial u} + \frac{\partial(\rho_1 u_1)}{\partial s} = 0, \qquad \frac{\partial \rho_2}{\partial u} + \frac{\partial(\rho_2 u_2)}{\partial s} = 0,$$

$$\frac{\partial(\rho_1 u_1)}{\partial u} + \frac{\partial}{\partial s}\left[\rho_1 u_1^2 - (\rho_1^2 + \rho_2^2) - \rho_1 \frac{\partial^2}{\partial s^2}(\log \rho_1)\right] = 0,$$

$$\frac{\partial(\rho_2 u_2)}{\partial u} + \frac{\partial}{\partial s}\left[\rho_2 u_2^2 - (\rho_1^2 + \rho_2^2) - \rho_2 \frac{\partial^2}{\partial s^2}(\log \rho_2)\right] = 0.$$

5. HF system is gauge equivalent to Manakov system

The vector nonlinear Schrödinger equation is associated with type **A.III** symmetric space $\mathrm{SU(n+1)/S(U(1)\otimes U(n))}$. The special case $n = 2$ of such symmetric space is associated with the famous Manakov system [26].

Let us first fix the notations and the normalizations of the basis of $\mathfrak{g}$. By Δ_+ (Δ_-) we shall denote the set of positive (negative) roots of the algebra with respect to some ordering in the root space. By $\{E_\alpha, H_i\}$, $\alpha \in \Delta$, $i = 1, \dots, r$ we denote the Cartan-Weyl basis of $\mathfrak{g}$ with the standard commutation relations.[27] Here H_i are Cartan generators dual to the basis vectors e_i in the root space. The root system is invariant under the action of the Weyl group $\mathfrak{W}(\mathfrak{g})$ of the simple Lie algebra $\mathfrak{g}$ [27].

Let us now consider the gauge equivalent systems. The notion of gauge equivalence allows us to associate with the vector nonlinear Schrödinger equation an equivalent equation solvable by the ISM for the gauge equivalent linear problem [28]:

$$\begin{aligned} \tilde{L}\tilde{\psi}(x,t,\lambda) &= \left(i\frac{d}{dx} - \lambda \mathcal{S}(x,t) \right) \tilde{\psi}(x,t,\lambda) = 0, \\ \tilde{M}\tilde{\psi}(x,t,\lambda) &= \left(i\frac{d}{dt} - \lambda^2 \mathcal{S} - \lambda \mathcal{S}_x \mathcal{S}(x,t) \right) \tilde{\psi}(x,t,\lambda) = 0, \end{aligned} \tag{43}$$

where

$$\begin{aligned} &\tilde{\psi}(x,t,\lambda) = \psi_0^{-1}\psi(x,t,\lambda), \quad \mathcal{S}(x,t) = \sum_{\alpha=1}^{r} (S_\alpha E_\alpha + S_\alpha^* E_{-\alpha}) + \sum_{j=1}^{r} S_j H_j, \\ &\mathcal{S}(x,t) = \mathrm{Ad}_{\hat{\psi}_0} J \equiv \psi_0^{-1} J \psi_0(x,t), \qquad J = \sum_{s=1}^{n} H_s, \end{aligned} \tag{44}$$

and $\psi_0 = \psi(x,t,0)$ is the Jost solution at $\lambda = 0$. The zero-curvature condition $[\tilde{L}, \tilde{M}] = 0$ is equivalent to $i\mathcal{S}_t - [\mathcal{S}, \mathcal{S}_{xx}] = 0$ with $\mathcal{S}^2 = I_n$.

6. Conclusions

In this paper the Manakov model is interpreted as two moving interacting curves. We derive new extended Da Rios system and obtain the soliton, one-, and two-phase periodic solution of two thin vortex filaments in an incompressible inviscid fluid. The solution was explicitly given in terms of Weierstrass and Jacobian elliptic functions.

Acknowledgements

The present work is supported by the National Science Foundation of Bulgaria, contract No F-1410. The work of one of us GGG is supported also by the Bulgarian National Scientific Foundation Young Scientists scholarship for the project "Solitons, differential geometry and biophysical models".

References

1. L.P. Eisenhart, *A treatise on the differential geometry of curves and surfaces*, Dover, New York, 1960.
2. W.K. Schief and C. Rogers, *Binormal motion of curves of constant curvature and torsion. Generation of soliton surfaces*, Proc. R. Soc. Lond. A, **455** (1999) 3163-3188;
   C. Rogers and W.K. Schief, *Bäcklund and Darboux Transformations: Geometry and Modern Applications in Soliton Theory*, Cambridge Univ. Press, Cambridge (2002).
3. H. Hasimoto, *A soliton on a vortex filament*, J. Fluid Mech. **51** (1972) 477–485.
4. G.L. Lamb, Jr., *Solitons on moving space curves*, J. Math. Phys. **18** (1977) 1654–1661.
5. J. Langer and R. Perline, *Poisson geometry of the filament equation*, J. Nonlinear Sci. 1991, **1**, N 1, (1991) 71–93.
6. A. Doliwa and P.M. Santini, *An elementary geometric characterization of the integrable motions of a curve*, Phys. Lett. A **185**, (1994) 373–384.
7. M. Lakshmanan, *Rigid body motions, space curves, prolongation structures, fiber bundles, and solitons*, J. Math. Phys. **20**, (1979) 1667–1672.
8. K. Nakayama, *Motion of curves in hyperboloid in the Minkowski space*, J. Phys. Soc. Jpn. **67**, (1998) 3031–3037.
9. A.S. Fokas and I.M. Gelfand, *Surfaces on Lie groups, on Lie algebras, and their integrability*, Comm. Math. Phys. **177** (1996), 203–220;
   A.S. Fokas, I.M. Gelfand, F. Finkel and Q.M. Liu, *A formula for constructing infinitely many surfaces on Lie algebras and integrable equations*, Selecta Math. (New ser.) **6** (2000) 347–375.
10. R.K. Dodd, *Soliton immersions*, Commun. Math. Phys. **197** (1998) 641–665.
11. A. Sym, *Soliton surfaces and their applications (soliton geometry from spectral problems)*, In: Geometrical aspects of the Einstein equations and integrable systems. (Lect. Notes Phys. **239**), Berlin-Heidelberg-New York: Springer, 1985, pp. 154–231.
12. K. Pohlmeyer, *Integrable Hamiltonian systems and interactions through quadratic constraints*, Commun. Math. Phys. **46** (1976) 207–221.
13. J. Cieśliński, *A generalized formula for integrable classes of surfaces in Lie algebras*, J. Math. Phys. **38** (1997) 4255–4272.
14. K. Chou and C. Qu, *Integrable equations arising from motions of plane curves*, Physica D **162** (2002) 9–33.
15. J. Langer and R. Perline, *Geometric Realizations of Fordy-Kulish Nonlinear*

Schrödinger Systems, Pacific J. Math. **195** (2000) 157–178.

16. H.J. Shin, *Vortex filament motion under the localized induction approximation in terms of Weierstrass elliptic functions*, Phys. Rev. **E 65**, (2002) 036317.
17. A. Calini and T. Ivey, *Connecting geometry, topology and spectra for finite-gap NLS potentials*, Physica D **152-153** (2001), 9–19;
    ______, *Finite-gap solutions of the Vortex Filament Equation: Genus one solutions and symmetric solutions*, Journal of Nonlinear Science **15**, (2005), 321–361.
18. L.S. Da Rios, *Sul moto d'un liquido indefinito con un filetto vorticoso*, Rend. Circ. Mat. Palermo **22** (1906) 117–135.
19. R. Balakrishnan and R. Dandoloff, *The Schrödinger equation as a moving curve*, Phys. Lett. A **260** (1999) 62–67.
20. R. Balakrishnan, A.R. Bishop and R. Dandoloff, *Geometric phase in the classical continuous antiferromagnetic Heisenberg spin chain*, Phys.Rev. Lett. **64** (1990) 2107–2110.
21. V. Enolskii and N. Kostov, *On the geometry of elliptic solitons*, Acta Applicandae Math. **36** (1994) 57–86.
22. M. Florjanzyk and R. Tremblay, *Periodic and solitary waves in bimodal optical fibers*, Phys. Lett. A **25** (1989) 34–36.
23. N. Kostov and M. Uzunov, *New kinds of periodical waves in birefringent optical fibers*, Opt. Commun. **89** (1992) 389–392.
24. P.L. Christiansen, J.C. Eilbeck, V.Z. Enolskii and N.A. Kostov, *Quasi-periodic solutions of the coupled nonlinear Schrodinger equations*, Proc. Royal Soc. London A **451** (1995) 685–700.
25. P. Christiansen, J. Eilbeck, V. Enolskii and N. Kostov, *Quasi-periodic and periodic solutions for Manakov type systems of coupled nonlinear Schrodinger equations*, Proc. Royal Soc. London A. **456** (2000) 2263–2281.
26. S.V. Manakov, *On the theory of two-dimensional stationary self-focusing of electromagnetic waves*, Sov. Phys. JETP **38** (1974) 248–253.
27. S. Helgasson, *Differential geometry, Lie groups and symmetric spaces*, AMS Providence, 2001.
28. A.P. Fordy and P.P. Kulish, *Nonlinear Schrodinger Equations and Simple Lie Algebras*, Commun. Math. Phys. **89** (1983) 427–443.

CLUSTER SETS AND PERIODICITY IN SOME STRUCTURE FRACTALS

J. ŁAWRYNOWICZ*

Institute of Physics, University of Łódź
Pomorska 149/153, PL-90-236 Łódź, Poland
Institute of Mathematics, Polish Academy of Sciences
Łódź Branch, Banacha 22, PL-90-238 Łódź, Poland
E-mail: jlawryno@uni.lodz.pl

S. MARCHIAFAVA

Dipartimento di Matematica "Guido Castelnuovo"
Università di Roma I "La Sapienza"
Piazzale Aldo Moro, 2, I-00-185 Roma, Italia
E-mail: marchiaf@mat.uniroma1.it

M. NOWAK-KĘPCZYK

High School of Business
Kolejowa 22, PL-26-600 Radom, Poland
E-mail: gosianmk@poczta.onet.pl

It is well known that starting with real structure, the Cayley-Dickson process gives complex, quaternionic, and octonionic (Cayley) structures related to the Adolf Hurwitz composition formula for dimensions $p = 2, 4$ and 8, respectively, but the procedure fails for $p = 16$ in the sense that the composition formula involves no more a triple of quadratic forms of the same dimension; the other two dimensions are $n = 2^7$. Instead, Ławrynowicz and Suzuki (2001) have considered graded fractal bundles of the flower type related to complex and Pauli structures and, in relation to the iteraton process $p \to p + 2 \to p + 4 \to \cdots$, they have constructed 2^4-dimensional "bipetals" for $p = 9$ and 2^7-dimensional "bisepals" for $p = 13$. The objects constructed appear to have an interesting property of periodicity related to the gradating function on the fractal diagonal interpreted as the "pistil" and a family of pairs of segments parallel to the diagonal and equidistant from it, interpreted as the "stamens".

2000 Mathematics Subject Classification: 81R25, 32L25, 53A50, 15A66.

*Research of the author partially supported by the State Committee for Scientific Research grant PB1 P03A 001 26 (Sections 1–3 of the paper), and partially by the grant of the University of Łódź no. 505/692 (Sections 4–5).

The present paper aims at an effective, explicit determination of the periods and expressing them in terms of complex and quaternionic structures, thus showing the quaternionic background of that periodicity. The proof of the Periodicity Theorem is given in the case where the index of the generator of the algebra in question does not exeed the order of the initial algebra, but, in contrast to our earlier paper [8], in the degenerated situations: with period 1, or involving perpendicular lower or upper convergence. In contrast to earlier results, the fractal bundle flower structure, in particular petals, bipetals, the ovary, ovules, pistils, and stamens are not introduced *ab initio*; they are quoted *a posteriori*, when they are fully motivated.

Keywords: Clifford algebra; Quaternion; Billinear form; Quadratic form.

1. Introduction and statement of the periodicity theorem

Given generators $A_1^1, A_2^1, \ldots, A_{2p-1}^1$ of a Clifford algebra $Cl_{2p-1}(\mathbb{C})$, $p = 2, 3, \ldots$, in particular the generators

$$\sigma_1 = \begin{pmatrix} 0 & 1 \\ 1 & 0 \end{pmatrix}, \quad \sigma_2 = \begin{pmatrix} 0 & -i \\ i & 0 \end{pmatrix}, \quad \sigma_3 = \begin{pmatrix} 1 & 0 \\ 0 & -1 \end{pmatrix}$$

of the Pauli algebra [19,20], consider the sequence

$$\begin{aligned} A_\alpha^{q+1} &= \sigma_3 \otimes A_\alpha^q \equiv \begin{pmatrix} A_\alpha^q & 0 \\ 0 & -A_\alpha^q \end{pmatrix}, \quad \alpha = 1, 2, \ldots, 2p+2q-3; \\ A_{2p+2q-2}^{q+1} &= \sigma_1 \otimes I_{p,q} \equiv \begin{pmatrix} 0 & I_{p,q} \\ I_{p,q} & 0 \end{pmatrix}, \\ A_{2p+2q-1}^{q+1} &= -\sigma_2 \otimes I_{p,q} \equiv \begin{pmatrix} 0 & iI_{p,q} \\ -iI_{p,q} & 0 \end{pmatrix}, \end{aligned} \tag{1}$$

of generators of Clifford algebras $Cl_{2p+2q-1}(\mathbb{C})$ $q = 1, 2, \ldots$, and the sequence of corresponding systems of closed squares Q_q^α of diameter 1, centered at the origin of $\mathbb{C}$, where $I_{p,q} = I_{2^{p+q-2}}$, the unit matrix of order 2^{p+q-2}. It is convenient to start with q always from 1, i.e., to shift q for $\alpha \geq 2p$ correspondingly.

Within a closed square Q_q^α consider its diameter

$$L_\infty = \left[\frac{1}{2\sqrt{2}}(-1+i);\ \frac{1}{2\sqrt{2}}(1-i)\right]$$

and two segments, symmetric and equidistant with respect to L_∞:

$$\begin{aligned} L_h^- &= \left[\frac{1}{2\sqrt{2}}(-1+i-i\varepsilon_q^r);\ \frac{1}{2\sqrt{2}}(1-\varepsilon_q^r-i)\right], \\ L_h^+ &= \left[\frac{1}{2\sqrt{2}}(-1+\varepsilon_q^r+i);\ \frac{1}{2\sqrt{2}}(1-i+i\varepsilon_q^r)\right], \end{aligned}$$

where

$$\varepsilon_q^r = 1/2^h, \quad h = p + q - 1 - r, \quad r = \begin{cases} 2 & \text{for} \quad \alpha = 1, 2, \ldots, 2p-1; \\ \left[\frac{1}{2}\alpha\right] & \text{for} \quad \alpha = 2p, 2p+1, \ldots \end{cases}$$

and [ ] denotes the function "entier". Clearly, dist $(L_h^{\pm}, L_\infty) = 1/2^{h+2}$.

Consider then the sets: L_∞^0 of points

$$z = \frac{1}{2\sqrt{2}} \frac{m}{2^n}(1-i), \quad m = 0, \pm 1, \ldots, \pm(2^n - 1); \; n = 0, 1, \ldots, \text{ of } L_\infty,$$

L_h^0 of points

$$z_-^-(h) = \frac{1}{2\sqrt{2}} \frac{m}{2^n}(1-i) - \frac{1}{2\sqrt{2}}\varepsilon_q^r \text{ and } z_+^-(h) = \frac{1}{2\sqrt{2}} \frac{m}{2^n}(1-i) - \frac{i}{2\sqrt{2}}\varepsilon_q^r,$$
$$m = 0, \pm 1, \ldots, \pm(2^n - 1); \quad n = 0, 1, \ldots, \quad \text{of} \quad L_h^-,$$
$$z_+^+(h) = \frac{1}{2\sqrt{2}} \frac{m}{2^n}(1-i) + \frac{1}{2\sqrt{2}}\varepsilon_q^r \text{ and } z_-^+(h) = \frac{1}{2\sqrt{2}} \frac{m}{2^n}(1-i) + \frac{i}{2\sqrt{2}}\varepsilon_q^r,$$
$$m = 0, \pm 1, \ldots, \pm(2^n - 1); \quad n = 0, 1, \ldots, \quad \text{of} \quad L_h^+,$$

and the corresponding limit *cluster sets*.

Let

$$A_\alpha^q = (a_{\alpha j}^{qk}), \quad A_\alpha = (a_{\alpha j}^k), \quad j, k = 1, 2, \ldots, 2^{p+q-2}. \tag{2}$$

Let further

$$g_q^\alpha(a_{\alpha j}^{qk}; z) = a_{\alpha j}^{qk} \text{ if } g_q^\alpha(z) = a_{\alpha j}^{qk}; \quad g_q^\alpha(a_{\alpha j}^{qk}; z) = 0 \text{ if } g_q^\alpha(z) \neq a_{\alpha j}^{qk},$$

where g_q^α is the gradating function equal $a_{\alpha j}^{qk}$ on the closed square $Q_{qk}^{\alpha j}$ corresponding to the pair (j, k); we suppose that the original square is divided into 4^{p+q-2} squares with sides parallel to the sides of Q_q^α for $\alpha \leq 2p-1$, and into 4^{p+q-1} analogous squares for $\alpha \geq 2p$. We shall call the squares $Q_{qk}^{\alpha j}$ *basic squares* for Q_q^α.

Given $z \in L_\infty^0$, consider the sequences

$$g_1^\alpha(z), g_2^\alpha(z), \ldots \quad \text{for} \quad \alpha < 2p, \tag{3}$$
$$\hat{g}_1^\alpha(z_-^1), \hat{g}_2^\alpha(z_-^2), \ldots \quad \text{for} \quad \alpha < 2p, \tag{4}$$
$$\hat{g}_1^\alpha(z_+^1), \hat{g}_2^\alpha(z_+^2), \ldots \quad \text{for} \quad \alpha < 2p, \tag{5}$$

where

$$\hat{g}_q^\alpha(z_-^q) = (g_q^\alpha(z_-^-(h)), g_q^\alpha(z_-^+(h)), \quad \hat{g}_q^\alpha(z_+^q) = (g_q^\alpha(z_+^-(h)), g_q^\alpha(z_+^+(h)), \tag{6}$$

as well as

$$\hat{g}_1^{2r}(z_-^1), \hat{g}_2^{2r}(z_-^2), \ldots \quad \text{for} \quad 2r = \alpha \geq 2p, \tag{7}$$

$$\hat{g}_1^{2r}(z_+^1),\ \hat{g}_2^{2r}(z_+^2),\dots \quad \text{for} \quad 2r=\alpha\geq 2p, \tag{8}$$
$$\hat{g}_1^{2r+1}(z_-^1),\ \hat{g}_2^{2r+1}(z_-^2),\dots \quad \text{for} \quad 2r+1=\alpha> 2p, \tag{9}$$
$$\hat{g}_1^{2r+1}(z_+^1),\ \hat{g}_2^{2r+1}(z_+^2),\dots \quad \text{for} \quad 2r+1=\alpha> 2p, \tag{10}$$

with the notation (6), and

$$\hat{g}_1^{2r}(z_1^-),\ \hat{g}_2^{2r}(z_2^-),\dots \quad \text{for} \quad 2r=\alpha\geq 2p, \tag{11}$$
$$\hat{g}_1^{2r}(z_1^+),\ \hat{g}_2^{2r}(z_2^+),\dots \quad \text{for} \quad 2r=\alpha\geq 2p, \tag{12}$$
$$\hat{g}_1^{2r+1}(z_1^-),\ \hat{g}_2^{2r+1}(z_2^-),\dots \quad \text{for} \quad 2r+1=\alpha> 2p, \tag{13}$$
$$\hat{g}_1^{2r+1}(z_1^+),\ \hat{g}_2^{2r+1}(z_1^+),\dots \quad \text{for} \quad 2r+1=\alpha> 2p, \tag{14}$$

where

$$\hat{g}_q^\alpha(z_q^-)=(g_q^\alpha(z_-^-(h)),g_q^\alpha(z_+^-(h)),\quad \hat{g}_q^\alpha(z_q^+)=(g_q^\alpha(z_-^+(h)),g_q^\alpha(z_+^+(h)). \tag{15}$$

Denote by $\mathbf{1}$, $\mathbf{i}$, $\mathbf{j}$, $\mathbf{k}$ the matrices representing the orthogonal unit vectors of the algebra $\mathbb{H}$ of (real) quaternions:

$$\mathbf{1}=\begin{pmatrix}1&0\\0&1\end{pmatrix},\ \mathbf{i}=\begin{pmatrix}0&i\\i&0\end{pmatrix}=\mathbf{i}\sigma_\mathbf{1},\ \mathbf{j}=\begin{pmatrix}0&1\\-1&0\end{pmatrix}=\mathbf{i}\sigma_\mathbf{2},\ \mathbf{k}=\begin{pmatrix}i&0\\0&-i\end{pmatrix}=\mathbf{i}\sigma_\mathbf{3}. \tag{16}$$

Then (1) becomes

$$A_\alpha^{q+1}=\frac{1}{i}\mathbf{k}\otimes A_\alpha^q,\quad \alpha=1,2,\dots,2p+2q-3; \tag{17}$$
$$A_{2p+2q-2}^{q+1}=\frac{1}{i}\mathbf{i}\otimes\mathbf{1}^{\otimes(p+q-2)},\quad A_{2p+2q-1}^{q+1}=-\frac{1}{i}\mathbf{j}\otimes\mathbf{1}^{\otimes(p+q-2)}.$$

We have

Periodicity Theorem (quaternionic formulation). (i) *If* $a_{\alpha\lambda}^\lambda\neq 0$ *for* $\lambda=2^{p+q-2}$, *the sequences* (3) *are periodic of period* 2, *starting from some term. The periods are:*

$$\frac{1}{2}\eta(a_{\alpha\lambda}^\lambda-a_{\alpha1}^1)\mathbf{1}+\frac{1}{2i}\eta(a_{\alpha\lambda}^\lambda+a_{\alpha1}^1)\mathbf{k},\ \frac{1}{2}\eta(a_{\alpha\lambda}^\lambda+a_{\alpha1}^1)\mathbf{1}-\frac{1}{2i}\eta(a_{\alpha\lambda}^\lambda-a_{\alpha1}^1)\mathbf{k} \tag{18}$$

where $\eta=1$ *or* -1.

(ii) *If*

$$a_{\alpha\lambda}^\lambda=0\quad \textit{and}\quad a_{\alpha\lambda-1}^{\lambda-1}=a_{\alpha2}^2=a_{\alpha1}^1=0,\quad \textit{where}\quad \lambda=2^{p+q-2}, \tag{19}$$

the sequences (3) *are constant-valued, starting from some term; it amounts at*

$$-\frac{1}{2}\eta a_{\alpha1}^1\mathbf{1}+\frac{1}{2i}\eta a_{\alpha1}^1\mathbf{k},\quad \textit{where}\quad \eta=1\ \textit{or}\ -1. \tag{20}$$

(iii) *If* (19) *holds, the sequences* (4) *are periodic of period* 2, *starting from some term. The periods are:*

$$\frac{1}{2}\eta\left(a_{\alpha\lambda}^{\lambda-1}+a_{\alpha,\lambda-1}^{\lambda}\right)\mathbf{j}+\frac{1}{2i}\eta\left(a_{\alpha\lambda}^{\lambda-1}-a_{\alpha,\lambda-1}^{\lambda}\right)\mathbf{i},\tag{21}$$
$$-\frac{1}{2}\eta\left(a_{\alpha\lambda}^{\lambda-1}+a_{\alpha,\lambda-1}^{\lambda}\right)\mathbf{j}-\frac{1}{2i}\eta\left(a_{\alpha\lambda}^{\lambda-1}-a_{\alpha,\lambda-1}^{\lambda}\right)\mathbf{i},$$

where $\eta = 1$ *or* $\eta = -1$.

(iv) *If* (19) *holds, the sequences* (5) *are constant-valued, starting from some term; it amounts at*

$$-\frac{1}{2}\eta\left(a_{\alpha2}^{1}+a_{\alpha1}^{2}\right)\mathbf{j}-\frac{1}{2i}\eta\left(a_{\alpha2}^{1}-a_{\alpha1}^{2}\right)\mathbf{i},\quad\text{where}\quad\eta=1\text{ or }-1.\tag{22}$$

(v) *The sequences* (7) *and* (9) *are periodic of period* 2, *starting from some term. The periods are:*

$$\left(\frac{1}{2}\mathbf{1}+\frac{1}{2i}\mathbf{k},\frac{1}{2}\mathbf{1}+\frac{1}{2i}\mathbf{k}\right),\quad\left(-\frac{1}{2}\mathbf{1}-\frac{1}{2i}\mathbf{k},-\frac{1}{2}\mathbf{1}-\frac{1}{2i}\mathbf{k}\right)\tag{23}$$

or

$$\left(-\frac{1}{2}\mathbf{1}-\frac{1}{2i}\mathbf{k},-\frac{1}{2}\mathbf{1}-\frac{1}{2i}\mathbf{k}\right),\quad\left(\frac{1}{2}\mathbf{1}+\frac{1}{2i}\mathbf{k},\frac{1}{2}\mathbf{1}+\frac{1}{2i}\mathbf{k}\right)\tag{24}$$

in the case of (7), *and*

$$\left(\frac{1}{2i}\mathbf{1}-\frac{1}{2}\mathbf{k},-\frac{1}{2i}\mathbf{1}+\frac{1}{2}\mathbf{k}\right),\quad\left(-\frac{1}{2i}\mathbf{1}+\frac{1}{2}\mathbf{k},\frac{1}{2i}\mathbf{1}-\frac{1}{2}\mathbf{k}\right)\tag{25}$$

or

$$\left(\frac{1}{2i}\mathbf{1}+\frac{1}{2}\mathbf{k},-\frac{1}{2i}\mathbf{1}-\frac{1}{2}\mathbf{k}\right),\quad\left(-\frac{1}{2i}\mathbf{1}-\frac{1}{2}\mathbf{k},\frac{1}{2i}\mathbf{1}+\frac{1}{2}\mathbf{k}\right)\tag{26}$$

in the case of (9).

(vi) *The sequences* (8) *and* (10) *are constant-valued, starting from some term; it amounts at*

$$\left(-\frac{1}{2}\mathbf{1}+\frac{1}{2i}\mathbf{k},-\frac{1}{2}\mathbf{1}+\frac{1}{2i}\mathbf{k}\right)\quad\text{or}\quad\left(\frac{1}{2}\mathbf{1}-\frac{1}{2i}\mathbf{k},\frac{1}{2}\mathbf{1}-\frac{1}{2i}\mathbf{k}\right)\tag{27}$$

in the case of (8), *and*

$$\left(\frac{1}{2i}\mathbf{1}-\frac{1}{2}\mathbf{k},-\frac{1}{2i}\mathbf{1}-\frac{1}{2}\mathbf{k}\right)\quad\text{or}\quad\left(-\frac{1}{2i}\mathbf{1}+\frac{1}{2}\mathbf{k},-\frac{1}{2i}\mathbf{1}-\frac{1}{2}\mathbf{k}\right)\tag{28}$$

in the case of (10).

(vii) *The sequences* (13) *and* (14) *are periodic of period* 2, *starting from some term. The periods are:*

$$\left(\frac{1}{2i}\mathbf{1} - \frac{1}{2}\mathbf{k}, -\frac{1}{2}\mathbf{1} - \frac{1}{2i}\mathbf{k}\right), \quad \left(-\frac{1}{2i}\mathbf{1} + \frac{1}{2}\mathbf{k}, -\frac{1}{2}\mathbf{1} - \frac{1}{2i}\mathbf{k}\right) \tag{29}$$

or

$$\left(-\frac{1}{2i}\mathbf{1} + \frac{1}{2}\mathbf{k}, \frac{1}{2}\mathbf{1} + \frac{1}{2i}\mathbf{k}\right), \quad \left(\frac{1}{2i}\mathbf{1} - \frac{1}{2}\mathbf{k}, \frac{1}{2}\mathbf{1} + \frac{1}{2i}\mathbf{k}\right) \tag{30}$$

in the case of (13), *and* (30) *or* (29) *in the case of* (14) (*given* $z \in L^0_\infty$, *the choices* (29) *and* (30) *are mutually correlated*).

(viii) *The sequences* (11) *and* (12) *are periodic of period* 2, *starting from some term. The periods are:*

$$\left(\frac{1}{2}\mathbf{1} + \frac{1}{2i}\mathbf{k}, -\frac{1}{2}\mathbf{1} + \frac{1}{2i}\mathbf{k}\right), \quad \left(-\frac{1}{2}\mathbf{1} - \frac{1}{2i}\mathbf{k}, -\frac{1}{2}\mathbf{1} + \frac{1}{2i}\mathbf{k}\right) \tag{31}$$

or

$$\left(-\frac{1}{2}\mathbf{1} - \frac{1}{2i}\mathbf{k}, \frac{1}{2}\mathbf{1} - \frac{1}{2i}\mathbf{k}\right), \quad \left(\frac{1}{2}\mathbf{1} + \frac{1}{2i}\mathbf{k}, \frac{1}{2}\mathbf{1} - \frac{1}{2i}\mathbf{k}\right). \tag{32}$$

The Periodicity Theorem is a consequence of Theorems 1–6 [18]. We do not repeat the assertions of these theorems that precise the starting terms of the periodicity — here we have no further contribution in that field. We are dealing only with the assertions that can be reformulated to involve quaternions explicitly. More precisely, the Periodicity Theorem in the quaternionic version has been announced in [8] with the proof of the assertion (i). Now we restrict ourselves to proving the assertions (ii)–(iv). Assertions (v)–(viii) will be proved in the forthcoming paper [9].

However, we can remove the second condition in (19) which is done in the Second and Third Periodicity Theorems in Sections 4 and 5, respectively.

2. Quaternions and octonions in the Cayley-Dickson construction

Let A be a finite-dimensional algebra over a field K equipped with an involution, namely the K-linear mapping: $\star : A \to A$ such that $(ab)^\star = b^\star a^\star$ for all $a, b \in A$ and with norm given by $\|a\|^2 = aa^\star$. Let us consider the Cartesian product $A \times A$ and define addition and multiplication in it by the formulae:

$$(a,b) + (c,d) = (a+c, b+d),$$
$$(a,b) \cdot (c,d) = (ac - db^\star, a^\star d + cb)$$

for all (a,b), $(c,d) \in A \times A$. The set $A' = A \times A$ with these operations is an algebra with involution:

$$(a,b)^\star = (a^\star, -b)$$

and norm

$$\|(a,b)\| = (a,b)(a,b)^\star.$$

Let us note that the algebra A is a subalgebra of A' via the embedding: $a \mapsto (a,0)$. The construction defined above is the so-called *Cayley-Dickson construction*. If we perform the Cayley-Dickson construction a number of times, say n times, starting from an arbitrary algebra A_0, we shall obtain a sequence of algebras:

$$A_0 \subset A_1 \subset \cdots \subset A_n$$

such that $\dim_K A_{j+1} = 2\dim_K A_j$ for $j = 0, 1, \ldots, n-1$.

Let us start the construction from $A_0 = \mathbb{R}$, the algebra of real numbers over itself. The mapping $\star$ is identity. It is easy to see $A_1 = \mathbb{C} = \mathbb{R} \oplus \mathbb{R}i$ is the field of complex numbers. Applying the construction to $\mathbb{C}$ we obtain $A_2 = \mathbb{C} \oplus \mathbb{C}j \cong \mathbb{H}$, the real quaternions, and consequently: $A_3 = \mathbb{H} \oplus \mathbb{H}\ell \cong \mathbb{O}$ which is the algebra of octonions. Here j, $j^2 = -1$, has to anticommute with i, and ℓ, $\ell^2 = -1$, has to anticommute with i, j, k; $(1, i, j, k)$ being the basis of $\mathbb{H}$. The next algebra after octonions would be the algebra of sedenions, which is no longer a division algebra, that is, it possesses zero divisors.

It seems obvious, that the process of Cayley-Dickson construction is worth continuing, because the bigger algebra, the better, we expect, might describe the world. However, there are two kinds of losses connected with this expansion.

Firstly, with every step of the Cayley-Dickson construction the properties of the new algebra are a little worse than these of the old one. For example: in $\mathbb{C}$, we have no order of elements which we had in $\mathbb{R}$. Obviously, lack of order refers to all further algebras in the construction. In $\mathbb{H}$ we have no commutativity, which we had in $\mathbb{C}$. In $\mathbb{O}$ we no longer have associativity of the multiplication.

The second inconvenience about bigger algebras is that, the objects considered become either too large or too small to be interesting. For example, the class of holomorphic functions over quaternions is very poor and reduces to right $\mathbb{H}$-linear functions $f(q) = aq + b$ where $a, b \in \mathbb{H}$. On the other hand, the power series of quaternion variable $q = w + ix + jy + kz$ gives a set which is too big to be interesting [18].

3. The case $\alpha \leq 2p - 1$, period 1. A fractal bundle of the flower type

For (19), the expressions (18) are still correct, but seem not so interesting: they reduce to (20), so the sequences (3) are constant-valued, starting form some term; in the cases of Pauli matrices σ_1 and σ_2 they reduce to 0 and give no information on the structure in question. Therefore we have to take into account sequences related to petals of the form

$$\begin{array}{cc} \boxed{} & \\ \boxed{u} & \boxed{v} \\ \boxed{} & \end{array}, \quad u \neq 0, \quad v \neq 0, \quad \text{where } \boxed{} \text{ stands for } \boxed{0}\,. \text{ cm} \tag{33}$$

This will be done in Sections 4 and 5 of the paper corresponding to Parts (iii) and (iv) of the Periodicity Theorem.

Having the most important examples of petals: (36), (37) in [8] and (33) above, we are in a position to understand the definition of a fractal bundle of the flower type and actually we need it [2,15].

Definition 3.1. Let K_0 be a compact metric space with metric ρ. Let $\sigma_j : K_0 \to K_0$ for $j = 1, 2, \ldots, N$ be contractible mappings with respect to ρ. The set K of the form

$$K := \bigcap_{n=1}^{\infty} K_n, \quad \text{where} \quad K_n = \bigcup_{j=1}^{N} \sigma_j(K_{n-1})$$

is called *fractal set of flower type*.

Moreover, in this case we have [5]:

$$K = \bigcup_{j=1}^{N} \sigma_j(K).$$

If, for fixed α, we now consider Q_α^n, $n = 1, 2, \ldots$, we obtain a graded (coloured) fractal, denoted Σ_α, which will be called *graded (coloured) Clifford-type fractal*. The corresponding sequence (Σ_α, $\alpha = 1, 2, \ldots$) will be called *graded (coloured) Clifford-type fractal bundle*.

This fractal bundle is well defined, which can be proved using the Cuntz algebra $\mathcal{O}(4)$ [1] and Kakutani dichotomy theorem [4].

4. The case: $\alpha \leq 2p - 1$, perpendicular lower convergence. Ovary and ovules

For (19), according to Theorem 2 [16], the sequences (4) are periodic of period 2 starting from some term determined in that theorem. The periods are:

$$\begin{pmatrix} 0 & \eta a^{\lambda}_{\alpha,\lambda-1} \\ \eta a^{\lambda-1}_{\alpha\lambda} & 0 \end{pmatrix}, \begin{pmatrix} 0 & -\eta a^{\lambda}_{\alpha,\lambda-1} \\ -\eta a^{\lambda-1}_{\alpha\lambda} & 0 \end{pmatrix}, \text{ where } \eta = 1 \text{ or } -1. \tag{34}$$

For an alternative direct proof we may observe that, in order to calculate

$$\begin{aligned}&(g^{\alpha}_{q}(z^{-}_{-}(p+q-3)), g^{\alpha}_{q}(z^{+}_{-}(p+q-3)))\\ &= (g^{\alpha}_{q}(\frac{1}{2\sqrt{2}}\frac{m}{2^n}(1-i) - \frac{1}{2\sqrt{2}}\frac{1}{2^{p+q-3}}), g^{\alpha}_{q}(\frac{1}{2\sqrt{2}}\frac{m}{2^n}(1-i) - \frac{i}{2\sqrt{2}}\frac{1}{2^{p+q-3}})),\\ &\qquad m = 0, \pm 1, \ldots, \pm(2^n - 1); \quad n = 0, 1, \ldots,\end{aligned}$$

we have to look for

$$(g^{\alpha}_{q-s}(z^{-}_{-}(p+q-s-3)), g^{\alpha}_{q-s}(z^{+}_{-}(p+q-s-3))), \quad s = 1, 2, \ldots.$$

To this end, in the case of Clifford-type fractal $\Sigma_2 = (Q^2_q)$, $p = 2$ and $A^1_2 = \sigma_2$, we consider the table, where rows numbered with positive integers, corresponding to odd numbers, represent $\{z^{-}_{-}(p+q-3)\} \cup \{z^{+}_{-}(p+q-3)\}$ in the q-th iteration step, and columns represent the configuration related to $s_m = m/2^n$ (in Fig. 1 (a) we take $n = 11$). The configuration consists of two values of the generating function if z belongs to the interior of the basic square $Q^{\alpha j}_{qk}$ corresponding to a pair (j, k); of four values if z is on the boundary but not in a vertex; finally, of eight values if z is in a vertex (of course, it may happen, that some of these values coincide). In the case of σ_2 we have the following possibilities for the two values corresponding to the direction perpendicular to L_∞ starting from some term of (4):

$$\boxed{i}\,\boxed{-i} \quad \text{or} \quad \boxed{-i}\,\boxed{i}. \tag{35}$$

As far as the direction of L_∞ corresponding to $z \in L_\infty$ is concerned, we have only one possibility:

$$\begin{matrix}\boxed{0}\\ \boxed{0}\end{matrix}, \text{ denoted by } \begin{matrix}\boxed{}\\ \boxed{}\end{matrix}. \tag{36}$$

For each n the first period is indicated with help of the upper part of a bigger square; this upper part containing four small squares $\boxed{x}$ and four small rectangles $\boxed{}$.

In the case of Σ_3 for $p = 3$ and $\alpha = 1, 2, \ldots, 5$, we consider an analogous table (in Fig. 2 (b) we take again $n = 11$). Let $\boxed{5}$, $\boxed{6}$, $\boxed{7}$, $\boxed{8}$ represent the values $a^1_{\alpha 2}$, $a^3_{\alpha 4}$, $a^2_{\alpha 1}$, $a^4_{\alpha 3}$, respectively. Then we have the following possibilities for the two values corresponding to the direction perpendicular

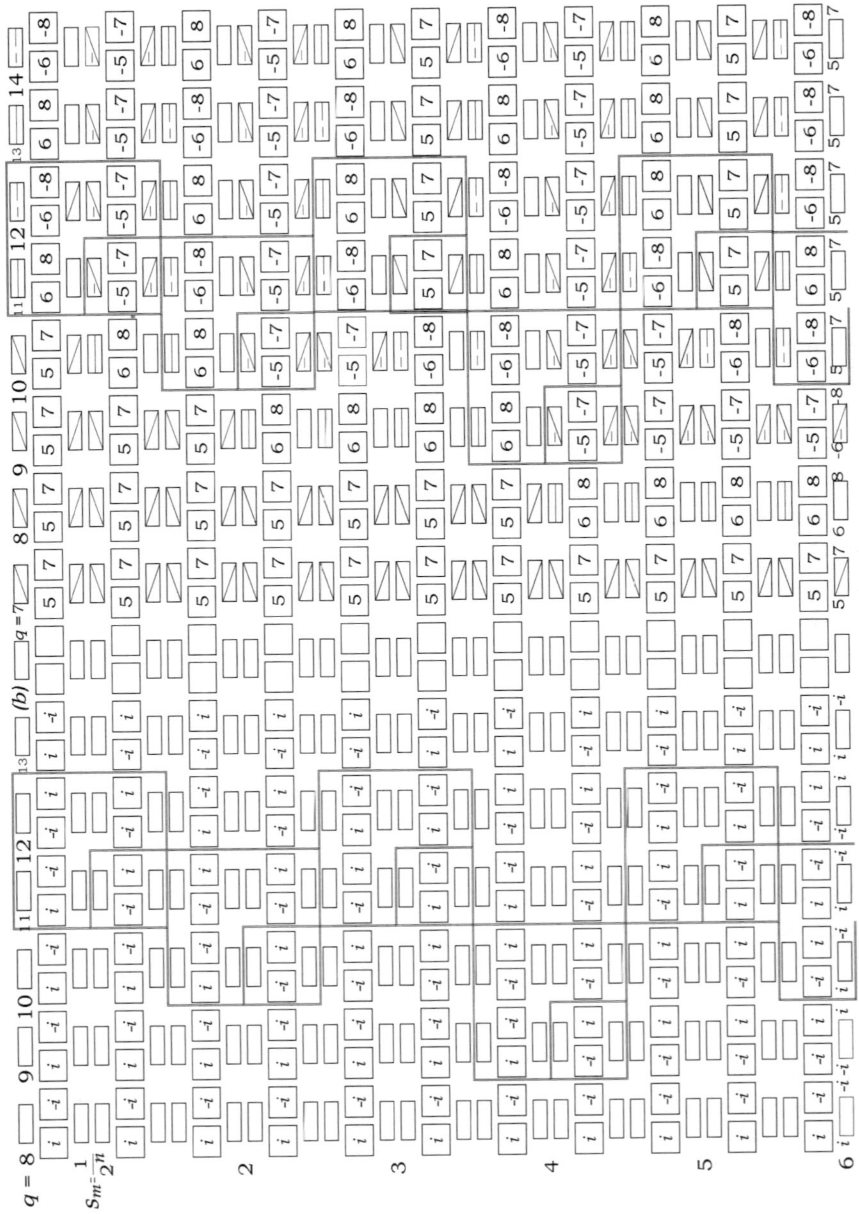

Fig. 1. Checking the periodicity in construction of a graded Clifford-type fractal: (a) Σ_2 for $p = 2$, (b) Σ_α for $p = 3$, $\alpha = 1, 2, \ldots, 5$; $a^\lambda_{\alpha\lambda} = 0$.

to L_∞:

$$\boxed{5}\,\boxed{7}\,,\quad \boxed{6}\,\boxed{8}\,,\quad \boxed{-5}\,\boxed{-7}\,,\quad \boxed{-6}\,\boxed{-8}\,.$$

As far as the direction of L_∞ corresponding to $z \in L_\infty$ is concerned, we have only one possibility (36). If, however, we remove the second condition in (19), the number of possibilities increases (cf. Fig. 2):

 with $\boxed{5}\,\boxed{7}$,  with $\boxed{x}\,\boxed{y}$,  with $\boxed{6}\,\boxed{8}$,

 with $\boxed{-5}\,\boxed{-7}$, 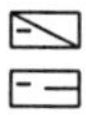 with $\boxed{-x}\,\boxed{-y}$, 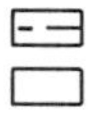 with $\boxed{-6}\,\boxed{-8}$.

For each n the first period is indicated with help of the upper part of a bigger square; these upper parts containing four squares and four rectangles form the collection:

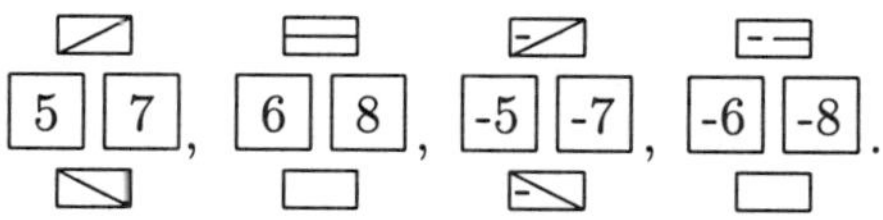

Then we can proceed by induction with respect to p, considering Σ_α, $\alpha = 1, 2, \ldots, 2p-1$. We observe that, by (16) the matrices (34) can be expressed in the quaternionic form (21) as desired.

If we remove the second condition from (19):

$$a^\lambda_{\alpha\lambda} = 0 \quad \text{for} \quad \lambda = 2^{p+q-2}, \tag{37}$$

then the periods (34) have a more general form

$$\begin{pmatrix} \eta a^{\lambda-1}_{\alpha,\lambda-1} & \eta a^{\lambda}_{\alpha,\lambda-1} \\ \eta a^{\lambda-1}_{\alpha\lambda} & 0 \end{pmatrix}, \begin{pmatrix} -\eta a^{\lambda-1}_{\alpha,\lambda-1} & -\eta a^{\lambda}_{\alpha,\lambda-1} \\ -\eta a^{\lambda-1}_{\alpha\lambda} & 0 \end{pmatrix}, \text{ where } \eta = 1 \text{ or } -1, \tag{38}$$

and the quaternionic form, instead of (21) reads:

$$\begin{gathered} \frac{1}{2}\eta a^{\lambda-1}_{\alpha,\lambda-1}\mathbf{1} + \frac{1}{2i}\eta(a^{\lambda-1}_{\alpha\lambda} - a^{\lambda}_{\alpha,\lambda-1})\mathbf{i} + \frac{1}{2}\eta(a^{\lambda-1}_{\alpha\lambda} + a^{\lambda}_{\alpha,\lambda-1})\mathbf{j} + \frac{1}{2}\eta a^{\lambda-1}_{\alpha,\lambda-1}\mathbf{k}, \\ -\frac{1}{2}\eta a^{\lambda-1}_{\alpha,\lambda-1}\mathbf{1} - \frac{1}{2i}\eta(a^{\lambda-1}_{\alpha\lambda} - a^{\lambda}_{\alpha,\lambda-1})\mathbf{i} - \frac{1}{2}\eta(a^{\lambda-1}_{\alpha\lambda} + a^{\lambda}_{\alpha,\lambda-1})\mathbf{j} - \frac{1}{2}\eta a^{\lambda-1}_{\alpha,\lambda-1}\mathbf{k}. \end{gathered} \tag{39}$$

Therefore we have arrived at the following generalization of Periodicity Theorem (iii):

Second Periodicity Theorem (quaternionic formulation). (i) *If* (37) *holds, the sequences* (4) *are periodic of period* 2, *starting from some term. The periods are* (39), *where* $\eta = 1$ *or* -1.

(ii) *The periodicity of the sequences* (4) *always starts from their* n*-th term or earlier whenever* m *in the definition of* $z \in L^0_\infty$ *is odd. If* m *is even, of the form* $m = \mu \cdot 2^\nu$, *where* μ *is odd, the periodicity of* (4) *starts from the* $(n-\nu)$*-th term or earlier. In particular, such a situation appears in the cases of Pauli matrices* σ_1 *and* σ_2, *where the periodicity, for* m *odd, starts exactly from the* n*-th term. For* m *even the periodicity starts exactly from the* $(n-\nu)$*-th term.*

The statement (ii) of the above theorem was not proved here, but the proof is merely a repetition of those of Theorems 1 and 2 [16]. (Because of importance of σ_1, Fig. 1 (a) is repeated as Fig. 3 (a) with $a^1_{22} = i$ and $a^2_{21} = -i$ formally replaced by $a^1_{12} = 1$ and $a^2_{11} = 1$, respectively. Of course, distribution of signs + and − in the table is now entirely different. It seems worth-while to draw also Fig. 3 (b) being a counterpart of Fig. 1 (b), where $\boxed{6}$ and $\boxed{8}$ are formally replaced by $\boxed{5}$ and $\boxed{7}$, respectively.)

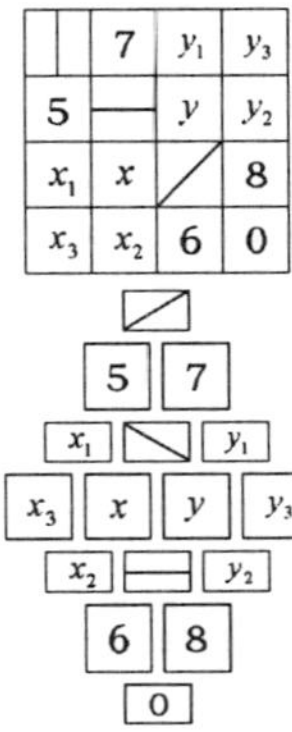

Figure 2: Conventions concerning $a^k_{\alpha j}$ for $p = 3$, $\alpha = 1, 2, \ldots, 5$; $a^\lambda_{\alpha\lambda} = 0$.

The above discussion fully motivates distinguishing the structures

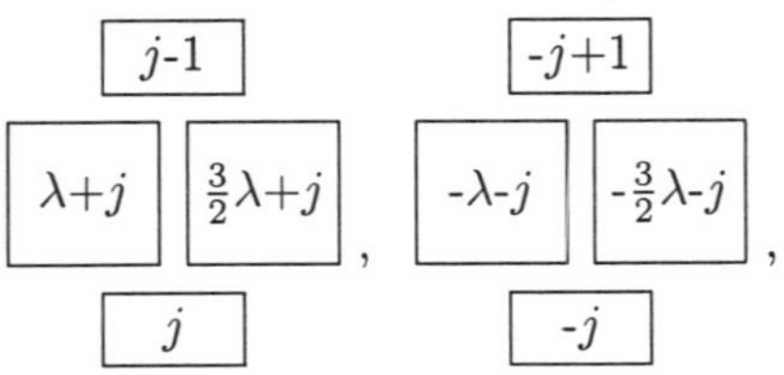

$$j = 1, 2, \ldots, \frac{1}{2}\lambda - 1, \tag{40}$$

where $\lambda = 2^{p+q-2}$, and

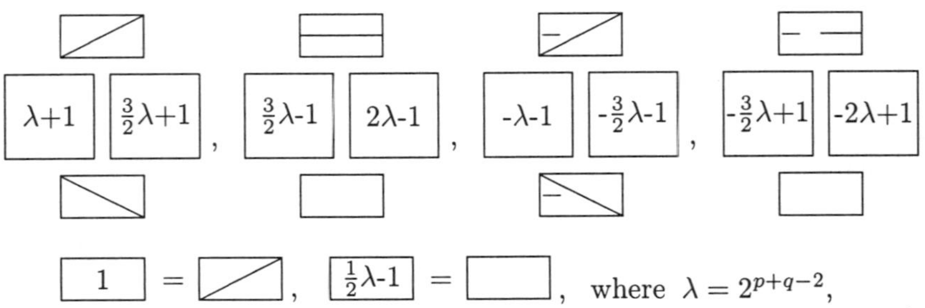

where $\lambda = 2^{p+q-2}$, (41)

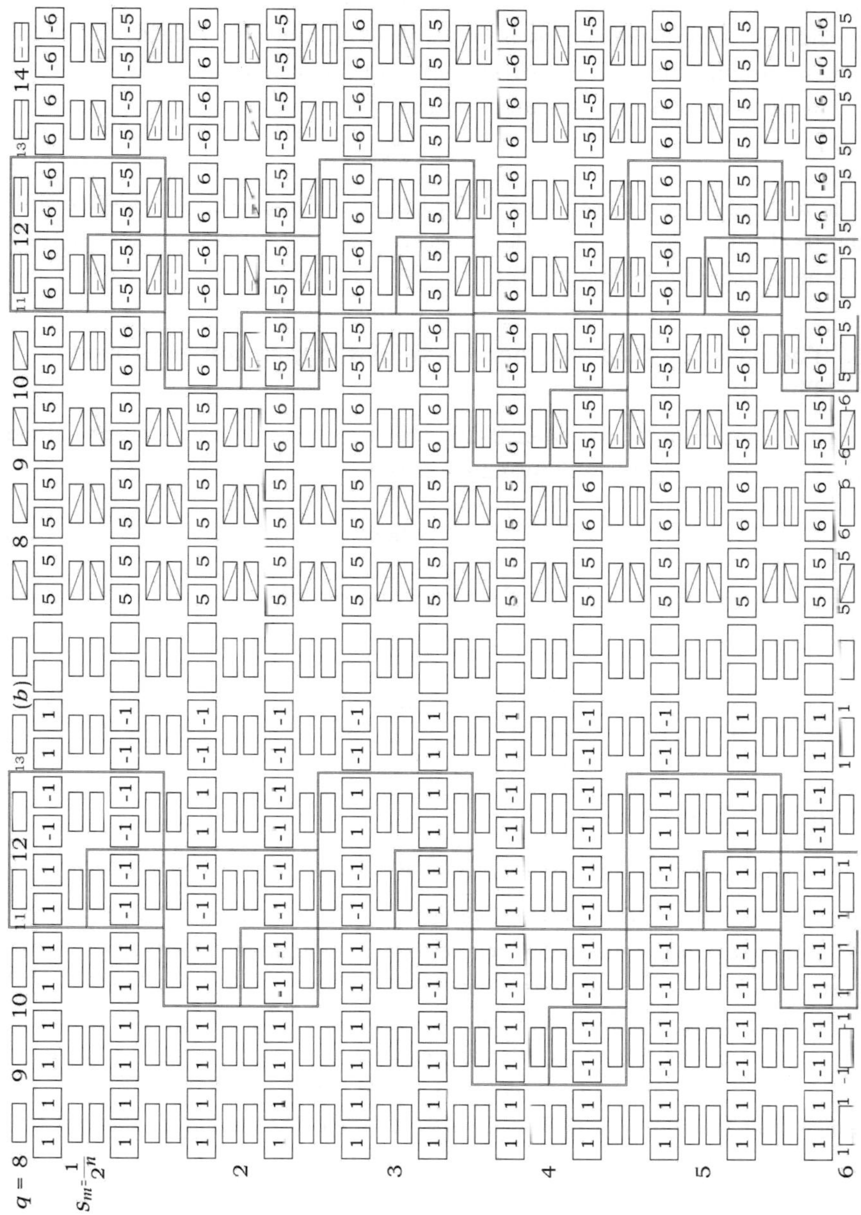

Fig. 3. Checking the periodicity in construction of a graded Clifford-type fractal: (a) Σ_1 for $p = 2$, (b) Σ_α for $p = 3$, $\alpha = 1, 2, \ldots, 5$; $a^{\lambda}_{\alpha\lambda} = 0$.

which are petals again. Because of the periodicity, any petal (40) has to be considered with one of the intrinsic petals (41), that is to consider bipetals, in this case ordered pairs of petals (40) and intrinsic petals (41).

In the case of a graded fractal bundle

$$(\Sigma_1, \Sigma_2, \ldots, \Sigma_{2p-1}), \tag{42}$$

it is natural to call the initial Clifford algebra $Cl_{2p-1}(\mathbb{C})$ — the *ovary* of the fractal bundle (42), and the generators $A^1_\alpha = A_\alpha$, $\alpha = 1, 2, \ldots, 2p-1$ — the *ovules* of the bundle (42).

5. The case: $\alpha \leq 2p-1$, perpendicular upper convergence. The pistil and stamens

For (19), according to Theorem 3 [16], the sequences (5) are constant-valued starting from some term determined in that theorem; it amounts at:

$$\begin{pmatrix} 0 & -\eta a^2_{\alpha 1} \\ -\eta a^1_{\alpha 2} & 0 \end{pmatrix}, \quad \text{where } \eta = 1 \text{ or } -1. \tag{43}$$

For an alternative, direct proof we may observe that, in order to calculate

$$\begin{gathered}(g^\alpha_q(z^-_+(p+q-3)),\, g^\alpha_q(z^+_+(p+q-3))) \\ = (g^\alpha_q(\frac{1}{2\sqrt{2}}\frac{m}{2^n}(1-i) - \frac{i}{2\sqrt{2}}\frac{1}{2^{p+q-3}}),\, g^\alpha_q(\frac{1}{2\sqrt{2}}\frac{m}{2^n}(1-i) + \frac{1}{2\sqrt{2}}\frac{1}{2^{p+q-3}})), \\ m = 0, \pm 1, \ldots, \pm(2^n-1); \quad n = 0, 1, \ldots,\end{gathered}$$

we have to look for

$$(g^\alpha_{q-s}(z^-_+(p+q-s-3)),\, g^\alpha_{q-s}(z^+_+(p+q-s-3))), \quad s = 1, 2, \ldots .$$

To this end, in the case of the Clifford-type fractal $\Sigma_2 = (Q^2_q)$, $p = 2$ and $A^1_2 = \sigma_2$, we consider the table, where rows numbered with positive integers, corresponding to even numbers, represent $\{z^-_+(p+q-3)\} \cup \{z^+_+(p+q-3)\}$ in the q-th iteration step, and columns represent the configuration related to $s_m = m/2^n$ (in Fig. 1 (a) we take $n = 11$).

Again, the configuration consists of two, four or eight values of the generating function (of course, it may happen, that some of these values coincide). In the case of σ_2 we have the two possibilities (35) for two values corresponding to the direction perpendicular to L_∞. As far as the direction of L_∞ corresponding to $z \in L_\infty$ is concerned, we have only one possibility (36). For each n the beginning of constancy of the sequences (5) is indicated with the help of a medium-size square, containing two small squares $\boxed{x}$ and two small rectangles $\square$. If we consider together the perpendicular lower and upper convergences related to $z \in L_\infty$, we have to state that for each n the first period is indicated with help of the upper and lower parts of a bigger square; the both parts containing eight small squares $\boxed{x}$ and eight small rectangles ▭.

In the case of Σ_α, for $p = 3$ and $\alpha = 1, 2, \ldots, 5$, we continue considering an analogous table (in Fig. 1 (b) we take again $n = 11$). A discussion, completely analogous to that of the preceding section, leads to the same conclusion, but it is interesting to observe that, as far as the periodicity is concerned,

intrinsic petals [6] [8], [-6] [-8] correspond to the perpendicular lower convergence,

intrinsic petals [5] [7], [-5] [-7] correspond to the perpendicular upper convergence.

We observe that, by (16), the matrices (43) can be expressed in the quaternionic form (22), as desired.

If we replace (19) by a less restrictive condition (37), then the one-element periods (38) have a more general form

$$\begin{pmatrix} -\eta a^1_{\alpha 1} & -\eta a^2_{\alpha 1} \\ -\eta a^1_{\alpha 2} & -\eta a^2_{\alpha 2} \end{pmatrix}, \quad \text{where } \eta = 1 \text{ or } -1, \tag{44}$$

and the quaternionic form, instead of (22), reads:

$$-\frac{1}{2}\eta\left(a^1_{\alpha 1} + a^2_{\alpha 2}\right)\mathbf{1} - \frac{1}{2i}\eta\left(a^1_{\alpha 2} - a^2_{\alpha 1}\right)\mathbf{i} - \frac{1}{2}\eta\left(a^1_{\alpha 2} + a^2_{\alpha 1}\right)\mathbf{j} - \frac{1}{2i}\eta\left(a^1_{\alpha 1} - a^2_{\alpha 2}\right)\mathbf{k}. \tag{45}$$

Therefore we have arrived at the following generalization of Periodicity Theorem (iv):

Third Periodicity Theorem (quaternionic formulation). (i) *If* (42) *holds, the sequences* (5) *are constant-valued, starting from some term. It amounts at* (45), *where* $\eta = 1$ *or* -1.

(ii) *The constancy of the sequences* (5) *always starts from their* n*-th term or earlier whenever* m *in the definition of* $z \in L^0_\infty$ *is odd. If* m *is even, of the form* $m = \mu \cdot 2^\nu$, *where* μ *is odd, the constancy of* (5) *starts from the* $(n - \nu)$*-th term or earlier. In particular, such a situation appears in the cases of Pauli matrices* σ_1 *and* σ_2, *where the constancy, for* m *odd, starts* *exactly from the n-th term. For m even the periodicity starts exactly from the $(n-\nu)$-th term.*

The statement (ii) of the above theorem was not proved here, but the proof is merely a repetition of those of Theorems 1 and 3 [16] (because of importance of σ_1, it is worthwhile to fill up the blank rows in Figs 4 (a) and (b) corresponding to the perpendicular upper convergence). Here we are dealing again with petals of the form (40) together with intrinsic petals of the form (41), that is with the corresponding bipetals.

Taking into account the role of sets L^0_{p+q-3}, $q = 1, 2, \ldots$, and L^0_∞ in a graded fractal bundle (42), expressed by the considerations of Sections 4 and 5, and — in particular — in the Second and Third Periodicity Theorems, we are let to call them the *stamens* of the fractal bundle (42) in case of the sets L^0_{p+q-3}, $q = 1, 2, \ldots$, and the *pistil* of that bundle in case of the set L^0_∞.

The research including the rest of proof of the Periodicity Theorem, will be continued in the forthcoming paper [9]. It is worthwhile to stress the relationship of fractal bundles of algebraic structure with studies of the Hurwitz problem [3], especially on its geometrical aspects introduced and studied in other papers [6,7,10-15,17].

References

1. J. Cuntz, *Simple C^*-algebras generated by isometries*, Comm. Math. Phys. **57** (1977), 173–185.
2. K.J. Falconer, *Fractal Geometry. Mathematical Foundations and Applications*, Wiley, Chichester 1990.
3. A. Hurwitz, *Über die Komposition der quadratischen Formen*, Math. Ann. **88** (1923), 1–25; reprinted in: A. Hurwitz, *Mathematische Werke II*, Birkhäuser Verlag, Basel 1933, 641–666.
4. S. Kakutani, *On equivalence of infinite product measures*, Ann. of Math. **49** (1948), 214–224.
5. J. Kigami, *Analysis on fractals* (Cambridge Tracts in Mathematics 143), Cambridge Univ. Press, Cambridge 2001.
6. R. Kovacheva, J. Ławrynowicz, and M. Nowak-Kępczyk, *Critical dimension* 13 *in approximation related to fractals of algebraic structure*, Bull. Soc. Sci. Lettres Łódź **52** Sér. Rech. Déform. **37** (2002) 77–102.
7. J. Ławrynowicz, *Type-changing transformations of pseudo-euclidean Hurwitz pairs, Clifford analysis, and particle lifetimes*, in: Clifford Algebras and Their Applications in Mathematical Physics. Ed. by V. Dietrich, K. Habetha, and G. Jank, Kluwer Academic, Dordrecht 1998, pp. 217–226.
8. J. Ławrynowicz, S. Marchiafava, M. Nowak-Kępczyk, *Quaternionic background of the periodicity of petal and sepal structures in some fractal of the*

flower type, in: Proceedings of the 5th ISAAC Congress, Catania, July 15-20, 2005. Ed. by H. Begehr and F. Nicolosi, World Scientific, Singapore, to appear.

9. ———, *Periodicity theorem for structure fractals in quaternionic formulation*, Internat. J. of Geom. Meth. in Modern Phys. **3** (2006), 1167-1197.
10. J. Ławrynowicz, M. Nowak-Kępczyk, and Osamu Suzuki, *A duality theorem for inoculated graded fractal bundles vs. Cuntz algebras and their central extensions*, in: C^*-Algebras in Geometry and Physics, Ed. by B. Bojarski, Birkhäuser Verlag, Basel, to appear.
11. J. Ławrynowicz and J. Rembieliński, *Pseudo-euclidean Hurwitz pairs and generalized Fueter equations*, (a) Inst. of Math. Polish Acad. Sci. Preprint no. 355 (1985), ii+10 pp., (b) in: Clifford Algebras and Their Applications in Mathematical Physics. Ed. by J.S.R. Chisholm and A.K. Common (NATO-ASI Series C: Mathematical and Physical Sciences 183), Reidel, Dordrecht 1986, pp. 39–48.
12. ———, *Pseudo-euclidean Hurwitz pairs and the Kałuża-Klein theories*, (a) Inst. of Phys. Univ. of Łódź Preprint no. 86-8 (1986), 28 pp., (b) J. Phys. A: Math. Gen. **20** (1987), 5831–5848.
13. ———, *On the composition of nondegenerate quadratic forms with an arbitrary index*, (a) Inst. of Math. Polish Acad. of Sci. Preprint no. 369 (1986), ii+29 pp., (b) Ann. Fac. Sci. Toulouse Math. (5) **10** (1989), 141–168 [due to a printing error in vol. **10** the whole article was reprinted in vol. **11** (1990), no. 1, of the same journal, pp. 141–168].
14. J. Ławrynowicz, J. Rembieliński and O. Suzuki, *An introduction to pseudotwistors: Spinor solutions vs. harmonic forms and cohomology groups*, Progress in Physics **18** (2000), 393–423.
15. J. Ławrynowicz and J. Rembieliński, *An introduction to pseudotwistors: Basic constructions*, in: Quaternionic Structures in Mathematics and Physics. Ed. by S. Marchiafava, P. Piccinni and M. Pontecorvo, World Scientific, Singapore 2001, pp. 241–252.
16. ———, *Periodicity theorems for graded fractal bundles related to Clifford structures*, Internat. J. of Pure and Appl. Math., to appear.
17. J. Ławrynowicz, J. Rembieliński and F.L. Castillo Alvarado, *Basic properties and applications of graded fractal bundles related to Clifford structures*, in: Clifford Algebras and Their Applications in Mathematical Physics, Ed. by P. Angelès, Birkhäuser Verlag, Basel, to appear.
18. P. Lounesto, *Clifford Algebras and Spinors*, (London Math. Soc. Lecture Notes Series 239), Cambridge Univ. Press, 1997; 2nd ed. (vol. 286), ibid. 2001.
19. W. Pauli, *Zur Quantenmechanik des magnetischen Elektrons*, Z. Phys. **42** (1927), 601–623.
20. ———, *Contributions mathématiques à la théorie des matrices de Dirac*, Ann. Inst. H. Poincaré **6** (1936), 109–136.

SECTIONAL CURVATURES OF SOME HOMOGENEOUS REAL HYPERSURFACES IN A COMPLEX PROJECTIVE SPACE

S. MAEDA† and M. KIMURA‡

Department of Mathematics, Shimane University
Matsue 690-8504, Japan
† *E-mail: smaeda@riko.shimane-u.ac.jp*
‡ *E-mail: mkimura@riko.shimane-u.ac.jp*

In this note, we pose two pinching problems on sectional curvatures of real hypersurfaces in a complex projective space.

Keywords: Sectional curvatures; Complex projective spaces; Homogeneous real hypersurfaces; type A hypersurfaces; type B hypersurfaces.

1. Introduction

Let $P_n(\mathbb{C})$ be an n-dimensional complex projective space with Fubini-Study metric of constant holomorphic sectional curvature 4 and M a real hypersurface of $P_n(\mathbb{C})$. Nice examples of real hypersurfaces in $P_n(\mathbb{C})$ are *homogeneous* real hypersurfaces, namely they are given as orbits under subgroups of the projective unitary group $PU(n+1)$. It is well-known that a homogeneous real hypersurface in $P_n(\mathbb{C})$ is congruent to one of the six model spaces of type A_1, A_2, B, C, D and E. By virtue of Takagi's result [3] we know that these homogeneous real hypersurfaces are realized as tubes of constant radius over certain compact Hermitian symmetric spaces of rank 1 or rank 2 (see Theorem T).

In the study of real hypersurfaces in $P_n(\mathbb{C})$, many differential geometers are interested in the following two problems:

(1) Give a characterization of homogeneous real hypersurfaces.
(2) Construct non-homogeneous *nice* real hypersurfaces in $P_n(\mathbb{C})$ and characterize such hypersurfaces.

2000 *Mathematics Subject Classification*. Primary 53B25, 53C40.

‡The second author is partially supported by Grant-in-Aid for Scientific Research (C) (No. 17540078), Japan Society for the Promotion of Science.

From the viewpoint of Problem (1) it is important to compute various geometric invariants of homogeneous real hypersurfaces. In this context, motivated by Kon's result (Theorem K), for all type A hypersurfaces and some type B hypersurfaces we compute their sectional curvatures (Proposition 1). Here type A means either type A_1 or type A_2. In the last section we pose two open problems on sectional curvatures of real hypersurfaces in $P_n(\mathbb{C})$ (Problems 1 and 2).

2. Preliminaries

Let M be an orientable real hypersurface (with Riemannian metric $\langle\ ,\ \rangle$) of $P_n(\mathbb{C})$ and $\mathcal{N}$ a unit normal vector field on M. The Riemannian connections $\widetilde{\nabla}$ in $P_n(\mathbb{C})$ and ∇ in M are related by the following formulas for any vector fields X and Y on M:

$$\widetilde{\nabla}_X Y = \nabla_X Y + \langle AX, Y\rangle \mathcal{N} \quad \text{and} \quad \widetilde{\nabla}_X \mathcal{N} = -AX,$$

where A is the shape operator of M in $P_n(\mathbb{C})$. An eigenvector X of the shape operator A is called a *principal curvature vector*. Also an eigenvalue λ of A is called a *principal curvature*. In the following, we denote by V_λ the eigenspace of A associated with eigenvalue λ. It is known that M has an *almost contact metric structure* induced from the Kähler structure J of $P_n(\mathbb{C})$. That is, we can define a tensor field ϕ of type (1,1), a vector field ξ and a 1-form η on M by

$$\langle \phi X, Y\rangle = \langle JX, Y\rangle \quad \text{and} \quad \langle \xi, X\rangle = \eta(X) = \langle JX, \mathcal{N}\rangle.$$

So we have

$$\phi^2 X = -X + \eta(X)\xi, \quad \langle \xi, \xi\rangle = 1, \quad \phi\xi = 0.$$

Let R denote the curvature tensor of M. Then we have the following Gauss equation:

$$\begin{aligned}\langle R(X,Y)Z, W\rangle = {} & \langle Y, Z\rangle\langle X, W\rangle - \langle X, Z\rangle\langle Y, W\rangle + \langle \phi Y, Z\rangle\langle \phi X, W\rangle \\ & - \langle \phi X, Z\rangle\langle \phi Y, W\rangle - 2\langle \phi X, Y\rangle\langle \phi Z, W\rangle \\ & + \langle AY, Z\rangle\langle AX, W\rangle - \langle AX, Z\rangle\langle AY, W\rangle.\end{aligned}$$

Hence, for each orthonormal pair of vectors X, Y of M the sectional curvature $K(X,Y) = \langle R(X,Y)Y, X\rangle$ is expressed as:

$$K(X,Y) = 1 + 3\langle \phi X, Y\rangle^2 + \langle AX, X\rangle\langle AY, Y\rangle - \langle AX, Y\rangle^2. \tag{2.1}$$

We here recall the classification theorem of homogeneous real hypersurfaces in $P_n(\mathbb{C})$ due to Takagi [3].

Theorem T. *Let M be a homogeneous real hypersurface of $P_n(\mathbb{C})$. Then M is congruent to one of the following:*

(A_1) *a geodesic sphere of radius r, where $0 < r < \pi/2$,*

(A_2) *a tube of radius r over totally geodesic $P_k(\mathbb{C})$ $(1 \leqq k \leqq n-2)$, where $0 < r < \pi/2$,*

(B) *a tube of radius r over a complex hyperquadric Q_{n-1}, where $0 < r < \pi/4$,*

(C) *a tube of radius r over $P_1(\mathbb{C}) \times P_{\frac{n-1}{2}}(\mathbb{C})$, where $0 < r < \pi/4$ and n is odd,*

(D) *a tube of radius r over a complex Grassmann $G_{2,5}(\mathbb{C})$, where $0 < r < \pi/4$ and $n = 9$,*

(E) *a tube of radius r over a Hermitian symmetric space $SO(10)/U(5)$, where $0 < r < \pi/4$ and $n = 15$.*

It is known that a geodesic sphere of radius r (with $0 < r < \pi/2$) is congruent to a tube of radius $(\pi/2) - r$ over a complex hyperplane $P_{n-1}(\mathbb{C})$ in $P_n(\mathbb{C})$.

The numbers of distinct principal curvatures of these real hypersurfaces are 2, 3, 3, 5, 5, 5, respectively. These principal curvatures are as follows.

	(A_1)	(A_2)	(B)	(C, D, E)
λ_1	$\cot r$	$\cot r$	$\cot(r - \pi/4)$	$\cot(r - \pi/4)$
λ_2	—	$-\tan r$	$\cot(r + \pi/4)$	$\cot(r + \pi/4)$
λ_3	—	—	—	$\cot r$
λ_4	—	—	—	$-\tan r$
α	$2\cot(2r)$	$2\cot(2r)$	$2\cot(2r)$	$2\cot(2r)$

Here λ_i denotes a principal curvature of a principal curvature vector orthogonal to the characteristic vector ξ (with principal curvature α) of M.

The multiplicity $m(\mu)$ of each principal curvature μ of a homogeneous real hypersurface is as follows.

	(A_1)	(A_2)	(B)	(C)	(D)	(E)
$m(\lambda_1)$	$2n-2$	$2n-2k-2$	$n-1$	2	4	6
$m(\lambda_2)$	—	$2k$	$n-1$	2	4	6
$m(\lambda_3)$	—	—	—	$n-3$	4	8
$m(\lambda_4)$	—	—	—	$n-3$	4	8
$m(\alpha)$	1	1	1	1	1	1

Remark 1. Direct computation yields that a homogeneous real hypersurface M (which is a tube of radius r) is minimal in one of the following cases:

(A$_1$) $\cot r = \dfrac{1}{\sqrt{2n-1}}$,

(A$_2$) $\cot r = \sqrt{\dfrac{2k+1}{2n-2k-1}}$,

(B) $\cot r = \sqrt{n} + \sqrt{n-1}$,

(C) $\cot r = \dfrac{\sqrt{n}+\sqrt{2}}{\sqrt{n-2}}$,

(D) $\cot r = \sqrt{5}$,

(E) $\cot r = \dfrac{\sqrt{15}+\sqrt{6}}{3}$.

3. Sectional curvatures of hypersurfaces of type A or type B

By elementary computation we have the following lemma which is a key in our discussion.

Lemma. *Let V be an n-dimensional vector space (with inner product $\langle\ ,\ \rangle$) and $A : V \to V$ is a linear mapping. We denote by $\lambda_1, \ldots, \lambda_n$ eigenvalues of A. Then for each orthonormal pair of vectors $X, Y \in V$ we have*

$$\min_{1 \leqq i \neq j \leqq n} \lambda_i\lambda_j \leqq \langle AX, X\rangle\langle AY, Y\rangle - \langle AX, Y\rangle^2 \leqq \max_{1 \leqq i \neq j \leqq n} \lambda_i\lambda_j.$$

Our aim here is to prove the following:

Proposition 1. *The sectional curvature K of a type* A *hypersurface or a type* B *hypersurface takes the following maximum value and minimum value.*

(A$_1$) $\cot^2 r \leqq K \leqq 4 + \cot^2 r$.

(A$_2$) $0 \leqq K \leqq 4 + \max\{\cot^2 r, \tan^2 r\}$.

(B) $-\dfrac{\cot^2 + \cot r + 1}{\cot r} \leqq K$ *for each $n (\geqq 2)$. In particular, when $n = 2$, K satisfies* $-\dfrac{\cot^2 r + \cot r + 1}{\cot r} \leqq K \leqq \max\left\{3, \dfrac{\cot^2 r - \cot r + 1}{\cot r}\right\}$.

Proof. We denote by X, Y a pair of orthonormal vectors on our real hypersurface M. We shall consider the case of type (A$_1$). It follows from (2.1) and our Lemma that

$$\begin{aligned} K(X,Y) &\leqq 4 + \langle AX, X\rangle\langle AY, Y\rangle - \langle AX, Y\rangle^2 \\ &\leqq 4 + \cot^2 r \\ &= K(X_1, \phi X_1) \quad \text{for} \quad \forall X_1 \in V_{\cot r} \end{aligned}$$

as well as

$$\begin{aligned} K(X,Y) &\geqq 1 + \langle AX, X\rangle\langle AY, Y\rangle - \langle AX, Y\rangle^2 \\ &\geqq 1 + \cot r \cdot 2\cot(2r) = \cot^2 r \\ &= K(X_1, \xi) \quad \text{for} \quad \forall X_1 \in V_{\cot r}. \end{aligned}$$

Case of type (A_2). Note that $\lambda_2 < \alpha < \lambda_1$. Then we have similarly

$$\begin{aligned} K(X,Y) &\leqq 4 + \langle AX, X\rangle\langle AY, Y\rangle - \langle AX, Y\rangle^2 \\ &\leqq 4 + \max\{\lambda_1^2, \lambda_2^2\} = 4 + \max\{\cot^2 r, \tan^2 r\} \\ &= \max\{K(X_1, \phi X_1), K(X_2, \phi_2)\} \quad \text{for} \quad \forall X_1 \in V_{\lambda_1},\ \forall X_2 \in V_{\lambda_2} \end{aligned}$$

and

$$\begin{aligned} K(X,Y) &\geqq 1 + \langle AX, X\rangle\langle AY, Y\rangle - \langle AX, Y\rangle^2 \\ &\geqq 1 + \lambda_1\lambda_2 = 0 \\ &= K(X_1, X_2) \quad \text{for} \quad \forall X_1 \in V_{\lambda_1},\ \forall X_2 \in V_{\lambda_2}. \end{aligned}$$

Case of type (B). Our three distinct principal curvatures λ_1, λ_2, α satisfy $\lambda_2 > 0$, $\lambda_1 < 0$ and $\alpha > \lambda_2 > \lambda_1$. Hence by the same computation as above we see that the minimum value of K is expressed as:

$$\begin{aligned} K(X,Y) &\geqq 1 + \langle AX, X\rangle\langle AY, Y\rangle - \langle AX, Y\rangle^2 \\ &\geqq 1 + \lambda_1\alpha = -\frac{\cot^2 r + \cot r + 1}{\cot r} \\ &= K(X_1, \xi) \quad \text{for} \quad \forall X_1 \in V_{\lambda_1}. \end{aligned}$$

In the rest of proof we shall compute the maximum value of K in the case of $n = 2$. That is, we consider a homogeneous real hypersurface of type B in $P_2(\mathbb{C})$. We take unit vectors $e_1 \in V_{\lambda_1}$, $e_2 \in V_{\lambda_2}$. Needless to say, $e_2 = \pm\phi e_1$. For the vectors X, Y we set

$$X = a_1e_1 + a_2e_2 + a_3\xi, \quad Y = b_1e_1 + b_2e_2 + b_3\xi$$

and

$$c_1 = a_2b_3 - a_3b_2, \quad c_2 = a_3b_1 - a_1b_3, \quad c_3 = a_1b_2 - a_2b_1.$$

So we may easily find that $c_1^2 + c_2^2 + c_3^2 = 1$. Then by direct computation the sectional curvature $K(X,Y)$ can be expressed in terms of c_1, c_2 (with $0 \leqq c_1^2 + c_2^2 \leqq 1$) as:

$$K(c_1, c_2) = 3 - c_1^2 \frac{\cot^2 r - 4\cot r + 1}{\cot r} - c_2^2 \frac{\cot^2 r + 4\cot r + 1}{\cot r}.$$

Here, setting $c_1 = s\cos\theta$, $c_2 = s\sin\theta$ ($0 \leqq s \leqq 1$, $0 \leqq \theta < 2\pi$), we have the following quadratic function $K(s,\theta)$ with respect to s:

$$K(s,\theta) = 3 - 4s^2 + \frac{\cot^2 r + 1}{\cot r} s^2 \cos(2\theta).$$

In the following, our discussion is divided into two cases. One is the case of $1 < \cot r \leqq 2 + \sqrt{3}$, that is $\pi/12 \leqq r < \pi/4$. Then, because of $\dfrac{\cot^2 r + 1}{\cot r} \leqq 4$, we know that for each fixed θ_0 the function $K(s,\theta_0)$ is monotone decreasing with respect to s, so that $\max K = K(0,\theta) = K(e_1, e_2) = 3$. We here draw Figure 1 which is the graph of $K(s,\theta)$ ($0 \leqq s \leqq 1$, $0 \leqq \theta \leqq \pi$), when $r = \pi/4$.

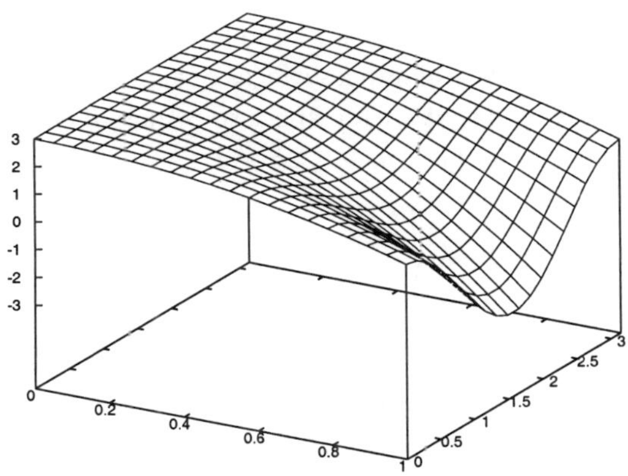

Fig. 1. The graph of $3 - 4s^2 + 2s^2\cos 2\theta$

The other is the case of $\cot r > 2 + \sqrt{3}$, that is $0 < r < \pi/12$. Then we have

$$K(s,\theta) \leqq K(s,0) = 3 + \frac{\cot^2 r - 4\cot r + 1}{\cot r} s^2 \leqq K(1,0)$$

and

$$\max K = K(e_2, \xi) = \frac{\cot^2 r - \cot r + 1}{\cot r}.$$

For example, when $\cot r = 5 + 2\sqrt{6}$, the graph of $K(s, \theta)$ $(0 \leqq s \leqq 1, 0 \leqq \theta \leqq \pi)$ is Figure 2.

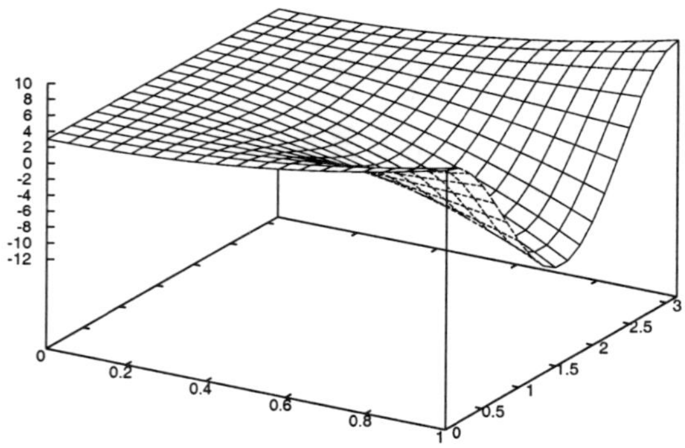

Fig. 2. The graph of $3 - 4s^2 + 10s^2 \cos 2\theta$

Thus we obtain the desirable maximum value. □

At the end of this section we pose the following conjecture on sectional curvatures of homogeneous real hypersurfaces of type B in $P_n(\mathbb{C})$, $n \geqq 3$. By direct computation we see that with respect to principal curvatures $\bar{\lambda}_i = \lambda_1$ $(i = 1, \cdots, n-1)$, $\bar{\lambda}_j = \lambda_2$ $(j = n, \cdots, 2n-2)$, $\bar{\lambda}_{2n-1} = \alpha$,

$$\max_{i \neq j} \bar{\lambda}_i \bar{\lambda}_j = \begin{cases} \lambda_2 \alpha & (0 < r \leqq r_0), \\ \lambda_1^2 & (r_0 \leqq r < \pi/4). \end{cases}$$

Here r_0 is given as follows. Let $t = \cot r$ with $0 < r < \pi/4$. Then $t_0 = \cot r_0$ is the unique solution of $t(1+t)^2 - (1-t)^4 = 0$, which is equivalent to $\lambda_2 \alpha = \lambda_1^2$, with $0 < t < 1$. Hence r_0 is expressed as:

$$r_0 = \cot^{-1} \left(\frac{5 + \sqrt{17}}{4} + \frac{1}{2} \sqrt{\frac{13 + 5\sqrt{17}}{2}} \right).$$

Conjecture. *For a homogeneous real hypersurface M of type B in $P_n(\mathbb{C})$, $n \geqq 3$ does the maximum value of the sectional curvature K of M take the following value?*

$$\max K = 1 + \begin{cases} \lambda_2 \alpha & (0 < r \leqq r_0), \\ \lambda_1^2 & (r_0 \leqq r < \pi/4). \end{cases}$$

4. Pinching problems on sectional curvature of real hypersurfaces

Proposition 1(B) gives us information on sectional curvatures of *some* plane section for other homogeneous real hypersurfaces in $P_n(\mathbb{C})$.

Proposition 2. *Let M be one of homogeneous real hypersurfaces of type* (B), (C), (D) *and* (E) *in* $P_n(\mathbb{C})$. *Then the sectional curvature K of M satisfies*

$$K(X,\xi) < 0 \quad \text{for each unit } X \in V_{\cot\left(r-\frac{\pi}{4}\right)}.$$

It is natural to pose the following two problems in connection with Propositions 1 and 2.

Problem 1. *Let M be a compact orientable real hypersurface of $P_n(\mathbb{C})$. If the sectional curvature K of M satisfies $0 \leqq K \leqq 5$, is M congruent to one of the following type A hypersurfaces?*

(1) *A geodesic sphere of radius r, where $\pi/4 \leqq r < \pi/2$ and $\cot^2 r \leqq K \leqq 4 + \cot^2 r$.*

(2) *A tube of radius $\pi/4$ over totally geodesic $P_k(\mathbb{C})$ $(1 \leqq k \leqq n-2)$, where $0 \leqq K \leqq 5$.*

Problem 2. *Let M be a compact orientable minimal real hypersurface of $P_n(\mathbb{C})$. If the sectional curvature K of M satisfies $0 \leqq K \leqq 5$, is M congruent to one of the following type A hypersurfaces?*

(1) *A geodesic sphere of radius r, where $\cot r = 1/\sqrt{2n-1}$ and $\dfrac{1}{2n-1} \leqq K \leqq 4 + \dfrac{1}{2n-1}$.*

(2) *A tube of radius $\pi/4$ over totally geodesic $P_k(\mathbb{C})$, where $2k = n-1$ and $0 \leqq K \leqq 5$.*

Finally we recall the following theorem due to Kon [1], which is related to Problem 2:

Theorem K. *Let M be a compact orientable minimal real hypersurface of $P_n(\mathbb{C})$. If the sectional curvature K satisfies $K \geqq \dfrac{1}{2n-1}$, then M is a geodesic sphere of radius r in $P_n(\mathbb{C})$, where $\cot r = 1/\sqrt{2n-1}$.*

References

1. M. Kon, *Real minimal hypersurfaces in a complex projective space*, Proc. Amer. Math. Soc. **79** (1980), 285–288.

2. R. Niebergall and P.J. Ryan, *Real hypersurfaces in complex space forms*, Tight and Taut Submanifolds (ed. T.E. Cecil and S.S. Chern), Math. Sci. Res. Inst. Publ. **32**, Cambridge Univ. Press, 1997, pp. 233–305.
3. R. Takagi, *On homogeneous real hypersurfaces in a complex projective space*, Osaka J. Math. **10** (1973), 495–506.

ON THREE-PARAMETRIC LIE GROUPS AS QUASI-KÄHLER MANIFOLDS WITH KILLING NORDEN METRIC

M. MANEV*, K. GRIBACHEV and D. MEKEROV

Faculty of Mathematics and Informatics, University of Plovdiv,
236 Bulgaria Blvd., Plovdiv, 4003, Bulgaria
**E-mail: mmanev@pu.acad.bg,*
http://www.fmi-plovdiv.org/

A 3-parametric family of 6-dimensional quasi-Kähler manifolds with Norden metric is constructed on a Lie group. This family is characterized geometrically. The condition for such a 6-manifold to be isotropic Kähler is given.

Keywords: Almost complex manifold; Norden metric; Quasi-Kähler manifold; Indefinite metric; Non-integrable almost complex structure; Lie group.

Introduction

It is a fundamental fact that on an almost complex manifold with Hermitian metric (almost Hermitian manifold), the action of the almost complex structure on the tangent space at each point of the manifold is isometry. There is another kind of metric, called a Norden metric or a B-metric on an almost complex manifold, such that the action of the almost complex structure is anti-isometry with respect to the metric. Such a manifold is called an almost complex manifold with Norden metric ([1]) or with B-metric ([2]). See also [5] for generalized B-manifolds. It is known that these manifolds are classified into eight classes ([1]).

The purpose of the present paper is to exhibit, by construction, almost complex structures with Norden metric on Lie groups as 6-manifolds, which are of a certain class, called quasi-Kähler manifold with Norden metric. It is proved that the constructed 6-manifold is isotropic Kählerian if and only if it is scalar flat or it has zero holomorphic sectional curvatures ([3]).

2000 *Mathematics Subject Classification.* Primary 53C15, 53C50; Secondary 32Q60, 53C55

1. Almost complex manifolds with Norden metric

1.1. *Preliminaries*

Let (M, J, g) be a $2n$-dimensional almost complex manifold with Norden metric, i. e. J is an almost complex structure and g is a metric on M such that

$$J^2X = -X, \qquad g(JX, JY) = -g(X, Y) \tag{1}$$

for all differentiable vector fields X, Y on M, i.e. $X, Y \in \mathfrak{X}(M)$.

The associated metric $\tilde{g}$ of g on M given by $\tilde{g}(X, Y) = g(X, JY)$ for all $X, Y \in \mathfrak{X}(M)$ is a Norden metric, too. Both metrics are necessarily of signature (n, n). The manifold $(M, J, \tilde{g})$ is an almost complex manifold with Norden metric, too.

Further, X, Y, Z, U (x, y, z, u, respectively) will stand for arbitrary differentiable vector fields on M (vectors in T_pM, $p \in M$, respectively).

The Levi-Civita connection of g is denoted by ∇. The tensor field F of type $(0, 3)$ on M is defined by

$$F(X, Y, Z) = g\big((\nabla_X J)Y, Z\big). \tag{2}$$

It has the following symmetries

$$F(X, Y, Z) = F(X, Z, Y) = F(X, JY, JZ). \tag{3}$$

Further, let $\{e_i\}$ $(i = 1, 2, \ldots, 2n)$ be an arbitrary basis of T_pM at a point p of M. The components of the inverse matrix of g are denoted by g^{ij} with respect to the basis $\{e_i\}$.

The Lie form θ associated with F is defined by

$$\theta(z) = g^{ij}F(e_i, e_j, z). \tag{4}$$

A classification of the considered manifolds with respect to F is given in Ref. 1. Eight classes of almost complex manifolds with Norden metric are characterized there according to the properties of F. The three basic classes are given as follows

$$\begin{aligned}
\mathcal{W}_1 : F(x, y, z) = \frac{1}{4n}\big\{&g(x, y)\theta(z) + g(x, z)\theta(y)\\
&+ g(x, Jy)\theta(Jz) + g(x, Jz)\theta(Jy)\big\};\\
\mathcal{W}_2 : \mathop{\mathfrak{S}}_{x,y,z} &F(x, y, Jz) = 0, \quad \theta = 0;\\
\mathcal{W}_3 : \mathop{\mathfrak{S}}_{x,y,z} &F(x, y, z) = 0,
\end{aligned} \tag{5}$$

where $\mathfrak{S}$ is the cyclic sum by three arguments.

The special class $\mathcal{W}_0$ of the Kähler manifolds with Norden metric belonging to any other class is determined by the condition $F = 0$.

1.2. *Curvature properties*

Let R be the curvature tensor field of ∇ defined by

$$R(X,Y)Z = \nabla_X \nabla_Y Z - \nabla_Y \nabla_X Z - \nabla_{[X,Y]} Z. \tag{6}$$

The corresponding tensor field of type $(0,4)$ is determined as follows

$$R(X,Y,Z,U) = g(R(X,Y)Z,U). \tag{7}$$

The Ricci tensor ρ and the scalar curvature τ are defined as usual by

$$\rho(y,z) = g^{ij} R(e_i, y, z, e_j), \qquad \tau = g^{ij} \rho(e_i, e_j). \tag{8}$$

Let $\alpha = \{x, y\}$ be a non-degenerate 2-plane (i.e. $\pi_1(x,y,y,x) = g(x,x)$ $g(y,y) - g(x,y)^2 \neq 0$) spanned by vectors $x, y \in T_pM$, $p \in M$. Then, it is known, the sectional curvature of α is defined by the following equation

$$k(\alpha) = k(x,y) = \frac{R(x,y,y,x)}{\pi_1(x,y,y,x)}. \tag{9}$$

The basic sectional curvatures in T_pM with an almost complex structure and a Norden metric g are

- *holomorphic sectional curvatures* if $J\alpha = \alpha$;
- *totally real sectional curvatures* if $J\alpha \perp \alpha$ with respect to g.

1.3. *Isotropic Kähler manifolds*

The square norm $\|\nabla J\|$ of ∇J is defined in [3] by

$$\|\nabla J\| = g^{ij} g^{kl} g\big((\nabla_{e_i} J)\, e_k, \big(\nabla_{e_j} J\big)\, e_l\big). \tag{10}$$

Having in mind the definition (2) of the tensor F and the properties (3), we obtain the following equation for the square norm of ∇J

$$\|\nabla J\| = g^{ij} g^{kl} g^{pq} F_{ikp} F_{jlq}, \tag{11}$$

where $F_{ikp} = F(e_i, e_k, e_p)$.

Definition 1.1 ([6]). *An almost complex manifold with Norden metric satisfying the condition $\|\nabla J\| = 0$ is called an* isotropic Kähler manifold with Norden metric.

Remark 1.1. It is clear, if a manifold belongs to the class $\mathcal{W}_0$, then it is isotropic Kählerian but the inverse statement is not always true.

2. Lie groups as quasi-Kähler manifolds with Killing Norden metric

The only class of the three basic classes, where the almost complex structure is not integrable, is the class $\mathcal{W}_3$ — the class of the *quasi-Kähler manifolds with Norden metric*.

Let us remark that the definitional condition from (5) implies the vanishing of the Lie form θ for the class $\mathcal{W}_3$.

Let V be a $2n$-dimensional vector space and consider the structure of the Lie algebra defined by the brackets $[E_i, E_j] = C_{ij}^k E_k$, where $\{E_1, E_2, \ldots, E_{2n}\}$ is a basis of V and $C_{ij}^k \in \mathbb{R}$.

Let G be the associated connected Lie group and $\{X_1, X_2, \ldots, X_{2n}\}$ be a global basis of left invariant vector fields induced by the basis of V. Then the Jacobi identity has the form

$$\underset{X_i,X_j,X_k}{\mathfrak{S}} \big[[X_i, X_j], X_k\big] = 0. \tag{12}$$

Next we define an almost complex structure by the conditions

$$JX_i = X_{n+i}, \quad JX_{n+i} = -X_i, \qquad i \in \{1, 2, \ldots, n\}. \tag{13}$$

Let us consider the left invariant metric defined by the following way

$$\begin{aligned} &g(X_i, X_i) = -g(X_{n+i}, X_{n+i}) = 1, \qquad i \in \{1, 2, \ldots, n\} \\ &g(X_j, X_k) = 0, \qquad j \neq k \in \{1, 2, \ldots, 2n\}. \end{aligned} \tag{14}$$

The introduced metric is a Norden metric because of (13).

In this way, the induced $2n$-dimensional manifold (G, J, g) is an almost complex manifold with Norden metric, in short *almost Norden manifold*.

The condition the Norden metric g be a Killing metric of the Lie group G with the corresponding Lie algebra $\mathfrak{g}$ is $g(\operatorname{ad} X(Y), Z) = -g(Y, \operatorname{ad} X(Z))$, where $X, Y, Z \in \mathfrak{g}$ and $\operatorname{ad} X(Y) = [X, Y]$. It is equivalent to the condition the metric g to be an invariant metric, i.e.

$$g\left([X, Y], Z\right) + g\left([X, Z], Y\right) = 0. \tag{15}$$

Theorem 2.1. *If (G, J, g) is an almost Norden manifold with a Killing metric g, then it is:*

(i) *a $\mathcal{W}_3$-manifold;*
(ii) *a locally symmetric manifold.*

Proof. (i) Let ∇ be the Levi-Civita connection of g. Then the condition (15) implies consecutively

$$\nabla_{X_i} X_j = \frac{1}{2}[X_i, X_j], \quad i,j \in \{1,2,\ldots,2n\}, \tag{16}$$

$$F(X_i, X_j, X_k) = \frac{1}{2}\Big\{g\big([X_i, JX_j], X_k\big) - g\big([X_i, X_j], JX_k\big)\Big\}. \tag{17}$$

According to (15) the last equation implies $\mathfrak{S}_{X_i,X_j,X_k} F(X_i, X_j, X_k) = 0$, i.e. the manifold belongs to the class $\mathcal{W}_3$.

(ii) The following form of the curvature tensor is given in [4]

$$R(X_i, X_j, X_k, X_l) = -\frac{1}{4}g\big(\big[[X_i, X_j], X_k\big], X_l\big).$$

Using the condition (15) for a Killing metric, we obtain

$$R(X_i, X_j, X_k, X_l) = -\frac{1}{4}g\big([X_i, X_j], [X_k, X_l]\big). \tag{18}$$

According to the constancy of the component R_{ijks} and (16) and (18), we get the covariant derivative of the tensor R of type $(0,4)$ as follows

$$\begin{aligned}(\nabla_{X_i} R)&(X_j, X_k, X_l, X_m)\\ &= \frac{1}{8}\Big\{g\big(\big[[X_i, X_j], X_k\big] - \big[[X_i, X_k], X_j\big], [X_l, X_m]\big)\\ &\quad + g\big(\big[[X_i, X_l], X_m\big] - \big[[X_i, X_m], X_l\big], [X_j, X_k]\big)\Big\}.\end{aligned} \tag{19}$$

We apply the the Jacobi identity (12) to the double commutators. Then the equation (19) gets the form

$$\begin{aligned}&(\nabla_{X_i} R)(X_j, X_k, X_l, X_m)\\ &= -\frac{1}{8}\Big\{g\big(\big[X_i, [X_j, X_k]\big], [X_l, X_m]\big) + g\big(\big[X_i, [X_l, X_m]\big], [X_j, X_k]\big)\big)\Big\}.\end{aligned} \tag{20}$$

Since g is a Killing metric, then applying (15) to (20) we obtain the identity $\nabla R = 0$, i.e. the manifold is locally symmetric. □

3. The Lie group as a 6-dimensional $\mathcal{W}_3$-manifold

Let (G, J, g) be a 6-dimensional almost Norden manifold with Killing metric g. Having in mind Theorem 2.1 we assert that (G, J, g) is a $\mathcal{W}_3$-manifold. Let the commutators have the following decomposition

$$[X_i, X_j] = \gamma_{ij}^k X_k, \quad \gamma_{ij}^k \in \mathbb{R}, \qquad i,j,k \in \{1,2,\ldots,6\}. \tag{21}$$

Further we consider the special case when the following two conditions are satisfied

$$g([X_i, X_j], [X_k, X_l]) = 0, \qquad g([X_i, JX_i], [X_i, JX_i]) = 0 \tag{22}$$

for all different indices i, j, k, l in $\{1, 2, \dots, 6\}$. In other words, the commutators of the different basis vectors are mutually orthogonal and moreover the commutators of the holomorphic sectional bases are isotropic vectors with respect to the Norden metric g.

According to the condition (15) for a Killing metric g, the Jacobi identity (12) and the condition (22), the equations (21) take the form given in Table 1.

Table 1. The Lie brackets with 3 parameters.

	X_1	X_2	X_3	X_4	X_5	X_6
$[X_2, X_3]$					λ_1	λ_2
$[X_3, X_1]$				λ_1		λ_3
$[X_1, X_2]$				λ_2	λ_3	
$[X_5, X_6]$		$-\lambda_1$	$-\lambda_2$			
$[X_6, X_4]$	$-\lambda_1$		$-\lambda_3$			
$[X_4, X_5]$	$-\lambda_2$	$-\lambda_3$				
$[X_1, X_4]$		λ_2	$-\lambda_1$		λ_2	$-\lambda_1$
$[X_1, X_5]$		λ_3		$-\lambda_2$		
$[X_1, X_6]$			$-\lambda_3$	λ_1		
$[X_2, X_4]$	$-\lambda_2$				λ_3	
$[X_2, X_5]$	$-\lambda_3$		λ_1	$-\lambda_3$		λ_1
$[X_2, X_6]$			λ_2		$-\lambda_1$	
$[X_3, X_4]$	λ_1					$-\lambda_3$
$[X_3, X_5]$		$-\lambda_1$				λ_2
$[X_3, X_6]$	λ_3	$-\lambda_2$		λ_3	$-\lambda_2$	

The Lie groups G thus obtained are of a family which is characterized by three real parameters λ_i ($i = 1, 2, 3$). Therefore, for the manifold (G, J, g) constructed above, we establish the truthfulness of the following

Theorem 3.1. *Let (G, J, g) be a 6-dimensional almost Norden manifold, where G is a connected Lie group with corresponding Lie algebra $\mathfrak{g}$ determined by the global basis of left invariant vector fields $\{X_1, X_2, \dots, X_6\}$; J is an almost complex structure defined by* (13) *and g is an invariant Norden metric determined by* (14) *and* (15). *Then (G, J, g) is a quasi-Kähler manifold with Norden metric if and only if G belongs to the 3-parametric family of Lie groups determined by Table 1.*

Let us remark, the Killing form $B(X, Y) = \operatorname{tr}(\operatorname{ad} X \operatorname{ad} Y)$, $X, Y \in \mathfrak{g}$, on the constructed Lie algebra $\mathfrak{g}$ has the following form

$$B = 4\left(\begin{array}{c|c} L & -L \\ \hline -L & L \end{array}\right), \qquad L = \begin{pmatrix} \lambda_3^2 & -\lambda_2\lambda_3 & -\lambda_1\lambda_3 \\ -\lambda_2\lambda_3 & \lambda_2^2 & -\lambda_1\lambda_2 \\ -\lambda_1\lambda_3 & -\lambda_1\lambda_2 & \lambda_1^2 \end{pmatrix}.$$

Obviously, it is degenerate.

4. Geometric characteristics of the constructed manifold

Let (G, J, g) be the 6-dimensional $\mathcal{W}_3$-manifold introduced in the previous section.

4.1. *The components of the tensor F*

Then by direct calculations, having in mind (2), (13), (14), (15), (16), (17) and Table 1, we obtain the nonzero components of the tensor F as follows

$$\begin{aligned}\lambda_1 &= 2F_{116} = 2F_{161} = -2F_{134} = -2F_{143} = 2F_{223} = 2F_{232} \\ &= 2F_{256} = 2F_{265} = -F_{322} = -F_{355} = 2F_{413} = 2F_{431} \\ &= 2F_{446} = 2F_{464} = -2F_{526} = -2F_{562} = 2F_{535} = 2F_{553} \\ &= -F_{611} = -F_{644} = -2F_{113} = -2F_{131} = -2F_{146} = -2F_{164} \\ &= -2F_{226} = -2F_{262} = 2F_{235} = 2F_{253} = F_{311} = F_{344} = 2F_{416} \\ &= 2F_{461} = -2F_{434} = -2F_{443} = -2F_{523} = -2F_{532} = -2F_{556} \\ &= -2F_{565} = F_{622} = F_{655},\end{aligned} \tag{23}$$

$$\begin{aligned}\lambda_2 &= -2F_{115} = -2F_{151} = 2F_{124} = 2F_{142} = F_{233} = F_{266} = -2F_{323} \\ &= -2F_{332} = -2F_{356} = -2F_{365} = -2F_{412} = -2F_{421} = -2F_{445} \\ &= -2F_{454} = F_{511} = F_{544} = -2F_{626} = -2F_{662} = 2F_{635} = 2F_{653} \\ &= 2F_{112} = 2F_{121} = 2F_{145} = 2F_{154} = -F_{211} = -F_{244} = -2F_{326} \\ &= -2F_{362} = 2F_{335} = 2F_{353} = -2F_{415} = -2F_{451} = 2F_{424} \\ &= 2F_{442} = -F_{533} = -F_{566} = 2F_{623} = 2F_{632} = 2F_{656} = 2F_{665},\end{aligned} \tag{24}$$

$$\begin{aligned}\lambda_3 &= -F_{133} = -F_{166} = -2F_{215} = -2F_{251} = 2F_{224} = 2F_{242} = 2F_{313} \\ &= 2F_{331} = 2F_{346} = 2F_{364} = -F_{422} = -F_{455} = 2F_{512} = 2F_{521} \\ &= 2F_{545} = 2F_{554} = 2F_{616} = 2F_{661} = -2F_{634} = -2F_{643} = F_{122} \\ &= F_{155} = -2F_{212} = -2F_{221} = -2F_{245} = -2F_{254} = 2F_{316} \\ &= 2F_{361} = -2F_{334} = -2F_{343} = F_{433} = F_{466} = -2F_{515} = -2F_{551} \\ &= 2F_{524} = 2F_{542} = -2F_{613} = -2F_{631} = -2F_{646} = -2F_{664},\end{aligned} \tag{25}$$

where $F_{ijk} = F(X_i, X_j, X_k)$.

4.2. *The square norm of* ∇J

According to (14) and (23)–(25), from (11) we obtain that the square norm of ∇J is zero, i.e. $\|\nabla J\| = 0$. Then we have the following

Proposition 4.1. *The manifold* (G, J, g) *is isotropic Kählerian.*

4.3. *The components of R*

Let R be the curvature tensor of type (0,4) determined by (7) and (6) on (G, J, g). We denote its components by $R_{ijks} = R(X_i, X_j, X_k, X_s)$; $i, j, k, s \in \{1, 2, \ldots, 6\}$. Using (16), (12), (18) and Table 1 we get the nonzero components of R as follows

$$-R_{1221} = R_{4554} = \frac{1}{4}\left(\lambda_2^2 + \lambda_3^2\right), \quad -R_{1551} = R_{2442} = \frac{1}{4}\left(\lambda_2^2 - \lambda_3^2\right),$$
$$-R_{1331} = R_{4664} = \frac{1}{4}\left(\lambda_1^2 + \lambda_3^2\right), \quad -R_{1661} = R_{3443} = \frac{1}{4}\left(\lambda_1^2 - \lambda_3^2\right),$$
$$-R_{2332} = R_{5665} = \frac{1}{4}\left(\lambda_1^2 + \lambda_2^2\right), \quad -R_{2662} = R_{3553} = \frac{1}{4}\left(\lambda_1^2 - \lambda_2^2\right),$$

$$R_{1361} = R_{2362} = -R_{4364} = -R_{5365} = \frac{1}{4}\lambda_1^2,$$
$$R_{1251} = R_{3253} = -R_{4254} = -R_{6256} = \frac{1}{4}\lambda_2^2,$$
$$R_{2142} = R_{3143} = -R_{5145} = -R_{6146} = \frac{1}{4}\lambda_3^2,$$

$$\begin{aligned} R_{1561} &= R_{2562} = R_{3563} = -R_{4564} = -R_{1261} = -R_{3263} \\ &= R_{4264} = R_{5265} = -R_{1351} = -R_{2352} = R_{4354} \\ &= R_{6356} = R_{1231} = -R_{4234} = -R_{5235} = -R_{6236} = \frac{1}{4}\lambda_1\lambda_2, \end{aligned}$$

$$\begin{aligned} -R_{1341} &= -R_{2342} = R_{5345} = R_{6346} = R_{2132} = -R_{4134} \\ &= -R_{5135} = -R_{6136} == R_{1461} = R_{2462} = R_{3463} \\ &= -R_{5465} = -R_{2162} = -R_{3163} = R_{4164} = R_{5165} = \frac{1}{4}\lambda_1\lambda_3, \end{aligned}$$

$$\begin{aligned} R_{3123} &= -R_{4124} = -R_{5125} = -R_{6126} = -R_{1241} = -R_{3243} \\ &= R_{5245} = R_{6246} == -R_{2152} = -R_{3153} = R_{4154} \\ &= R_{6156} = R_{1451} = R_{2452} = R_{3453} = -R_{6456} = \frac{1}{4}\lambda_2\lambda_3. \end{aligned}$$

4.4. *The components of ρ and the value of τ*

Having in mind (8) and the components of R, we obtain the components $\rho_{ij} = \rho(X_i, X_j)$ $(i, j = 1, 2, \dots, 6)$ of the Ricci tensor ρ and the the value of the scalar curvature τ as follows

$$
\begin{aligned}
&\rho_{11} = \rho_{44} = -\rho_{14} = -\lambda_3^2, \quad \rho_{12} = -\rho_{15} = -\rho_{24} = \rho_{45} = \lambda_2\lambda_3,\\
&\rho_{22} = \rho_{55} = -\rho_{25} = -\lambda_2^2, \quad \rho_{13} = -\rho_{16} = -\rho_{34} = \rho_{46} = \lambda_1\lambda_3, \qquad (26)\\
&\rho_{33} = \rho_{66} = -\rho_{36} = -\lambda_1^2, \quad \rho_{23} = -\rho_{26} = -\rho_{35} = \rho_{56} = \lambda_1\lambda_2,
\end{aligned}
$$

$$\tau = 0. \qquad (27)$$

The last equation implies immediately

Proposition 4.2. *The manifold (G, J, g) is scalar flat.*

4.5. *The sectional curvatures*

Let us consider the characteristic 2-planes α_{ij} spanned by the basis vectors $\{X_i, X_j\}$ at an arbitrary point of the manifold:

- holomorphic 2-planes — α_{14}, α_{25}, α_{36};
- pairs of totally real 2-planes — $(\alpha_{12}, \alpha_{45})$; $(\alpha_{13}, \alpha_{46})$; $(\alpha_{15}, \alpha_{24})$; $(\alpha_{16}, \alpha_{34})$; $(\alpha_{23}, \alpha_{56})$; $(\alpha_{26}, \alpha_{35})$.

Then, using (9), (14) and the components of R, we obtain the corresponding sectional curvatures

$$k(\alpha_{14}) = k(\alpha_{25}) = k(\alpha_{36}) = 0;$$

$$
\begin{aligned}
-k(\alpha_{12}) = k(\alpha_{45}) = \frac{1}{4}\left(\lambda_2^2 + \lambda_3^2\right), \quad k(\alpha_{15}) = -k(\alpha_{24}) = \frac{1}{4}\left(\lambda_2^2 - \lambda_3^2\right),\\
-k(\alpha_{13}) = k(\alpha_{46}) = \frac{1}{4}\left(\lambda_1^2 + \lambda_3^2\right), \quad k(\alpha_{16}) = -k(\alpha_{34}) = \frac{1}{4}\left(\lambda_1^2 - \lambda_3^2\right),\\
-k(\alpha_{23}) = k(\alpha_{56}) = \frac{1}{4}\left(\lambda_1^2 + \lambda_2^2\right), \quad k(\alpha_{26}) = -k(\alpha_{35}) = \frac{1}{4}\left(\lambda_1^2 - \lambda_2^2\right).
\end{aligned}
$$

Therefore we have the following

Proposition 4.3. *The manifold (G, J, g) has zero holomorphic sectional curvatures.*

4.6. *The isotropic-Kählerian property*

Having in mind Propositions 4.1–4.3 and Theorem 2.1, we give the following characteristics of the constructed manifold

Theorem 4.1. *The manifold (G, J, g) constructed as an $\mathcal{W}_3$-manifold with Killing metric in Theorem 3.1:*

(i) *is isotropic Kählerian;*
(ii) *is scalar flat;*
(iii) *is locally symmetric;*
(iv) *has zero holomorphic sectional curvatures.*

References

1. G. Ganchev and A. Borisov, *Note on the Almost Complex Manifolds with a Norden Metric*, Compt. Rend. Acad. Bulg. Sci. **39** (1986), 31–34.
2. G. Ganchev, K. Gribachev and V. Mihova, *B-Connections and their Conformal Invariants on Conformally Kahler Manifolds with B-Metric*, Publ. Inst. Math. (Beograd) (N.S.) **42** (1987), 107–121.
3. E. García-Río and Y. Matsushita, *Isotropic Kähler Structures on Engel 4-Manifolds*, J. Geom. Phys. **33** (2000), 288–294.
4. K. Gribachev, M. Manev and D. Mekerov, *A Lie group as a 4-dimensional quasi-Kahler manifold with Norden metric*, JP J. Geom. Topol. **6** (2006), 55–68.
5. K.I. Gribachev, D.G. Mekerov and G.D. Djelepov, *Generalized B-manifolds*, Compt. Rend. Acad. Bulg. Sci. **38** (1985), 299–302.
6. D. Mekerov and M. Manev, *On the geometry of quasi-Kähler manifolds with Norden metric*, Nihonkai Math. J. (2006).

APPLICATION OF UNSCENTED AND EXTENDED KALMAN FILTERING FOR ESTIMATING QUATERNION MOTION

V. MARKOVA

Institute of Control and System Research
159 Russki str., Plovdiv, 4000, Bulgaria
e-mail: markovavanya@yahoo.com

In this paper, we present a study comparing the performance of unscented and extended Kalman filtering for improving model for trajectory interpolation. Specifically, we examine orientation motion with quaternions, which are critical for appreciating direction. Our experimental results and analysis indicate that unscented Kalman filtering performs equivalently with extended Kalman filtering.

Keywords: Extended Kalman filtering; Unscented Kalman filtering; Trajectory interpolation; Quaternions.

1. Introduction

Trajectory interpolation algorithms represent an important component of any 2D/3D Short Term Dispersion simulation system. Without these algorithms, the systems must be limited to use the observed points only to compute and predict the dispersion of the pollutant.

The industrial short term model accepts hourly meteorological data records to define the conditions for plume rise, transport, diffusion and deposition and uses the steady-state Gaussian plume equation for a continue elevated sources. For a steady-state Gaussian plume, the hourly concentration is given by [2].

$$\mu = \frac{Q * K * V * D}{2 * \pi * u_s * \sigma_y * \sigma_z} * \exp{(-0.5 * \frac{y}{\sigma_y})^2}. \tag{1}$$

The Kalman filter is a set of mathematical equations that provides an efficient computational (recursive) means to estimate the state of a process, in a way that minimizes the mean of the squared error. The filter is very powerful in a several aspects: it supports estimations of past, present and even future states, and it can do so even when the precise nature of the modeled system is unknown.

The extended Kalman filter (EKF) provides this modification by linearizing all nonlinear models (i.e., process and measurement models) so the traditional KF can be applied [1]. Unfortunately, the EKF has two important potential drawbacks. First, the derivation of the Jacobean matrices, the linear approximation to the nonlinear functions, can be complex causing implementation difficulties. Second, these linearizations can lead to filter instability if the time step intervals are not sufficiently small [3].

To address these limitations, Julier and Uhlmann developed the unscented Kalman filter UKF [3]. The UKF operates on the premise that it is easier to approximate a Gaussian distribution than it is to approximate an arbitrary nonlinear function. Instead of linearizing using Jacobian matrices, the UKF using a deterministic sampling approach to capture the mean and covariance estimates with a minimal set of sample points. The UKF is a powerful nonlinear estimation technique and has been shown to be a superior alternative to the EKF in a variety of applications including parameter estimation for time series modeling [3], and neural network training [3].

The aim of the paper is to provide a practical application to the discrete Kalman filter. This application includes a description and some discussion of the basic and extended Kalman filter and example,results. The main purpose in this article is developing of the algorithm for comparison of UKF and EKF effects in moving trajectory of virtual point source of air pollution. This trajectory is described by means of the quaternion geometry and about this reason is defined measure for correlation of the quaternion curves. We describe the results of an experimental study which examines the estimation accuracy of the EKF and UKF on of virtual point source of air pollution represented with quaternions. Quaternions are a common way to represent rotations in tracking, robotics, and mechanical engineering because they are compact and avoid gumball lock [4]. The results of our study indicate that, although the EKF and UKF have equivalent performance, the additional computational overhead of the UKF and the quasilinear nature of the quaternion dynamics makes the EKF a more appropriate choice for orientation estimation.

2. Kalman's filters - short background

The Extended Kalman gain is computed [3] using

$$K_k = P_k^- H_k^T (H_k P_k^- H_k^T + R)^{-1} \tag{2}$$

where R is the measurement noise covariance, and the measurement matrix is calculated using

$$H_{k,[i,j]} = \frac{dh_{(i)}}{dx_{(j)}}(x_k^-) \tag{3}$$

a Jacobian matrix that linearizes around the non-linear measurement function h. In our case,h is quaternion normalisation

$$h = \frac{q}{\sqrt{q_x^2 + q_y^2 + q_z^2 + q_w^2}} \tag{4}$$

for the quaternion in $\hat{x}_k^-$. Then, we find the a priory estimate of the error covariance Matrix

$$P_k^- = B_k P_{k-1} B_k^T + Q_k \tag{5}$$

The basic premise behind the Unscented Kalman filter is it is easier to approximate a Gaussian distribution than it is to approximate an arbitrary nonlinear function. Instead of linearizing using Jacobian matrices, the UKF uses a deterministic sampling approach to capture the mean and covariance estimates with a minimal set of sample points. As with the EKF, we present an algorithmic description of the UKF omitting some theoretical considerations:

$$K_k = P_{\hat{x}_k \hat{z}_k} P_{\hat{x}_k \hat{z}_k}^{-1} \tag{6}$$

$$P_{\hat{x}_k \hat{z}_k} = \sum_i^{2L} W_i^c [(Z_k)_i - \hat{z}_k^-][(Z_k)_i - \hat{z}_k^-]^T + R \tag{7}$$

$$P_{\hat{x}_k \hat{z}_k}^{-1} = \sum_i^{2L} W_i^c [(X_k)_i - \hat{x}_k^-][(Z_k)_i - \hat{z}_k^-]^T \tag{8}$$

where W are weights defined by

$$W_0^c = \frac{\lambda}{(L-\lambda)} + (1 + \alpha^2 + \beta) \tag{9}$$

$$W_i^c = \frac{1}{(L-\lambda)} \tag{10}$$

$$i = 1, \ldots, 2L. \tag{11}$$

To compute the correction step, we fist must transform the columns of X_k through the measurement function

$$(Z_k)_i = h((X_k)_i), \quad i = 0, \dots, 2L.$$

Once Xk is computed, we perform the prediction step by first propagating each column of $Xk - 1$ through time. With $(X_k)_i$ calculated the a priory state estimate is

$$x_k^{-1} = \sum_i^{2L} W_i^m [(X_k)_i \tag{12}$$

3. The Problem of Trajectory interpolation

The list of dispersion models is long and including different models. We are interesting by Industrial Source Models (ISM) [2]. This is a US EPA multi-source Gaussian model capable of predicting both long-term (annual mean) and short-term (down to 1-hour mean) concentrations arising from point, area and volume sources. Gravitational settling of particles can be accounted for using a dry deposition algorithm; wet deposition and depletion due to rainfall can also be treated. Effects of buildings can be considered (using the BREEZEWAKE/BPIP facility). The model has urban and rural dispersion coefficients, and percentile concentrations can be calculated using the- PERCENT post-processor if sequential meteorological data are used. ISM can handle up to 1000 sources and 10,000 receptors. It is widely used in the USA for the evaluation of industrial sources and has been updated over the years to remain compatible with PC systems.

Indeed, we investigated the dispersion of concentrations from point air source, as trajectories of virtual pollutant sources and estimate the real dispersion field as well ideal symmetricall field.

The process model we use is an orientation/angular velocity (OV) model defined by

$$f = \frac{dq}{dt} = \frac{1}{2} qw, \tag{13}$$

where q is the current quaternion and w is a pure vector quaternion representing angular velocity. We use a single EKF/UKF, where the state vector at time k is defined by

$$\hat{x}_k = [q_x, q_y, q_z, q_w, w_0, w_1, w_2]^T. \tag{14}$$

Given the state vector at step $k-1$, we first perform the prediction step by finding the a priory state estimate $\hat{x}_k^-$ by integrating equation (1) through

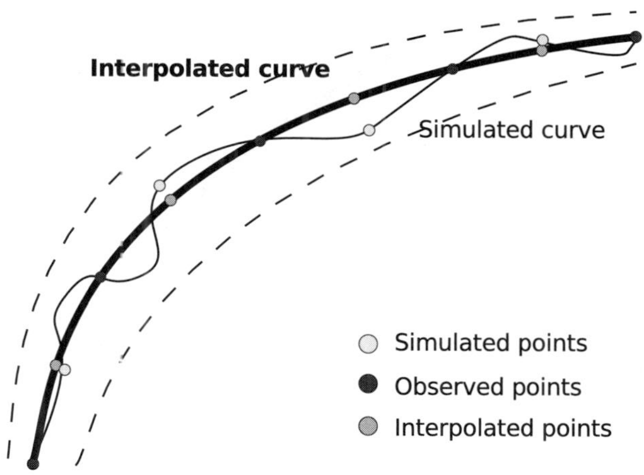

Fig. 1. The pollution trajectory curve.

time by Δt. Because, the real curve have fluctuation, which are not presented with interpolation curve, we use a small Monte Carlo simulation on each experiment since we have random Gaussian noise added to the motion trajectory of ideal point pollutant, which is used to simulate jittery tracking data (Figure 1). After that we use Kalman's filters for additionally smoothing the curve. Applying the EKF and UKF for accomplishing to the hybrid model for simulation of the pollution dispersion in the course of time, and at the end we make comparison of the results.

4. Experimental Study

The Kalman filter is a set of mathematical equations that use information from multiple sources; it uses a predictor/corrector mechanism to find an optimal estimate in the sense that it minimises the estimated error covariance. In other words, the filter uses an underlying process model to make an estimate of the current system state and then corrects the estimate using any available sensor measurements.

Then, after the correction is made, we use the process model to make a prediction.

For orientation prediction, we use extended Kalman filtering since the standard Kalman filter is a linear estimator and orientation is non-linear in nature.

The data set was tested with different prediction times giving us four

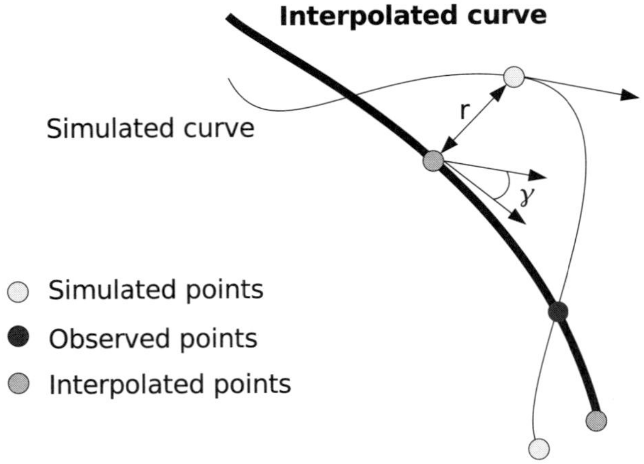

Fig. 2. The speed of the wind at the observed and predicted point

different test sets. The predictors are evaluated using root mean square error (RMSE). Algorithm running times are also calculated by grouping the UKF and EKF predictors. We can calculate the root mean square error (RMS) for each step and take the Monte Carlo simulation runs. Let we assume that the speed of the wind at the observed and predicted point is identical

$$q = \cos(\gamma) + r * \sin(\gamma),$$

then for truth and estimated quaternions q_{t_i} and q_{e_i} (Figure 2). RMS is defined by

$$RMS_q = \sqrt{(\frac{1}{2}\sum_{i=0}^{n-1} e_i^2)}. \tag{15}$$

where the RMS_q is in $n-1$ degrees and

$$RMS_e = \sqrt{(\frac{1}{2}\sum_{i=0}^{n-1} \parallel e_i^2 \parallel)}. \tag{16}$$

$$e_i = \frac{4}{\pi}\arccos((q_{t_i}(q_{e_i}^{-1})w). \tag{17}$$

To compare the performance of the EKF and UKF algorithms described above in the paper, we conducted an experiment to determine which filtering algorithm is preferable for improving direction of plume flux trajectory of air pollution. First, the data sets were used in our study to represent common orientation dynamics fluxes in trajectory applications. We use a small Monte Carlo simulation on each test scenario since we have random Gaussian noise added to the motion signals, which is used to simulate jittery tracking data.

To determine how well the EKF and UKF algorithms are performing, we need comparison data. Comparing estimated output with reported user orientations is problematic syntheses records have noise and small distortions associated with them. Thus, any comparison with the recorded data would count tracking error with the estimation error.

5. Conclusion

In this paper, we have presented an experiment which compares extended and unscented Kalman filtering for simulation of the pollution dispersion in the course of time. The results of our study indicate that, although the EKF and UKF have equivalent performance, the additional computational overhead of the UKF and the quasilinear nature of the quaternion dynamics makes the EKF a more appropriate choice for orientation estimation. Our results indicate that, although the EKF and UKF have roughly the same accuracy, the computational overhead of the UKF,the simplicity of the Jacobian matrix calculations, and the quasilinear nature of the quaternion dynamics makes the EKF a better choice for the task of improving trajectory interpolation of noisy quaternion signals in virtual pollutant point.

References

1. T.L. Chia, D. Simon and H.J. Chizeck, in *Kalman filtering with statistical state constrains*, J. Control and Intelligent Systems, (2006).
2. http://www.uwe ac.uk/aqm/centre/model.html
3. G. Welch and G. Bishop, in *An introduction to the Kalman filter*, Technical TR-95041, Dep. Computer Science, Uni of North Carolina at Chapel Hill, 1995.
4. S. Ken, *Animating Rotations with Quaternion Curves*, ASM Press 245–254 (1985).

COUNTEREXAMPLES OF COMPACT TYPE TO THE GOLDBERG CONJECTURE AND VARIOUS VERSION OF THE CONJECTURE

Y. MATSUSHITA

Section of Mathematics, School of Engineering
University of Shiga Prefecture,
Hikone 522-8533, Japan
E-mail: yasuo@mech.usp.ac.jp

We review a newly discovered counterexample to the Goldberg Conjecture of compact type. Such a counterexample is constructed on an eight-dimensional Walker manifold of neutral signature. The original Goldberg conjecture is, of course, reviewed, and counterexamples to some variants of the conjecture are also exhibited. Analogues of the Goldberg conjecture for almost Norden manifolds and for almost para-Hermitian manifolds are proposed. Finally, some work on the odd-dimensional version of the Goldberg conjecture will be briefly reviewed.

1. The Goldberg conjecture

There is a famous conjecture, posed in 1969 by Goldberg [11], which states that the almost complex structure of a compact almost Kähler-Einstein Riemannian manifold is integrable.

Let (M, J, g) be an almost Hermitian manifold, with J an almost complex structure and g a J-invariant Riemannian metric, i.e., $g(JX, JY) = g(X, Y)$. Here, by an almost complex structure we mean a linear endomorphism of the tangent bundle TM of M, satisfying

$$J^2 = -1. \tag{1}$$

The pair (J, g) of an almost Hermitian structure defines a fundamental 2-form Ω by

$$\Omega(X, Y) = g(JX, Y). \tag{2}$$

Then, Goldberg's conjecture states that an almost Hermitian manifold (M, J, g) must be Kähler if the following three conditions are imposed:

(GC1) the manifold M is compact
(GC2) the Riemannian metric g is Einstein
(GC3) the fundamental 2-form Ω is symplectic.

Note that an almost Hermitian manifold (M, J, g) is called *almost Kähler* if Ω is symplectic (GC3).

Despite many papers by various authors concerning the Goldberg conjecture, there are only two papers which obtained substantial results to the original Goldberg conjecture. Sekigawa is the author of these two papers.

Theorem 1 (Sekigawa [19,20]). *Let $M = (M, J, g)$ be an almost Hermitian manifold, which satisfies the three conditions* (GC1), (GC2) *and* (GC3). *If the scalar curvature of M is nonnegative, then M must be Kähler.*

This partially affirmative result was proven for 4-dimensional manifolds in [19], and two years later for arbitrary even-dimensional manifolds in [20]. It should be noted that no progress has been made on the original conjecture, other than Sekigawa's theorem.

2. Noncompact type for almost Hermitian manifolds

Removing the condition (GC1), Alekseevsky [1] deduced the existence of counterexamples in the noncompact case, namely, certain quaternion spaces of rank > 1. Nurowski and Przanowski [18] exhibited an explicit counterexample in the noncompact case in four dimensions. Further counterexamples in the noncompact case have recently been reported by Apostolov, Draghici and Moroianu [2]. (The assertion in [18] that Sekigawa has established Goldberg's conjecture in four dimensions is erroneous, [21].)

3. Noncompact type for almost pseudo-Hermitian manifolds

Haze has demonstrated counterexamples by constructing a neutral Ricci-flat metric on an open subset of $\mathbb{R}^4$ which is almost-Kähler with respect to a non-integrable almost complex structure. In [14], Haze's metric is shown to be an instance of a Walker metric. By a Walker manifold, we mean a pseudo-Riemannian n-manifold which admits a field of parallel null r-planes, with $r \leq \dfrac{n}{2}$ [22]. The corresponding metric is called a Walker metric. When $n = 2k$ and $r = k$, such a metric is of neutral signature. Especially for four-dimensional Walker manifolds, see [10,13,14,6,9]. For a survey on the neutral 4-manifolds, see e.g., [15,17].

We now exhibit Haze's counterexamples. Let (x^1, x^2, x^3, x^4) be the coordinates on $\mathbb{R}^4$, and consider the metric on it defined by

$$g = \begin{bmatrix} 0 & 0 & 1 & 0 \\ 0 & 0 & 0 & 1 \\ 1 & 0 & a(x^1, x^2, x^3, x^4) & 0 \\ 0 & 1 & 0 & -a(x^1, x^2, x^3, x^4) \end{bmatrix}. \tag{3}$$

For the metric, the Einstein equation (cf. [14, Appendix D]) consists of the following PDE's for the function a:

$$\begin{aligned} a_{12} = 0, \qquad & a\,a_{11} - 2\,a_{24} - (a_2)^2 = 0, \\ a\,a_{14} - a_{23} + a_1 a_2 = 0, \qquad & a\,a_{11} - 2a_{13} + (a_1)^2 = 0, \end{aligned} \tag{4}$$

where $a_1 = \partial a/\partial x^1$, $a_{12} = \partial^2 a/\partial x^1 \partial x^2$, etc. If a is independent of x^2 and x^4, and if a contains x^1 only linearly, the first three PDE's hold trivially, and the last one reduces to: $2a_{13} - (a_1)^2 = 0$. We see that $a = -\dfrac{2x^1}{x^3}$ is a solution to the PDE, and therefore the metric

$$g = \begin{bmatrix} 0 & 0 & 1 & 0 \\ 0 & 0 & 0 & 1 \\ 1 & 0 & -\dfrac{2\,x^1}{x^3} & 0 \\ 0 & 1 & 0 & \dfrac{2\,x^1}{x^3} \end{bmatrix}, \tag{5}$$

is Einstein on the coordinate patch $x^3 > 0$ (or $x^3 < 0$). Thus, the second condition (GC2) holds. Note that this is a Ricci flat metric.

We know that this metric admits a proper almost complex structure as follows (cf. [14, (3)]):

$$J\partial_1 = \partial_2, \qquad J\partial_2 = -\partial_1, \qquad J\partial_3 = a\partial_2 + \partial_4, \qquad J\partial_4 = a\partial_1 - \partial_3. \tag{6}$$

From [14, (5)], together with the neutral metric (3), we have the fundamental 2-form as follows:

$$\Omega = dx^1 \wedge dx^4 - dx^2 \wedge dx^3. \tag{7}$$

Ω is clearly symplectic [14, Theorem 2]. Therefore, the condition (GC3) holds.

For the Einstein metric (5), the proper almost complex structure J in (6) becomes

$$J\partial_1 = \partial_2, \quad J\partial_2 = -\partial_1, \quad J\partial_3 = -\frac{x^1}{x^3}\,\partial_2 + \partial_4, \quad J\partial_4 = -\frac{x^1}{x^3}\,\partial_1 - \partial_3. \tag{8}$$

Then, the integrability condition of J, given in [14, Theorem 3], becomes

$$a_1 - b_1 - 2c_2 = 2a_1 = -\frac{4}{x^3} \neq 0, \qquad a_2 - b_2 + 2c_1 = 2a_2 = 0. \tag{9}$$

Thus, J does not satisfy (9), and hence it cannot be integrable.

Finally, the domain $x^3 > 0$ (or $x^3 < 0$) in $\mathbb{R}^4$, where the Einstein metric g in (5) and the nonintegrable J in (8) exist, must be noncompact. We thus obtain a counterexample of noncompact and neutral type to the Goldberg conjecture, as discovered by Haze.

Remark. In a recent paper [9], we exhibited several families of counterexamples of noncompact and neutral type, which are constructed also on Walker 4-manifolds. Haze's counterexample is of course contained in one of these families.

4. Counterexamples of compact type for almost pseudo-Hermitian Walker 8-manifolds

This section is the main part of the present note. The counterexamples reported in [16] are the topic of the present section, since these are the first ones of *compact type*.

By considering 8-dimensional Walker manifolds, we are able to exhibit almost Kähler-Einstein neutral structures, which are not Kähler, initially on $\mathbb{R}^8$ and then on an 8-torus. Thus, the neutral-signature version of Goldberg's conjecture fails.

Let (M, g, D) be an 8-dimensional Walker manifold, where g is a metric of neutral signature and D a field of parallel null 4-planes. From Walker's theorem [22], there is, locally, a system of coordinates $(x^1, \dots, x^8)$ so that g takes the canonical form

$$g = \left[g_{ij}\right] = \begin{bmatrix} 0 & I_4 \\ I_4 & B \end{bmatrix}, \tag{10}$$

where I_4 is the unit 4×4 matrix and B is a 4×4 symmetric matrix whose entries are functions of the coordinates $(x^1, \dots, x^8)$. Note that g is of neutral signature $(++++----)$, and that $D = \operatorname{span}\left\{\partial_1, \dots, \partial_4\right\}$ $(\partial_i = \partial/\partial x^i)$. In this article, we consider the specific Walker metrics on $\mathbb{R}^8$ with B of the form:

$$B = \begin{bmatrix} p & 0 & 0 & 0 \\ 0 & 0 & 0 & 0 \\ 0 & 0 & r & 0 \\ 0 & 0 & 0 & 0 \end{bmatrix}, \tag{11}$$

where p, r are arbitrary functions of the coordinates:

$$p = p(x^1, \ldots, x^8), \quad r = r(x^1, \ldots, x^8). \tag{12}$$

We can construct various g-orthogonal almost complex structures J (abbr. almost complex structures) on a Walker 8-manifold M with the metric g as in (10), (11) so that (M, g, J) is almost (neutral) Hermitian. The following J is one of the simplest examples of such an almost complex structure.

$$\begin{gathered} J\partial_1 = \partial_3, \quad J\partial_2 = \partial_4, \quad J\partial_3 = -\partial_1, \quad J\partial_4 = -\partial_2, \\ J\partial_5 = \frac{1}{2}(p - r)\partial_3 + \partial_7, \quad J\partial_6 = \partial_8, \\ J\partial_7 = \frac{1}{2}(p - r)\partial_1 - \partial_5, \quad J\partial_8 = -\partial_6. \end{gathered} \tag{13}$$

If we write the above action as $J\partial_i = \sum_{j=1}^{8} J_i^j\, \partial_j$, then we can read off the nonzero components J_i^j as follows:

$$\begin{gathered} J_1^3 = -J_3^1 = J_2^4 = -J_4^2 = J_5^7 = -J_7^5 = J_6^8 = -J_8^6 = 1, \\ J_5^3 = J_7^1 = \frac{1}{2}(p - r). \end{gathered} \tag{14}$$

Associated with the almost Hermitian structure (g, J) is the Kähler form Ω, given by $\Omega(X, Y) = g(JX, Y)$ for arbitrary vector fields X, Y. The coordinate expression for the Kähler form is:

$$\begin{aligned} \Omega &= \sum_{i<j} \Omega(\partial_i, \partial_j)\, dx^i \wedge dx^j \\ &= dx^1 \wedge dx^7 + dx^2 \wedge dx^8 - dx^3 \wedge dx^5 - dx^4 \wedge dx^6 + \frac{1}{2}\,(p + r)\, dx^5 \wedge dx^7. \end{aligned} \tag{15}$$

It is easy to see that Ω is nondegenerate by noting that

$$\Omega^4 = \Omega \wedge \Omega \wedge \Omega \wedge \Omega = 24 dx^1 \wedge dx^2 \wedge dx^3 \wedge dx^4 \wedge dx^5 \wedge dx^6 \wedge dx^7 \wedge dx^8. \tag{16}$$

The differential of Ω is easily computed:

$$\begin{aligned} d\Omega = &\frac{1}{2}(p_1 + r_1)\, dx^1 \wedge dx^5 \wedge dx^7 + \frac{1}{2}(p_2 + r_2)\, dx^2 \wedge dx^5 \wedge dx^7 \\ &+ \frac{1}{2}(p_3 + r_3)\, dx^3 \wedge dx^5 \wedge dx^7 + \frac{1}{2}(p_4 + r_4)\, dx^4 \wedge dx^5 \wedge dx^7 \\ &- \frac{1}{2}(p_6 + r_6)\, dx^5 \wedge dx^6 \wedge dx^7 + \frac{1}{2}(p_8 + r_8)\, dx^5 \wedge dx^7 \wedge dx^8, \end{aligned} \tag{17}$$

where $p_i = \partial p/\partial x^i$ and $r_i = \partial r/\partial x^i$. From this expression, we have the following

Proposition 2. *Ω is symplectic if and only if the following PDE's hold.*

$$\begin{aligned} &p_1 + r_1 = 0, \quad p_2 + r_2 = 0, \quad p_3 + r_3 = 0, \\ &p_4 + r_4 = 0, \quad p_6 + r_6 = 0, \quad p_8 + r_8 = 0. \end{aligned} \tag{18}$$

Corollary 3. *Ω is symplectic if and only if*

$$r = -p + \xi, \tag{19}$$

where ξ is any function of x^5 and x^7 alone.

The almost complex structure J is integrable if and only if the torsion of J (Nijenhuis tensor) vanishes, i.e., the components

$$N^i_{jk} = 2\sum_{h=1}^{8}\left(J^h_j \frac{\partial J^i_k}{\partial x^h} - J^h_k \frac{\partial J^i_j}{\partial x^h} - J^i_h \frac{\partial J^h_k}{\partial x^j} + J^i_h \frac{\partial J^h_j}{\partial x^k}\right) \tag{20}$$

all vanish (cf. [12, p.124]), with J^j_i as in (14). Since $N^i_{jk} = -N^i_{kj}$, we need only consider N^i_{jk} $(j < k)$. By explicit calculation, the nonzero components of the Nijenhuis tensor are as follows:

$$\begin{aligned} &N^1_{15} = -N^1_{37} = -N^3_{17} = -N^3_{35} = p_1 - r_1, \quad N^3_{57} = -\frac{1}{2}(p - r)(p_1 - r_1), \\ &N^1_{25} = -N^1_{47} = -N^3_{27} = -N^3_{45} = p_2 - r_2, \quad N^1_{57} = \frac{1}{2}(p - r)(p_3 - r_3) \\ &N^1_{17} = N^1_{35} = N^3_{15} = -N^3_{37} = p_3 - r_3, \\ &N^1_{27} = N^1_{45} = N^3_{25} = -N^3_{47} = p_4 - r_4, \\ &N^1_{56} = -N^1_{78} = -N^3_{58} = N^3_{67} = -p_6 + r_6, \\ &N^1_{58} = -N^1_{67} = N^3_{56} = -N^3_{78} = -p_8 + r_8. \end{aligned} \tag{21}$$

Proposition 4. *J is integrable if and only if the following PDE's hold.*

$$\begin{aligned} &p_1 - r_1 = 0, \quad p_2 - r_2 = 0, \quad p_3 - r_3 = 0, \\ &p_4 - r_4 = 0, \quad p_6 - r_6 = 0, \quad p_8 - r_8 = 0. \end{aligned} \tag{22}$$

Corollary 5. *J is integrable if and only if*

$$r = p + \eta, \tag{23}$$

where η is any function of x^5 and x^7 alone.

The almost Hermitian structure (g, J) is *Kähler* if Ω is a symplectic form and J is integrable.

Theorem 6. *The almost-Hermitian Walker 8-manifold (M, g, J), with g as in (10), (11) and J as in (13), is Kähler if and only if p and r are each arbitrary functions of (x^5, x^7) only.*

On the basis of the above results, we can derive conditions for the triple (g, J, Ω) to be an almost Kähler structure, which is not Kähler. For p and r in (19), put $f := \frac{1}{2}(p-r) = p - \frac{1}{2}\xi = -r + \frac{1}{2}\xi$.

Proposition 7. *The pair (g, J) is an almost Kähler structure on $\mathbb{R}^8$, which is not Kähler, if at least one of the derivatives f_1, f_2, f_3, f_4, f_6 and f_8 does not vanish.*

We now turn our attention to the Einstein conditions for the Walker metric (10), (11) with p and r given by (19). Now, as $p = f + \frac{1}{2}\xi$ and $r = -f + \frac{1}{2}\xi$, B in (11) is written explicitly as follows:

$$B = \begin{bmatrix} f + \frac{1}{2}\xi & 0 & 0 & 0 \\ 0 & 0 & 0 & 0 \\ 0 & 0 & -f + \frac{1}{2}\xi & 0 \\ 0 & 0 & 0 & 0 \end{bmatrix}, \tag{24}$$

where $f = f(x^1, \ldots, x^8)$ and $\xi = \xi(x^5, x^7)$.

Let R_{ij} and S denote the Ricci curvature and the scalar curvature of the metric (10) with B given as in (24). The Einstein tensor is defined by $G_{ij} = R_{ij} - \frac{1}{8}S\, g_{ij}$ and has nonzero components as follows:

$$\begin{aligned} &G_{25} = \frac{1}{2}\, f_{12}, \quad G_{17} = -G_{35} = -\frac{1}{2}\, f_{13}, \quad G_{45} = \frac{1}{2}\, f_{14}, \quad G_{56} = \frac{1}{2}\, f_{16}, \\ &G_{58} = \frac{1}{2}\, f_{18}, \quad G_{27} = -\frac{1}{2}\, f_{23}, \quad G_{47} = -\frac{1}{2}\, f_{34}, \quad G_{67} = -\frac{1}{2}\, f_{36}, \\ &G_{78} = -\frac{1}{2}\, f_{38}, \; G_{15} = \frac{1}{8}\,(3\, f_{11} + f_{33}), \; G_{26} = G_{48} = -\frac{1}{8}\,(f_{11} - f_{33}), \\ &G_{37} = -\frac{1}{8}\,(f_{11} + 3\, f_{33}), \quad G_{57} = \frac{1}{2}\,(f_{17} + f_1\, f_3 - f_{35}), \\ &G_{55} = -f_{26} - f_{37} - f_{48} + \frac{3}{8}\, f\,(f_{11} - f_{33}) + \frac{1}{8}\,\xi\,(3\, f_{11} + 5\, f_{33}) - \frac{1}{2}\, {f_3}^2, \\ &G_{77} = f_{15} + f_{26} + f_{48} - \frac{3}{8}\, f\,(f_{11} - f_{33}) - \frac{1}{8}\,\xi\,(5\, f_{11} + 3\, f_{33}) - \frac{1}{2}\, f_1^2. \end{aligned} \tag{25}$$

The metric g with B as in (24) is almost Kähler-Einstein if all the above components G_{ij} vanish ($G_{ij} = 0$: the Einstein equation). In the present note, we do not try to find the general solution to the Einstein equations but rather specific solutions which serve our purpose.

Since there are a lot of partial derivatives of f with respect to x^1 and x^3 among the nontrivial components of the Einstein tensor in (25), we assume that f does not depend on x^1 and x^3, i.e., $f = f(x^2, x^4, x^5, x^6, x^7, x^8)$. Then the Einstein equation $G_{ij} = 0$ drastically reduces to the following PDE's

$$G_{55} = -G_{77} = -f_{26} - f_{48} = 0. \tag{26}$$

It is easy to find solutions to these PDE's as follows.

Theorem 8. *Let f be a sum of four functions F^1, F^2, F^3 and F^4, each a function of four arguments as follows:*

$$\begin{aligned} f &= f(x^2, x^4, x^5, x^6, x^7, x^8) \\ &= F^1(x^2, x^4, x^5, x^7) + F^2(x^2, x^5, x^7, x^8) \\ &\qquad + F^3(x^4, x^5, x^6, x^7) + F^4(x^5, x^6, x^7, x^8), \end{aligned} \tag{27}$$

with the property that at least one of the derivatives f_2, f_4, f_6 and f_8 does not vanish. Then f gives a class of solutions to (26). Moreover, $p = f + \frac{1}{2}\xi$ and $r = -f + \frac{1}{2}\xi$ satisfy (19) but not the J integrability condition (4). It follows that the metric g as in (10) with B as in (24) and f as stated is an almost Kähler-Einstein neutral structure, but not a Kähler structure, on $\mathbb{R}^8$. Note that g is a Ricci flat metric.

These examples of almost Kähler-Einstein structures which are not Kähler are constructed on $\mathbb{R}^8$. Identifying a point $(x^1, \ldots, x^8)$ with a point $(\tilde{x}^1, \ldots, \tilde{x}^8)$ iff $\tilde{x}^1 = x^1 + 1, \ldots, \tilde{x}^8 = x^8 + 1$, we obtain an 8-torus T^8. Therefore, by choosing the functions $f(x^2, x^4, x^5, x^6, x^7, x^8)$ and $\xi(x^5, x^7)$ to be periodic with respect to their arguments, the metrics of the previous theorem descend to the 8-torus, thus giving examples of almost Kähler-Einstein neutral structures, which are not Kähler, on a compact manifold, which was our goal.

We end with an explicit example of such a neutral almost Kähler-Einstein structure, which is not Kähler, on an 8-torus: Put $f = f(x^8) = \sin 2\pi x^8$ and $\xi = \xi(x^7) = 2\sin 2\pi x^7$. Then the metric is

$$g = \begin{bmatrix} 0 & 0 & 0 & 0 & 1 & 0 & 0 & 0 \\ 0 & 0 & 0 & 0 & 0 & 1 & 0 & 0 \\ 0 & 0 & 0 & 0 & 0 & 0 & 1 & 0 \\ 0 & 0 & 0 & 0 & 0 & 0 & 0 & 1 \\ 1 & 0 & 0 & 0 & \sin 2\pi x^7 + \sin 2\pi x^8 & 0 & 0 & 0 \\ 0 & 1 & 0 & 0 & 0 & 0 & 0 & 0 \\ 0 & 0 & 1 & 0 & 0 & 0 & \sin 2\pi x^7 - \sin 2\pi x^8 & 0 \\ 0 & 0 & 0 & 1 & 0 & 0 & 0 & 0 \end{bmatrix}. \tag{28}$$

5. Almost Norden manifolds

An almost Norden manifold of dimension $2n$ is a triple (M, J, g) of a $2n$-manifold M, an almost complex structure J and a J-skew invariant metric g. That is, J and g satisfy the following

$$g(JX, JY) = -g(X, Y). \tag{29}$$

Note that such a Norden metric g is of neutral signature. For Norden 4-manifolds, see [4].

The pair (J, g) defines, as usual, a rank two tensor $\tilde{g}(X, Y) = g(JX, Y)$, but $\tilde{g}$ is symmetric (in fact another neutral metric), rather than a two-form. For an almost Norden manifold (M, J, g), the conditions (GC1) and (GC2) may, of course, still be imposed but, as the Norden pair (J, g) does not give rise to a two-form, the third condition (GC3) has no immediate meaning.

It is a fundamental fact, however, that an almost complex structure J on M solely gives a two-form on M (not unique). We denote it by Ω_J, and call it a J-compatible two-form. In terms of such a two-form Ω_J, we can replace the third condition (GC3) by

(GC3$'$) the J-compatible two-form Ω_J is symplectic.

Let $M = (M, J, g)$ be an almost Norden manifold, and choose a J-compatible two-form Ω_J on M. Then we can propose an *almost Norden version of Goldberg Conjecture* as follows: *If M is compact* (GC1) *and g is Einstein* (GC2), *and if a J-compatible two-form Ω_J is symplectic* (GC3$'$), *then J must be integrable.*

6. Almost para-Hermitian manifolds

An almost product structure P on a manifold M is a linear endomorphism of the tangent bundle TM of M, which satisfies $P^2 = 1$. We call the pair

(M, P) an almost product manifold. The almost product structure P induces a splitting TM into a sum $TM^+ \oplus TM^-$ of two eigen subbundles, where $TM^{\pm}$ are the ± 1 eigenspaces of P. An almost product structure P is called an almost paracomplex structure if $TM^{\pm}$ are of the same dimension $(= \frac{1}{2} \dim M)$. For paracomplex structures and related topics, see the excellent review articles [7,8]. It should be noted that an almost paracomplex structure is nothing but a $GL(n, \mathbb{R}) \times GL(n, \mathbb{R})$-structure, with $n = \frac{1}{2} \dim M$. Then the pair (M, P) with P an almost paracomplex structure is called an almost paracomplex manifold. An almost paracomplex structure P is called a paracomplex structure if the $GL(n, \mathbb{R}) \times GL(n, \mathbb{R})$-structure defined by P is integrable. (Note that an integrable almost product structure is called an *locally product manifold.*) The integrability condition of an almost paracomplex structure P is known as follows: An almost paracomplex structure P is the integrable if one of the following equivalent conditions holds

i) the two distributions $TM^{\pm}$ defined by P are involutive,
ii) the Nijenhuis tensor N of P, defined by

$$N(X, Y) = [PX, PY] - P[PX, Y] - P[X, PY] + [X, Y], \tag{30}$$

vanishes, and
iii) there exists a torsion free linear connection with respect to which P is parallel.

If an almost paracomplex manifold (M, P) admits a neutral metric g, and if g is P-skew invariant, i.e.,

$$g(PX, PY) = -g(X, Y), \tag{31}$$

then the triple (M, P, g) is called an almost para-Hermitian manifold. The para-Hermitian structure (P, g) defines a 2-form Ω as follows:

$$\Omega(X, Y) = g(PX, Y). \tag{32}$$

An almost para-Hermitian manifold (M, J, g) is called an *almost para-Kähler manifold* if $d\Omega = 0$. Moreover, an almost para-Hermitian manifold (M, J, g) is called a *para-Hermitian manifold* if P is integrable.

Then, we can consider an *almost para-Hermitian version of the Goldberg conjecture* as follows. *For an almost para-Hermitian manifold $M = (M, P, g)$, if the manifold M is compact, the neutral metric g is Einstein and the fundamental 2-form Ω is symplectic, then P is integrable.* Equivalently, an almost para-Kähler manifold $M = (M, P, g)$ is para-Kähler if the manifold M is compact and the neutral metric g is Einstein.

7. Odd-dimensional manifolds

There are some analogies between odd-dimensional manifolds and even-dimensional manifolds. In fact, it is often said that a Sasakian manifold of odd dimension corresponds to a Kähler manifold of even dimension, and a K-contact manifold of odd dimension to an almost Kähler manifold of even dimension.

even dimension		odd dimension
Kähler manifold	$\leftrightarrow$	Sasakian manifold
almost Kähler manifold	$\leftrightarrow$	K-contact manifold

Therefore, the *odd-dimensional version of the Goldberg conjecture* states that a compact Einstein K-contact manifold must be Sasakian. Boyer and Galicki [5] proved this affirmatively, and another proof of this fact is given by Apostolov, Draghici and Moroianu [3]. Such an odd-dimensional version of the conjecture is nothing but the contact structure version of the Goldberg conjecture.

References

1. D.V. Alekseevsky, *Quaternion Riemannian spaces with transitive reductive or solvable group of motions*, Functional Anal. Appl. **4** (1970), 321–322. (Translated from Funktsional'nyi Analiz i Ego Prilozheniya, **4** (4) October-December (1970), 68–69.)
2. V. Apostolov, T. Draghici and A. Moroianu, *A splitting theorem for Kähler manifolds whose Ricci tensors have constant eigenvalues*, Int. J. Math. **12** (2001), 769–789.
3. ———, *The odd-dimensional Goldberg conjecture*, Math. Nach. **279** (2006), 948–952.
4. A. Bonome, R. Castro, L. M. Hervella and Y. Matsushita, *Construction of Norden structures on neutral 4-manifolds*, JP Journal of Geometry and Topology **5** (2005), 121–140.
5. C. Boyer and K. Galicki, *Einstein manifolds and contact geometry*, Proc. Amer. Math. Soc. **129** (2001), 2419–2430.
6. M. Chaichi, E. García-Río and Y. Matsushita, *Curvature properties of four-dimensional Walker metrics*, Class. Quantum Grav. **22** (2005), 559–577.
7. V. Cruceanu, P. Fortuny and P.M. Gadea, *A survey on paracomplex geometry*, Rocky Mountain J. Math. **26** (1996), 83–115.
8. V. Cruceanu, P. M. Gadea and J. Muñoz-Masqué, *Para-Hermitian and para-Kähler manifolds*, Quaderni Inst. Mat. Univ. Messina **1** (1995), 1–72.
9. J. Davidov, J.C. Díaz-Ramos, E. García-Río, Y. Matsushita, O. Muskarov and R. Vázquez-Lorenzo, *Almost Kähler Walker 4-manifolds*, J. Geom. Phys. **57** (2007), 1075–1088.

10. R. Ghanam and G. Thompson, *The holonomy Lie algebras of neutral metrics in dimension four*, J. Math. Phys. **42** (2001), 2266–2284.
11. S.I. Goldberg, *Integrability of almost Kähler manifolds*, Proc. Amer. Math. Soc. **21** (1969), 96–100.
12. S. Kobayashi and K. Nomizu, *Foundations of differential geometry II*, John Wiley, New York, London, 1969.
13. Y. Matsushita, *Four-dimensional Walker metrics and symplectic structures*, J. Geom. Phys. **52** (2004), 89–99.
14. ———, *Walker 4-manifolds with proper almost complex structures*, J. Geom. Phys. **55** (2005), 385–398.
15. ———, *The existence of indefinite metrics of signature* (++−−) *and two kinds of almost complex structures in dimension four*, Proceedings of The Seventh International Workshop on Complex Structures and Vector Fields, Plovdiv, Bulgaria, August 31–September 4, 2004. World Scientific 2005, "Contemporary Aspects of Complex Analysis, Differential Geometry and Mathematical Physics," ed. K. Sekigawa, World Scientific Publishing, July 2005.
16. Y. Matsushita, S. Haze and P.R. Law, *Almost Kähler-Einstein structures on 8-dimensional Walker manifolds*, Monatshefte für Mathematik **150** (2007), 41–48.
17. Y. Matsushita and P.R. Law, *Hitchin-Thorpe-Type inequalities for pseudo-Riemannian 4-manifolds of metric signature* (++−−), Geom. Dedicata **87** (2001), 65–89.
18. P. Nurowski and M. Przanowski, *A four-dimensional example of a Ricci flat metric admitting almost-Kähler non-Kähler structure*, Classical Quantum Gravity **16** (1999), no. 3, L9–L13.
19. K. Sekigawa, *On some 4-dimensional compact Einstein almost Kähler manifolds*, Math. Ann. **271** (1985) 333–337.
20. ———, *On some compact Einstein almost Kähler manifolds*, J. Math. Soc. Japan **36** (1987) 677–684.
21. ———, private communications, December 2004.
22. A.G. Walker, *Canonical form for a riemannian space with a parallel field of null planes*, Q. J. Math. Oxford (2) **1** (1950), 69–79.

DISPERSION AND ASYMPTOTIC PROFILES FOR KIRCHHOFF EQUATIONS

T. MATSUYAMA

Department of Mathematics,
Tokai University,
Hiratsuka, Kanagawa 259-1292, Japan
E-mail: tokio@keyaki.cc.u-tokai.ac.jp

M. RUZHANSKY

Department of Mathematics,
Imperial College London,
180 Queen's gate, London SW7 2AZ, United Kingdom
E-mail: m.ruzhansky@imperial.ac.uk

The aim of this article is to describe asymptotic profiles for the Kirchhoff equation, and to establish time decay properties and dispersive estimates for Kirchhoff equations. For this purpose, the method of asymptotic integration is developed for the corresponding linear equations and representation formulae for their solutions are obtained. These formulae are analysed further to obtain the time decay rate of L^p–L^q norms of propagators for the corresponding Cauchy problems.

Keywords: Kirchhoff equation; Asymptotic profiles; Dispersive estimates.

1. Introduction

This article is devoted to several aspects of Kirchhoff equations or Kirchhoff systems, which were discussed in [12,13,15]. In particular, we will discuss the asymptotic profiles and dispersion properties, or time decay of L^p–L^q norms of propagators for some relevant classes of hyperbolic equations. These properties are well-known for the wave equations, but several aspects of Kirchhoff equations still remain far from being understood. The global well-posedness of Kirchhoff equations or Kirchhoff systems is known if the data is sufficiently small in some suitable Sobolev spaces of L^2 type (see [3–

2000 Mathematics Subject Classification: Primary 35L05; Secondary 35L10
The second author was supported by the Leverhulme Research Fellowship

6,8,10,11,26–28]). Up to now, if one takes any large data from these Sobolev spaces, the problem of the global well-posedness is still open.

In this article the asymptotics and the global well-posedness are discussed for small data. The first topic was developed in [15] by relating the problem to the asymptotic behaviour of the Bessel potentials (Theorem from [12] is the anoucement of [15]). More precisely, the first author proved that there exists a solution which is never asymptotically free. Here we say that $u = u(t,x)$ is asymptotically free if it is asymptotically convergent to some solution of the free wave equation as the time goes to $\pm\infty$. From the point of view of the scattering theory all solutions with data satisfying some fast decay conditions in space variables are asymptotically free (see [7,8,26]), while the result of [15] states that if the data satisfy the opposite condition to [7,8,26], then the scattering theory is not possible. This is stated more precisely in Theorem 2.2. For deriving these asymptotics, we need a delicate analysis of an oscillatory integral associated with Kirchhoff equation, which was introduced by Greenberg and Hu [8] in the one dimensional case (see also [4,5,26]), and we will develop an asymptotic expansion of this oscillatory integral.

For further investigations, for example, such as the nonlinear scattering theory, the second topic is very important. This means that there exists a scattering state for Kirchhoff equations or systems with nonlinear perturbations, which can be discussed in the standard way but is quite lengthy, hence we do not touch it (see e.g., [17]). Quite recently, the first author obtained the dispersive estimates for the Kirchhoff equation (see [13]), which will be introduced as Theorem 2.1. The essential point of the proof relies on the stationary phase method together with Littman's lemma.

Now let us give the precise formulation of Kirchhoff equations considered problems. In 1883 G. Kirchhoff proposed the equation

$$u_{tt} - \left(1 + \int_0^L u_x^2\,dx\right) u_{xx} = 0 \tag{1.1}$$

for $u = u(t,x)$ on $\mathbb{R}_t \times (0,L)$ (see [9]), which describes the nonlinear vibrations of one dimensional elastic strings having the natural length L. For simplicity, all the physical constants are normalised. Generalising the equation (1.1) to a multi-dimensional version, we can consider the Cauchy problem for $u = u(t,x)$ on $\mathbb{R}_t \times \mathbb{R}^n_x$:

$$\partial_t^2 u - \left(1 + \int_{\mathbb{R}^n} |\nabla u|^2\,dx\right) \Delta u = 0, \tag{1.2}$$

$$u(0,x) = f_0(x), \quad \partial_t u(0,x) = f_1(x), \tag{1.3}$$

where $\partial_t = \frac{\partial}{\partial t}$, $\nabla = \left(\frac{\partial}{\partial x_1}, \dots, \frac{\partial}{\partial x_n}\right)$ and Δ is the standard Laplacian in $\mathbb{R}^n$ defined by $\Delta = \sum_{j=1}^{n} \frac{\partial^2}{\partial x_j^2}$.

Higher order nonlinear equations of Kirchhoff type are also of great interest, and they can be viewed as dispersion relations for Kirchhoff systems. In particular, since higher order equations are influenced by the geometric properties of characteristics (cf. [22–24]), in such problem it is important to know how this phenomenon is affected by nonlinearities (this is contrary to the L^p–L^p estimates, see [19] for a survey of such results).

Thus, let us consider the following nonlinear equation

$$\widetilde{L}(t, D_t, D_x, \|\nabla u\|_{L^2}^2) = D_t^m u + \sum_{\substack{|\nu|+j=m \\ j\le m-1}} b_{\nu,j}\left(\|\nabla u(t,\cdot)\|_{L^2}^2\right) D_x^\nu D_t^j u = 0, \tag{1.4}$$

for $t \neq 0$, with the initial condition

$$D_t^k u(0,x) = f_k(x), \quad k = 0, 1, \dots, m-1, \quad x \in \mathbb{R}^n, \tag{1.5}$$

where $D_t = \frac{1}{i}\frac{\partial}{\partial t}$ and $D_x^\nu = \left(\frac{1}{i}\frac{\partial}{\partial x_1}\right)^{\nu_1} \cdots \left(\frac{1}{i}\frac{\partial}{\partial x_n}\right)^{\nu_n}$, $i = \sqrt{-1}$, for $\nu = (\nu_1, \dots, \nu_n)$. We will assume that the symbol of the differential operator $\widetilde{L}(t, D_t, D_x, \|\nabla u\|_{L^2}^2)$ has real and distinct roots $\widetilde{\varphi}_1(t,s;\xi), \dots, \widetilde{\varphi}_m(t,s;\xi)$ for $\xi \neq 0$ and $0 \le s \le \delta$ with $\delta > 0$, i.e.

$$\widetilde{L}(t,\tau,\xi,s) = (\tau - \widetilde{\varphi}_1(t,s;\xi)) \cdots (\tau - \widetilde{\varphi}_m(t,s;\xi)),$$

$$\inf_{\substack{|\xi|=1,\ t\in\mathbb{R}, s\in[0,\delta] \\ j\neq k}} |\widetilde{\varphi}_j(t,s;\xi) - \widetilde{\varphi}_k(t,s;\xi)| > 0.$$

The detail analysis of the Cauchy problem (1.4)–(1.5) will be done in [16], and we will consider only the Cauchy problem (1.2)–(1.3) of the second order in this article.

We conclude the introduction by fixing the notation used in this article. For $s \in \mathbb{R}$ and $1 \le p \le \infty$, let $\dot{L}_s^p = \dot{L}_s^p(\mathbb{R}^n)$ and $L_s^p = L_s^p(\mathbb{R}^n)$ be the Riesz and Bessel potential spaces with semi-norm or norm

$$\|u\|_{\dot{L}_s^p} = \|\mathcal{F}^{-1}[|\xi|^s \widehat{u}(\xi)]\|_{L^p(\mathbb{R}^n)} \equiv \||D|^s u\|_{L^p(\mathbb{R}^n)},$$

$$\|u\|_{L_s^p} = \|\mathcal{F}^{-1}[\langle\xi\rangle^s \widehat{u}(\xi)]\|_{L^p(\mathbb{R}^n)} \equiv \|\langle D\rangle^s u\|_{L^p(\mathbb{R}^n)},$$

respectively. Here $\widehat{\ }$ denotes the Fourier transform, $\mathcal{F}^{-1}$ is its inverse, and $\langle\xi\rangle = \sqrt{1+|\xi|^2}$. Throughout this article, we fix the notation as follows:

$$\dot{H}^s = \dot{L}_s^2, \quad H^s = L_s^2.$$

We also put, for $s \ge 1$,

$$\dot{X}^s(\mathbb{R}) = C(\mathbb{R}; \dot{H}^s) \cap C^1(\mathbb{R}; \dot{H}^{s-1}) \cap C^2(\mathbb{R}; \dot{H}^{s-2}).$$

Finally we shall denote by $\mathcal{S} = \mathcal{S}(\mathbb{R}^n)$ the Schwartz space on $\mathbb{R}^n$.

2. Results

In this section we survey the results of [13] and [15] on the Cauchy problem (1.2)–(1.3). In order to state these asymptotics for the solutions to Kirchhoff equation, we refer to a general theorem of Yamazaki (see [26]). For this purpose, let us introduce the set

$$Y_k := \{ \{\phi, \psi\} \in \dot{H}^{3/2} \times H^{1/2} ; |\{\phi, \psi\}|_{Y_k} < \infty \}, \quad k > 1,$$

where

$$\begin{aligned} |\{\phi, \psi\}|_{Y_k} := & \sup_{\tau \in \mathbb{R}} (1 + |\tau|)^k \left| \int_{\mathbb{R}^n} \mathrm{e}^{i\tau|\xi|} |\xi|^3 |\widehat{\phi}(\xi)|^2 \, d\xi \right| \\ & + \sup_{\tau \in \mathbb{R}} (1 + |\tau|)^k \left| \int_{\mathbb{R}^n} \mathrm{e}^{i\tau|\xi|} |\xi| |\widehat{\psi}(\xi)|^2 \, d\xi \right| \\ & + \sup_{\tau \in \mathbb{R}} (1 + |\tau|)^k \left| \int_{\mathbb{R}^n} \mathrm{e}^{i\tau|\xi|} |\xi|^2 \mathrm{Re} \left(\widehat{\phi}(\xi) \overline{\widehat{\psi}(\xi)} \right) d\xi \right| . \end{aligned}$$

Then we have the following:

Theorem A ([26]). *Let $n \geq 1$ and $s_0 \geq \frac{3}{2}$. If the data u_0, u_1 satisfy $u_0 \in \dot{H}^{s_0} \cap H^1$, $u_1 \in H^{s_0 - 1}$, and*

$$\delta_1 := \|\nabla u_0\|_{L^2}^2 + \|u_1\|_{L^2}^2 + |\{u_0, u_1\}|_{Y_k} \ll 1 \quad \text{for some } k > 1, \tag{2.1}$$

then the problem (1.2)–(1.3) has a unique solution $u(t,x) \in \dot{X}^{s_0}(\mathbb{R})$ having the following property: there exists a constant $c_{\pm\infty} \equiv c_{\pm\infty}(u_0, u_1) > 0$ such that

$$1 + \|\nabla u(t, \cdot)\|_{L^2}^2 = c_{\pm\infty}^2 + O\big(|t|^{-k+1}\big) \quad \text{as } t \to \pm\infty$$

Furthermore, if (2.1) holds with $k > 2$, then $c_{+\infty} = c_{-\infty} := c_\infty$ and each solution $u(t, x) \in \dot{X}^{s_0}(\mathbb{R})$ is asymptotically free in $\dot{H}^\sigma \times \dot{H}^{\sigma-1}$ for all $\sigma \in [1, s_0]$ as $t \to \pm\infty$, i.e., there exists a solution $v_\pm = v_\pm(t,x) \in \dot{X}^\sigma(\mathbb{R})$ of the equation

$$\big(\partial_t^2 - c_\infty^2 \Delta\big) v_\pm = 0 \quad \text{on } \mathbb{R} \times \mathbb{R}^n$$

such that

$$\|u(t, \cdot) - v_\pm(t, \cdot)\|_{\dot{H}^\sigma} + \|\partial_t u(t, \cdot) - \partial_t v_\pm(t, \cdot)\|_{\dot{H}^{\sigma-1}} \to 0 \quad (t \to \pm\infty).$$

The inclusions among the classes Y_k are as follows:

$$Y_k \subset Y_\ell \quad \text{if } k > \ell > 1, \quad \text{and} \quad \mathcal{S} \subset Y_k \quad \text{for all } k \in (1, n+1].$$

The latter inclusion can be shown by using the asymptotic expansion of oscillatory integral $I(\widetilde{\vartheta}(t), 0)$ which was proved in [15]. The definition of Y_k

is somewhat complicated. There are some examples of spaces contained in Y_k. For more details see [15].

Keeping in mind Theorem A, we have L^p–L^q estimates:

Theorem 2.1 ([13]). *Let $n \geq 2$ and let $1 < p \leq 2 \leq q < +\infty$ and $\frac{1}{p} + \frac{1}{q} = 1$. Then each solution $u(t,x)$ in Theorem A with $k = n+1$ has the following properties for all $\delta > 0$:*

$$\|\partial_t^j \partial_x^\alpha u(t,\cdot\,)\|_{L^q} \leq C(1+|t|)^{-(\frac{n-1}{2}-\delta)(\frac{1}{p}-\frac{1}{q})} \sum_{i=0,1} \|u_i\|_{L^p_{N_p+j+|\alpha|-i,p}}$$

where $N_p = \frac{3n+1}{2}\left(\frac{1}{p} - \frac{1}{q}\right)$, $j = 0, 1, 2$, and α is any multi-index.

Based on Theorem 2.1, we can develop the nonlinear scattering problems for the Kirchhof equation. But here, we want to exhibit the opposite phenomenon; for this, we will find the asymptotic profiles for the solutions to (1.2)–(1.3). Let us present the definitions of free and non-free waves.

Definition. (i) We say that $v_\pm = v_\pm(t,x) = \{v_+(t,x), v_-(t,x)\}$ is a *free wave* if it satisfies the equation

$$\left(\partial_t^2 - c_{\pm\infty}^2 \Delta\right) v_\pm = 0 \quad \text{on } \mathbb{R} \times \mathbb{R}^n.$$

(ii) Let $\sigma \geq 1$. We say that $v = v(t,x)$ is *asymptotically free* in $\dot{H}^\sigma \times \dot{H}^{\sigma-1}$ if it is asymptotically convergent to some free wave $v_\pm$ in $\dot{H}^\sigma \times \dot{H}^{\sigma-1}$, i.e.,

$$\|v(t,\cdot) - v_\pm(t,\cdot)\|_{\dot{H}^\sigma} + \|\partial_t v(t,\cdot) - \partial_t v_\pm(t,\cdot)\|_{\dot{H}^{\sigma-1}} \to 0 \quad (t \to \pm\infty).$$

(iii) Let $\sigma \geq 1$. We say that $w = w(t,x)$ is a *non-free wave* in $\dot{H}^\sigma \times \dot{H}^{\sigma-1}$ if it is not asymptotically free.

Theorem A states that each solution u of (1.2)–(1.3) with initial data satisfying (2.1) with $k > 2$, is asymptotically free. On the other hand, the next theorem states that the bound $k > 2$ is sharp. More precisely, we have the following:

Theorem 2.2 ([15]). *Assume that*

$$\textit{either } n \geq 2 \textit{ and } 1 < k \leq 2, \textit{ or } n = 1 \textit{ and } 1 < k < 2.$$

Then there exists a solution $u(t,x) \in \cap_{s\geq 1}\dot{X}^s(\mathbb{R})$ of (1.2)–(1.3) with data satisfying (2.1), which is a non-free wave in $\dot{H}^\sigma \times \dot{H}^{\sigma-1}$ for all $\sigma \geq 1$.

The proof of Theorems 2.1–2.2 relies on the representation formulae for the corresponding linear equation. In §3 we will introduce the representation formulae for more general strictly hyperbolic equations. Moreover, the argument of Theorem 2.2 is relating with the asymptotic behaviour of Bessel functions (see [1]).

3. Representation of solutions to linear Cauchy problems

In this section we introduce the representation formulae for more general equations than previously considered by using the asymptotic integration method along the argument of [16]. Let us consider the Cauchy problem for an mth order strictly hyperbolic equation with time-dependent coefficients, for function $u = u(t,x)$:

$$L(t, D_t, D_x)u \equiv D_t^m u + \sum_{\substack{|\nu|+j=m \\ j\le m-1}} a_{\nu,j}(t) D_x^\nu D_t^j u = 0, \quad t \neq 0, \tag{3.1}$$

with the initial condition

$$D_t^k u(0,x) = f_k(x) \in C_0^\infty(\mathbb{R}^n), \quad k = 0, 1, \cdots, m-1, \quad x \in \mathbb{R}^n. \tag{3.2}$$

Denoting by $\mathcal{B}^{m-1}(\mathbb{R})$ the space of all functions whose derivatives up to $(m-1)^{\text{th}}$ order are all bounded and continuous on $\mathbb{R}$, we assume that each $a_{\nu,j}(t)$ belongs to $\mathcal{B}^{m-1}(\mathbb{R})$ and satisfies

$$\partial_t^k a_{\nu,j}(t) \in L^1(\mathbb{R}) \quad \text{for all } \nu, j \text{ with } |\nu| + j = m, \text{ and } k = 1, \ldots, m-1. \tag{3.3}$$

Moreover, following the standard definition of equations of the regularly hyperbolic type (e.g. Mizohata [18]), we will assume that the symbol of the differential operator $L(t, D_t, D_x)$ has real and distinct roots $\varphi_1(t;\xi), \ldots, \varphi_m(t;\xi)$ for $\xi \neq 0$, and

$$L(t, \tau, \xi) = (\tau - \varphi_1(t;\xi)) \cdots (\tau - \varphi_m(t;\xi)), \tag{3.4}$$

$$\inf_{\substack{|\xi|=1,\ t\in\mathbb{R} \\ j\neq k}} |\varphi_j(t;\xi) - \varphi_k(t;\xi)| > 0. \tag{3.5}$$

By applying the Fourier transform on $\mathbb{R}_x^n$ to (3.1), we get

$$D_t^m v + \sum_{j=1}^m h_j(t;\xi) D_t^{m-j} v = 0, \tag{3.6}$$

where

$$h_j(t;\xi) = \sum_{|\nu|=j} a_{\nu,m-j}(t)\xi^\nu, \quad \xi \in \mathbb{R}^n.$$

This is the ordinary differential equation, homogeneous of mth order, with the parameter $\xi = (\xi_1, \ldots, \xi_n)$. As usual, the strict hyperbolicity means that the characteristic roots of (3.6) are real and can be written as $\varphi_1(t;\xi), \ldots, \varphi_m(t;\xi)$ satisfying (3.4)–(3.5). Notice that each $\varphi_\ell(t;\xi)$ has a homogeneous degree one with respect to ξ. In this section we will establish the representation formulae for solutions of the Cauchy problem (3.1) in the form of the oscillatory integrals. Let $\widehat{u}(t;\xi)$ be the solution of (3.6) with the intial data $\widehat{f}_k(\xi)$ $(k=0,\ldots,m-1)$. Let $v_k(t;\xi)$ be the solution of (3.6) with $(D_t^j v_k)(0;\xi)=\delta_k^j$ for $j,k=0,1,\ldots,m-1$. We set

$$W(t;\xi)=\begin{pmatrix} v_0(t;\xi) & v_1(t;\xi) & \cdots & v_{m-1}(t;\xi) \\ D_t v_0(t;\xi) & D_t v_1(t;\xi) & \cdots & D_t v_{m-1}(t;\xi) \\ \vdots & \vdots & & \vdots \\ D_t^{m-1} v_0(t;\xi) & D_t^{m-1} v_1(t;\xi) & \cdots & D_t^{m-1} v_{m-1}(t;\xi) \end{pmatrix}.$$

Hence, $W(t;\xi)$ is the fundamental matrix of (3.6). Defining

$$\vartheta_j(t;\xi)=\int_0^t \varphi_j(s;\xi)\,ds, \quad j=1,\ldots,m,$$

we introduce the matrix

$$Y(t;\xi)=\begin{pmatrix} e^{i\vartheta_1(t;\xi)} & \ldots & e^{i\vartheta_m(t;\xi)} \\ D_t e^{i\vartheta_1(t;\xi)} & \ldots & D_t e^{i\vartheta_m(t;\xi)} \\ \vdots & & \vdots \\ D_t^{m-1} e^{i\vartheta_1(t;\xi)} & \ldots & D_t^{m-1} e^{i\vartheta_m(t;\xi)} \end{pmatrix}.$$

Matrix $Y(t;\xi)$ is the fundamental matrix of a perturbed ordinary differential equation of (3.6):

$$(D_t-\varphi_1(t;\xi))\cdots(D_t-\varphi_m(t;\xi))\,w=0.$$

Then we can write this equation as

$$D_t^m w+\sum_{j=1}^m h_j(t;\xi)D_t^{m-j}w+\sum_{j=2}^m \widetilde{h}_j(t;\xi)D_t^{m-j}w=0, \tag{3.7}$$

where $\widetilde{h}_j(t;\xi)$ satisfies

$$\widetilde{h}_j(t;\xi)=\begin{cases} 0, \quad j=1, \\ \displaystyle\sum_{\substack{1\le|\nu|\le j-1 \\ (\nu_2,\ldots,\nu_j)\neq(0,\ldots,0)}} \widetilde{c}_{\nu_1\ldots\nu_j}\,\varphi_{\ell_1}^{\nu_1}(D_t\varphi_{\ell_2})^{\nu_2}\cdots(D_t^{m-j+1}\varphi_{\ell_j})^{\nu_j}, \\ \hfill j=2,\ldots,m, \end{cases}$$

with some constants $\widetilde{c}_{\nu_1\ldots\nu_j}\neq 0$. This means that each $e^{i\vartheta_\ell(t;\xi)}$ satisfies (3.7), and $e^{i\vartheta_1(t;\xi)},\ldots,e^{i\vartheta_m(t;\xi)}$ are linearly independent for $\xi\neq 0$ and $t\in\mathbb{R}$. It can be checked that the coefficient $\widetilde{h}_1(t;\xi)$ of $D_t^{m-1}w$ always vanishes for every m, by an induction argument on m. Then it follows from Proposition 2.4 of [16] (cf. [2,25]) that there exists the limit

$$\lim_{t\to\pm\infty} Y(t;\xi)^{-1}W(t;\xi) = L_\pm(\xi), \tag{3.8}$$

where we set

$$L_\pm(\xi) = \begin{pmatrix} \alpha^1_{0,\pm}(\xi) & \alpha^1_{1,\pm}(\xi) & \cdots & \alpha^1_{m-1,\pm}(\xi) \\ \alpha^2_{0,\pm}(\xi) & \alpha^2_{1,\pm}(\xi) & \cdots & \alpha^2_{m-1,\pm}(\xi) \\ \vdots & \vdots & & \vdots \\ \alpha^m_{0,\pm}(\xi) & \alpha^m_{1,\pm}(\xi) & \cdots & \alpha^m_{m-1,\pm}(\xi) \end{pmatrix}.$$

Furthermore, writing

$$\begin{aligned} R_\pm(t;\xi) &= Y(t;\xi)^{-1}W(t;\xi) - L_\pm(\xi) \\ &= \begin{pmatrix} \varepsilon^1_{0,\pm}(t;\xi) & \varepsilon^1_{1,\pm}(t;\xi) & \cdots & \varepsilon^1_{m-1,\pm}(t;\xi) \\ \varepsilon^2_{0,\pm}(t;\xi) & \varepsilon^2_{1,\pm}(t;\xi) & \cdots & \varepsilon^2_{m-1,\pm}(t;\xi) \\ \vdots & \vdots & & \vdots \\ \varepsilon^m_{0,\pm}(t;\xi) & \varepsilon^m_{1,\pm}(t;\xi) & \cdots & \varepsilon^m_{m-1,\pm}(t;\xi) \end{pmatrix}, \end{aligned} \tag{3.9}$$

we have

$$W(t;\xi) = Y(t;\xi)\left(L_\pm(\xi) + R_\pm(t;\xi)\right).$$

Thus we arrive at

$$\begin{aligned} D^\ell_t v_k(t;\xi) &= \sum_{j=1}^m \left(\alpha^j_{k,\pm}(\xi) + \varepsilon^j_{k,\pm}(t;\xi)\right) D^\ell_t e^{i\vartheta_j(t;\xi)} \\ &= \sum_{j=1}^m \left(\alpha^j_{k,\pm}(\xi) + \varepsilon^j_{k,\pm}(t;\xi)\right) p_\ell(\varphi_j(t;\xi)) e^{i\vartheta_j(t;\xi)} \end{aligned}$$

for $k, \ell = 0, \cdots, m-1$, where each $p_\ell(\varphi_j(t;\xi))$ is determined by the equation

$$D^\ell_t e^{i\vartheta_j(t;\xi)} = p_\ell(\varphi_j(t;\xi)) e^{i\vartheta_j(t;\xi)}.$$

We note that for the second order equations we have $m = 2$ and the next theorem covers the case of the wave equation as a special case, also improving the corresponding result in [13–15]. The result is as follows:

Theorem 3.1. *Assume that the characteristic roots* $\varphi_1(t;\xi), \dots, \varphi_m(t;\xi)$ *of* (3.6) *are real and distinct for all* $t \in \mathbb{R}$ *and for all* $\xi \in \mathbb{R}^n\backslash 0$, *and that they satisfy* (3.5). *Then there exists* $\alpha^j_{k,\pm}(\xi)$ *and* $\varepsilon^j_{k,\pm}(t;\xi)$ *determined by*

(3.8) and (3.9), respectively, such that the solution $u(t,x)$ of our problem (3.1)–(3.2) is represented by

$$D_t^\ell u(t,x)$$
$$= \sum_{k=0}^{m-1}\sum_{j=1}^{m}\mathcal{F}^{-1}\big[\big(\alpha^j_{k,\pm}(\xi)+\varepsilon^j_{k,\pm}(t;\xi)\big)p_\ell(\varphi_j(t;\xi))\mathrm{e}^{i\vartheta_j(t;\xi)}\widehat{f}_k(\xi)\big](x),\ t\gtrless 0,$$

for $\ell=0,\dots,m-1$, where

$$\big|\alpha^j_{k,\pm}(\xi)\big|\le c|\xi|^{-k},\qquad \big|\varepsilon^j_{k,\pm}(t;\xi)\big|\le c|\xi|^{-k}\int_{|t|}^{+\infty}\Psi(s)\,ds,$$

and $\Psi(t)$ is given by

$$\Psi(t)=\sum_{\substack{|\nu|+j=m\\ j\le m-2}}\big|\partial_t a_{\nu,j}(t)\big|\cdots\big|\partial_t^{m-j-1}a_{\nu,j}(t)\big|.$$

For the higher order derivatives of amplitude functions, we have, for $|\mu|\ge 1$,

$$\begin{aligned}
&\big|D^\mu_\xi\alpha^j_{k,\pm}(\xi)\big|\le c|\xi|^{-k},\\
&\big|D^\mu_\xi\varepsilon^j_{k,\pm}(t;\xi)\big|\le c\,\mathrm{e}^{\int_0^{|t|}(1+s)^{|\mu|}\Psi(s)\,ds}|\xi|^{-k}, && |\xi|\ge 1,\\
&\big|D^\mu_\xi\alpha^j_{k,\pm}(\xi)\big|\le c|\xi|^{-k-|\mu|},\\
&\big|D^\mu_\xi\varepsilon^j_{k,\pm}(t;\xi)\big|\le c\,\mathrm{e}^{\int_0^{|t|}(1+s)^{|\mu|}\Psi(s)\,ds}|\xi|^{-k-|\mu|}, && 0<|\xi|<1.
\end{aligned}$$

If we further assume that

$$(1+|t|)^{|\mu|}\partial_t^k a_{\nu,j}(t;\xi)\in L^1(\mathbb{R})$$

for some μ with $|\mu|\ge 1$, and for all ν, j, and $k=1,\dots,m-1$, then the bound of each $D^\mu_\xi\varepsilon^j_{k,\pm}(t;\xi)$ is uniform in t.

References

1. N. Aronszajn and K.T. Smith, *Theory of Bessel potentials. Part I*, Ann. Inst. Fourier, Grenoble **11** (1964), 385–475.
2. G. Ascoli, *Sulla forma asintotica degli integrali dell'equazione differenziale $y''+A(x)y=0$ in un caso notevole di stabilità*, Univ. Nac. Tucumán, Revista A **2** (1941), 131–140.
3. E. Callegari and R. Manfrin, *Global existence for nonlinear hyperbolic systems of Kirchhoff type*, J. Differential Equations **132** (1996), 239–274.
4. P. D'Ancona and S. Spagnolo, *A class of nonlinear hyperbolic problems with global solutions*, Arch. Rational Mech. Anal. **124** (1993), 201–219.
5. ———, *Nonlinear perturbations of the Kirchhoff equation*, Comm. Pure Appl. Math. **47** (1994), 1005–1029.

6. ———, *Kirichhoff type equations depending on a small parameter*, Chin. Ann. of Math. **16B** (1995), 413–430.
7. M. Ghisi, *Asymptotic behavior of Kirchhoff equation*, Ann. Mat. Pura Appl. **171** (1996), 293–312.
8. J.M. Greenberg and S.C. Hu, *The initial-value problem for a stretched string*, Quart. Appl. Math. **38** (1980), 289–311.
9. G. Kirchhoff, *Vorlesungen über Mechanik*, Teubner, 1883.
10. R. Manfrin, *On the global solvability of symmetric hyperbolic systems of Kirchhoff type*, Discrete Contin. Dynam. Systems **3** (1997), 91–106.
11. ———, *On the global solvability of Kirchhoff equation for non-analytic initial data*, J. Differential Equations **211** (2005), 38–60.
12. T. Matsuyama, *Asymptotic behaviours for Kirchhoff equation*, Proceedings of Isaac Congress 2006, in press.
13. ———, *L^p–L^q estimates for wave equations and the Kirchhoff equation*, submitted to Osaka J. Math. (2006).
14. ———, *Asymptotic behaviours for wave equations with time-dependent coefficients*, Annali dell'Università di Ferrara, Sec. VII — Sci. Math. **52** (2006), 383–393.
15. ———, *Asymptotic profiles for Kirchhoff equation*, Rend. Lincei Mat. Appl. **17** (2006), 377–395.
16. T. Matsuyama and M. Ruzhansky, *Asymptotic integration and dispersion for hyperbolic equations, with applications to Kirchhoff equations*, in preparation.
17. ———, *A remark on nonlinear perturbations of the Kirchhoff equation*, in preparation.
18. S. Mizohata, *The theory of partial differential equations*, Cambridge Univ. Press, 1973.
19. M. Ruzhansky, *Singularities of affine fibrations in the theory of regularity of Fourier integral operators*, Russian Math. Surveys **55** (2000), 93–161.
20. M. Ruzhansky and J. Smith, *Strichartz estimates for strictly hyperbolic equations with lower order terms*, preprint.
21. W. Rzymowski, *One-dimensional Kirchhoff equation*, Nonlinear Analysis **48** (2002), 209–221.
22. M. Sugimoto, *A priori estimates for higher order hyperbolic equations*, Math. Z. **215** (1994), 519–531.
23. ———, *Estimates for hyperbolic equations with non-convex characteristics*, Math. Z. **222** (1996), 521–531.
24. ———, *Estimates for hyperbolic equations of space dimension 3*, J. Funct. Anal. **160** (1998), 382–407.
25. A. Wintner, *Asymptotic integrations of adiabatic oscillator*, Amer. J. Math. **69** (1947), 251–272.
26. T. Yamazaki, *Scattering for a quasilinear hyperbolic equation of Kirchhoff type*, J. Differential Equations **143** (1998), 1–59.
27. ———, *Global solvability for the Kirchhoff equations in exterior domains of dimension larger than three*, Math. Methods Appl. Sci. **27** (2004), 1893–1916.
28. ———, *Global solvability for the Kirchhoff equations in exterior domains of dimension three*. J. Differential Equations **210** (2005), 290–316.

GEOMETRY OF STATISTICAL MANIFOLDS AND ITS GENERALIZATION

H. MATSUZOE

Department of Computer Science and Engineering,
Graduate School of Engineering,
Nagoya Institute of Technology,
Gokiso-cho, Showa-ku, Nagoya, 466-8555, Japan
E-mail: matsuzoe@nitech.ac.jp
http://venus.kyy.nitech.ac.jp/~matsuzoe/

A statistical manifold is a generalization of a Riemannian manifold with the Levi-Civita connection. On a statistical manifold, duality of affine connections naturally arises. In the present paper, basic properties of statistical manifolds and dual affine connections are given. In addition, generalizations of statistical manifolds and dual affine connections are given. This generalization turns out to be a generalization of Weyl manifolds.

Keywords: Statistical manifold; Dual connection; Information geometry, Affine differential geometry; Weyl manifold; Semi-Weyl manifold.

Introduction

Information geometry begun in 1980s, which studies natural geometrical structures of sets of probability distributions. Information geometry has rich applications in the field of mathematical sciences, then the study has been developing widely [1]. On the other hand, affine differential geometry studies hypersurfaces immersed into an affine space. Though affine differential geometry is a classical research topic in differential geometry, it has been redeveloping since 1980s [2].

A key fact is that duality of affine connections arises naturally, then information geometry and affine differential geometry have common geometric ideas. In particular, a statistical manifold is a natural manifold structure on a set of probability distributions in information geometry. Besides, a statistical manifold structure is induced from an equiaffine immersion in affine differential geometry.

In the present paper, basic formulas for dual connections and statistical manifolds are given. Then we consider a generalization of dual affine connections and statistical manifolds. We can see that this generalization is also regarded as a generalization of a Weyl manifold.

1. Dual connections

We assume that all the objects are smooth throughout this paper. In addition, we may assume that a manifold is simply connected since we discuss local geometric properties on a manifold.

At the beginning of this paper, we introduce the notion of dual affine connections.

Let (M, h) be a semi-Riemannian manifold, and let ∇ be an affine connection on M. We define another affine connection ∇^* by

$$Xh(Y,Z) = h(\nabla_X Y, Z) + h(Y, \nabla^*_X Z).$$

We call ∇^* the *dual connection* or the *conjugate connection* of ∇ with respect to h. It is easy to check that $(\nabla^*)^* = \nabla$.

Proposition 1.1. *Denote by R and R^* the curvature tensors of ∇ and ∇^*, respectively. Then the following formula holds.*

$$h(R(X,Y)Z,V) = -h(Z, R^*(X,Y)V)$$

Proof. From straightforward calculations, we obtain

$$\begin{aligned} h(\nabla_X \nabla_Y Z, V) &= XYh(Z,V) - h(\nabla_X Z, \nabla^*_Y V) - h(\nabla_Y Z, \nabla^*_X V) \\ &\quad - h(Z, \nabla^*_X \nabla^*_Y V), \\ h(\nabla_Y \nabla_X Z, V) &= YXh(Z,V) - h(\nabla_Y Z, \nabla^*_X V) - h(\nabla_X Z, \nabla^*_Y V) \\ &\quad - h(Z, \nabla^*_Y \nabla^*_X V), \\ h(\nabla_{[X,Y]} Z, V) &= [X,Y]h(Z,V) - h(Z, \nabla_{[X,Y]} V). \end{aligned}$$

These equations imply that $h(R(X,Y)Z,V) = -h(Z, R^*(X,Y)V)$. □

Proposition 1.2. *Set $\nabla^0 := (\nabla + \nabla^*)/2$. Then ∇^0 is a metric connection, namely, $\nabla^0 h = 0$.*

Proof. Since the metric h is symmetric, we have the following relation.

$$\begin{aligned} (\nabla^0_X h)(Y,Z) &= Xh(Y,Z) - h(\nabla^0_X Y, Z) - h(Y, \nabla^0_X Z) \\ &= Xh(Y,Z) - \frac{1}{2}h(\nabla_X Y + \nabla^*_X Y, Z) - \frac{1}{2}h(Y, \nabla_X Z + \nabla^*_X Z) \\ &= Xh(Y,Z) - \frac{1}{2}Xh(Y,Z) - \frac{1}{2}Xh(Z,Y) = 0. \end{aligned}$$

This implies that ∇^0 is a metric connection. □

We remark that ∇^0 is not the Levi-Civita connection of h in general. The connection ∇^0 may have a torsion.

We define a $(0,3)$-tensor field C, and a $(1,2)$-tensor field K by

$$C(X,Y,Z) := (\nabla_X h)(Y,Z),$$
$$K_X Y := \nabla_X Y - \nabla^0_X Y.$$

The $(0,3)$-tensor field C is called the *cubic form* of (M,∇,h) and the $(1,2)$-tensor field K is called the *difference tensor field*. (The tensor C is also called the *skewness* in statistics.)

Proposition 1.3. *Denote by C and K the cubic form and the difference tensor field of (M,∇,h), respectively. Then the following formulas hold.*

(i) $K_X Y = \nabla^0_X Y - \nabla^*_X Y = (\nabla_X Y - \nabla^*_X Y)/2.$

(ii) $C(X,Y,Z) = -2h(K_X Y, Z) = -2(Y, K_X Z).$

Proof. Since $\nabla = 2\nabla^0 - \nabla^*$, we obtain the formula (i).

From the definition of the dual connection, we have

$$\begin{aligned} h(K_X Y, Z) &= \frac{1}{2}h(\nabla_X Y - \nabla^*_X Y, Z) = \frac{1}{2}h(Y, \nabla_X Z - \nabla^*_X Z) \\ &= h(Y, K_X Z), \\ C(X,Y,Z) &= (\nabla_X h)(Y,Z) = Xh(Y,Z) - h(\nabla_X Y, Z) - h(Y, \nabla_X Z) \\ &= h(Y, \nabla^*_X Z) - h(Y, \nabla_X Z) \qquad (1) \\ &= -2h(Y, K_X Z). \end{aligned}$$

□

For the cubic form of the dual connection, we have the following relation.

Proposition 1.4. *Denote by C the cubic form for (M,∇,h), and by C^* for the cubic form for (M,∇^*,h). Then $C = -C^*$.*

Proof. From Equation (1) and the definition of the dual connection, we obtain

$$\begin{aligned} C(X,Y,Z) &= h(Y, \nabla^*_X Z) - h(Y, \nabla_X Z) \\ &= h(Y, \nabla^*_X Z) + h(\nabla^*_X Y, Z) - Xh(Y,X) \\ &= -(\nabla^*_X h)(Y,Z) \ = \ -C^*(X,Y,Z). \end{aligned}$$

□

For geometry of dual affine connections, similar formulas have been obtained in previous papers [3–6]. However, we remark that Propositions 1.2–1.4 hold even if ∇ has a torsion.

2. Statistical manifolds

In this section, we introduce the notion of statistical manifolds. Statistical manifolds was originally introduced by Lauritzen [4], then the definition was modified by Kurose [7]. Hence we show that the two definitions are equivalent.

Let (M, h) be a semi-Riemannian manifold, and let ∇ be an affine connection on M. Denote by ∇^* the dual connection of ∇ with respect to h.

Proposition 2.1. *If we assume two conditions from the followings, then the other conditions hold.*

(1) ∇ *is torsion-free.*
(2) ∇^* *is torsion-free.*
(3) $C = \nabla h$ *is totally symmetric.*
(4) $\nabla^0 = (\nabla + \nabla^*)/2$ *is the Levi-Civita connection with respect to* h.

Proof. Let us show (2) and (4) under the assumptions (1) and (3), for example.

From Equation (1), we have

$$(\nabla_X h)(Y, Z) = h(Y, \nabla^*_X Z) - h(Y, \nabla_X Z),$$
$$(\nabla_Z h)(Y, X) = h(Y, \nabla^*_Z X) - h(Y, \nabla_Z X).$$

Since ∇h is totally symmetric and ∇ is torsion-free, we obtain

$$h(Y, \nabla^*_X Z - \nabla^*_Z X) = h(Y, \nabla_X Z - \nabla_Z X) = h(Y, [X, Z]).$$

This implies that ∇^* is torsion-free.

The connection ∇^0 is torsion-free because

$$\begin{aligned} h(\nabla^0_X Y - \nabla^0_Y X, Z) &= \frac{1}{2}\{h(\nabla_X Y - \nabla_Y X, Z) + h(\nabla^*_X Y - \nabla^*_Y X, Z)\} \\ &= h([X, Y], Z). \end{aligned}$$

From Proposition 1.2, we obtain $\nabla^0 h = 0$. This implies that the connection ∇^0 is the Levi-Civita connection with respect to h. □

If ∇ and ∇^* are torsion-free, we can define a totally symmetric $(0,3)$-tensor field. Conversely, for a given totally symmetric $(0,3)$-tensor field, we can define mutually dual connections.

Proposition 2.2. *Let* (M, h) *be a semi-Riemannian manifold and let* C *be a totally symmetric* $(0,3)$*-tensor field. Set*

$$h(\nabla_X Y, Z) = h(\nabla^0_X Y, Z) - \frac{1}{2}C(X, Y, Z), \tag{2}$$

$$h(\nabla^*_X Y, Z) = h(\nabla^0_X Y, Z) + \frac{1}{2}C(X, Y, Z), \tag{3}$$

where ∇^0 is the Levi-Civita connection with respect to h. Then ∇ and ∇^* are torsion-free affine connections mutually dual with respect to h, and the derivatives ∇h and $\nabla^* h$ are totally symmetric, respectively.

Proof. From Equations (2) and (3), we obtain

$$h(\nabla_X Y, Z) + h(Y, \nabla^*_X Z) = h(\nabla^0_X Y, Z) + h(Y, \nabla^0_X Z) = Xh(Y, Z).$$

This implies that ∇ and ∇^* are mutually dual. The torsion tensor of ∇ is given by

$$\begin{aligned}
&h(\nabla_X Y - \nabla_Y X - [X, Y], Z) \\
&= h(\nabla^0_X Y - \nabla^0_Y X - [X, Y], Z) - \frac{1}{2}C(X, Y, Z) + \frac{1}{2}C(Y, X, Z) \\
&= h(T^0(X, Y), Z) = 0.
\end{aligned}$$

This implies that ∇ is torsion-free. We can show ∇^* is torsion-free from a similar calculation.

From Proposition 2.1, we obtain ∇h and $\nabla^* h$ are totally symmetric, respectively. □

Now we give the definitions of statistical manifold.

Definition 2.1. Let (M, h) be a semi-Riemannian manifold and let C be a totally symmetric $(0, 3)$-tensor field. We call the triplet (M, h, C) a *statistical manifold*.

This definition followed to Lauritzen's formulation [4]. This geometric structure naturally arises in geometric theory of a set of probability density functions [1,4].

Definition 2.2. Let (M, h) be a semi-Riemannian manifold, and let ∇ be a torsion-free affine connection on M. We call the triplet (M, ∇, h) a *statistical manifold* if ∇h is totally symmetric.

This definition followed to Kurose's formulation [7]. This geometric structure naturally arises in affine differential geometry. In fact, any nondegenerate equiaffine immersions induce a statistical manifold structure [2,7].

Remark 2.1. Lauritzen's definition and Kurose's definition are mutually equivalent. Proposition 2.2 shows that Kurose's definition is obtained from Lauritzen's definition. The converse is obtained from Proposition 2.1.

3. A generalization of statistical manifolds

In this section, we consider a generalization of dual connections and statistical manifolds.

Let (M, h) be a semi-Riemannian manifold, let ∇ be an affine connection on M, and let ω be a 1-form on M. We can define the *generalized dual connection* $\bar{\nabla}^*$ [5] by

$$Xh(Y, Z) = h(\nabla_X Y, Z) + h(Y, \bar{\nabla}^*_X Z) - \omega(X)h(Y, Z).$$

For the generalized dual connection, similar calculations in Proposition 1.2 show the following proposition.

Proposition 3.1. *Let ∇ be an affine connection on a semi-Riemannian manifold (M, h). Denote by $\bar{\nabla}^*$ the generalized dual connection of ∇ by a 1-form ω. Set $\bar{\nabla}^0 := (\nabla + \bar{\nabla}^*)/2$. Then the followings hold.*

(i) $\overline{(\bar{\nabla}^*)}^* = \nabla$.

(ii) $(\bar{\nabla}^0_X h)(Y, Z) + \omega(X)h(Y, Z) = 0$.

We consider geometry of generalized dual connections. At first, we define a $(0, 3)$-tensor field $\bar{C}$ by $\bar{C}(X, Y, Z) := (\nabla_X h)(Y, Z) + \omega(X)h(Y, Z)$. We call $\bar{C}$ the *generalized cubic form* or *generalized skewness tensor field*. The role of the generalized cubic form $\bar{C}$ is similar to a cubic form on a statistical manifold. For example, $\bar{C}$ gives the difference between the given connection ∇ and its generalized dual connection $\bar{\nabla}$:

$$\begin{aligned}\bar{C}(X, Y, Z) &= Xh(Y, Z) - h(\nabla_X Y, Z) - h(Y, \nabla_X Z) + \omega(X)h(Y, Z) \\ &= h(Y, \bar{\nabla}^*_X Z - \nabla_X Z).\end{aligned}$$

Let us recall the definition of a Weyl manifold. Suppose that (M, h) is a semi-Riemannian manifold, and ∇ is a torsion-free affine connection on M. We call the triplet (M, ∇, h) a *Weyl manifold* if there exists a 1-form ω such that $(\nabla_X h)(Y, Z) + \omega(X)h(Y, Z) = 0$. This implies that the generalized cubic form characterizes a Weyl manifold if the given connection ∇ is torsion-free.

Proposition 3.2. *Let (M, h) be a semi-Riemannian manifold, and let ∇ be an affine connection on M. Denote by $\bar{C}(X, Y, Z)$ the generalized cubic form on (M, ∇, h). If ∇ is torsion-free and $\bar{C}(X, Y, Z) = 0$, then (M, ∇, h) is a Weyl manifold.*

For generalized dual connections and generalized cubic forms, we have the following theorem.

Theorem 3.1. *Let (M, h) be a semi-Riemannian manifold, let ∇ be an affine connection on M, and let ω be a 1-form on M. Denote by $\bar{\nabla}^*$ the generalized dual connection of ∇ by ω. If we assume two conditions of the followings, then the others hold.*

(1) *∇ is torsion-free.*
(2) *$\bar{\nabla}^*$ is torsion-free.*
(3) *$\bar{C}(X, Y, Z) = (\nabla_X h)(Y, Z) + \omega(X)h(Y, Z)$ is totally symmetric.*
(4) *$\bar{\nabla}^0 = (\nabla + \bar{\nabla}^*)/2$ is a Weyl connection, namely, $(M, \bar{\nabla}^0, h)$ is a Weyl manifold.*

Proof. Suppose that the conditions (1) and (3) hold. Since $\bar{C}$ is totally symmetric, we obtain

$$h(Y, \bar{\nabla}^*_X Z - \bar{\nabla}^*_Z X) = h(Y, \nabla_X Z - \nabla_Z X).$$

This implies that $\bar{\nabla}^*$ is torsion-free since ∇ is torsion-free. Hence the condition (2) holds.

It is easy to show that $\bar{\nabla}^0$ is torsion-free since $\bar{\nabla}^0 = (\nabla + \bar{\nabla}^*)/2$. This implies that, from Proposition 3.1, $\bar{\nabla}^0$ is a Weyl connection. Hence the condition (4) holds.

In the case we assume the condition (3) in addition to (2) or (4), similar calculations hold.

Suppose that (1) and (2) hold. Then we obtain

$$\begin{aligned}\bar{C}(X, Y, Z) &= h(Y, \bar{\nabla}^*_X Z - \nabla_X Z)\\ &= h(Y, \bar{\nabla}^*_Z X - \nabla_Z X)\\ &= \bar{C}(Z, Y, X).\end{aligned}$$

In the second equality, we used the conditions ∇ and $\bar{\nabla}^*$ are torsion-free. Since the second and the third arguments of $\bar{C}$ are symmetric, the condition (3) holds. Then the condition (4) also holds from the first part of the proof.

In the case we assume two conditions from (1), (2) or (4), similar calculations hold. □

Let (M, h) be a semi-Riemannian manifold, and let ∇ be a torsion-free affine connection on M. We call the triplet (M, ∇, h) a *semi-Weyl manifold* [8] if there exists a 1-form ω such that $(\nabla_X h)(Y, Z) + \omega(X)h(Y, Z)$ is totally symmetric. In this case, the given connection ∇ is said to be *semi-compatible* [9].

A semi-Weyl manifold is regarded as a natural generalization of a Weyl manifold and a statistical manifold. Semi-Weyl manifolds arise naturally in affine hypersurface theory. Any nondegenerate affine hypersurface has a semi-Weyl manifold structure [8].

Acknowledgement

This research is supported by Grant-in-Aid for Encouragement of Young Scientists (B) No. 16740037, Japan Society for the Promotion of Science.

References

1. S. Amari and H. Nagaoka, *Methods of information geometry*, Amer. Math. Soc., Oxford University Press, Providence, RI, 2000.
2. K. Nomizu and T. Sasaki, *Affine differential geometry: geometry of affine immersions*, Cambridge Univ. Press, 1994.
3. F. Dillen, K. Nomizu and L. Vranken, Monatsh. Math. **109** (1990), 221.
4. S.L. Lauritzen, *Statistical manifolds*, in Differential Geometry in Statistical Inferences, IMS Lecture Notes Monograph Series 10, Institute of Mathematical Statistics, Hayward California, 1987.
5. K. Nomizu, *Affine connections and their use*, in Geometry and Topology of Submanifolds VII, ed. F. Dillen, World Scientific, 1995.
6. K. Nomizu and U. Simon, *Notes on conjugate connections*, in Geometry and Topology of Submanifolds IV, ed. F. Dillen and L. Verstraelen, World Scientific, 1992.
7. T. Kurose, Tôhoku Math. J. **46** (1994), 427.
8. H. Matsuzoe, Kyushu J. Math. **55** (2001), 107.
9. S. Ivanov, J. Geom. **53** (1995), 89.

ON SOME NON-INTEGRABLE ALMOST CONTACT MANIFOLDS WITH NORDEN METRIC OF DIMENSION 5*

G. NAKOVA

Department of Algebra and Geometry,
Faculty of Pedagogics, University of Veliko Tarnovo,
1 Theodosij Tirnovsky Str., 5000 Veliko Tarnovo, Bulgaria
E-mail: gnakova@yahoo.com

A Lie group as a 5-dimensional almost contact manifold with Norden metric is considered. By using the complexification of the real Lie algebra this manifold is constructed. The obtained manifold is non-integrable and belongs to one of the basic classes of almost contact manifolds with Norden metric. Curvature properties of the constructed manifold are given.

Keywords: Norden metric; Almost contact manifold; Partially integrable; Integrable and non-integrable almost contact structure, Lie group.

Introduction

The almost contact manifolds with Norden metric are classified [2] into eleven basic classes $\mathcal{F}_i$ $(i = 1, 2, \ldots, 11)$. Examples of some classes of these manifolds are obtained. Examples of manifolds of the classes $\mathcal{F}_6$, $\mathcal{F}_7$, $\mathcal{F}_8$, $\mathcal{F}_9$, $\mathcal{F}_{10}$ have not been found yet. In this paper we construct an almost contact structure with Norden metric on Lie groups as 5-dimensional manifolds, which belong to the class $\mathcal{F}_9$.

1. Almost contact manifolds with Norden metric

1.1. *Preliminaries*

Let $(M, \varphi, \xi, \eta, g)$ be a $(2n+1)$-dimensional almost contact manifold with Norden metric, i.e. (φ, ξ, η) is an almost contact structure [1] and g is a

*Partially supported by Scientific researches fund of "St. Cyril and St. Methodius" University of Veliko Tarnovo under contract RD620-02006.

metric [2] on M such that

$$\varphi^2 X = -id + \eta \otimes \xi, \qquad \eta(\xi) = 1, \\ g(\varphi X, \varphi Y) = -g(X,Y) + \eta(X)\eta(Y), \tag{1}$$

where id denotes the identity transformation and X, Y are differentiable vector fields on M, i.e. $X, Y \in \mathfrak{X}(M)$.

The tensor $\tilde{g}$ given by $\tilde{g}(X,Y) = g(X, \varphi Y) + \eta(X)\eta(Y)$ is a Norden metric, too. Both metrics g and $\tilde{g}$ are indefinite of signature $(n+1, n)$.

Let $(V, \varphi, \xi, \eta, g)$ be a $(2n+1)$-dimensional vector space with almost contact structure (φ, ξ, η) and Norden metric g. For an arbitrary $X \in V$ we have $\varphi^2 X = -X + \eta(X)\xi \Longleftrightarrow X = -\varphi^2 X + \eta(X)\xi$. Hence V admits a decomposition into a direct sum of vector subspaces:

$$V = D \oplus \{\xi\}, \tag{2}$$

where $D = \operatorname{Ker}\eta$, $\{\xi\} = (\operatorname{Im}\eta)\xi$. Then for an arbitrary $X \in V$ it follows $X = x + \eta(X)\xi$, where $x \in D$, $\eta(X)\xi \in \{\xi\}$. Denoting the restrictions of g and φ on D with the same letters we obtain the almost complex vector space (D, φ, g) of dimension $2n$ with a complex structure φ and Norden metric g.

Let ∇ be the Levi-Civita connection of the metric g. The tensor field F of type $(0,3)$ on M is defined by

$$F(X,Y,Z) = g((\nabla_X \varphi)Y, Z). \tag{3}$$

It has the following symmetries

$$F(X,Y,Z) = F(X,Z,Y), \\ F(X,\varphi Y,\varphi Z) = F(X,Y,Z) - \eta(Y)F(X,\xi,Z) - \eta(Z)F(X,Y,\xi). \tag{4}$$

A classification of the almost contact manifolds with Norden metric with respect to the tensor F is given in [2] and [11] basic classes $\mathcal{F}_i$ $(i = 1, 2, \dots, 11)$ are obtained.

1.2. *Curvature properties*

Let R be the curvature tensor field of ∇ defined by

$$R(X,Y)Z = \nabla_X \nabla_Y Z - \nabla_Y \nabla_X Z - \nabla_{[X,Y]} Z. \tag{5}$$

The corresponding tensor field of type $(0,4)$ is determined as follows

$$R(X,Y,Z,W) = g(R(X,Y)Z, W).$$

The Ricci-type tensors ρ, ρ^*, $\tilde{\rho}$ and scalar curvatures τ, τ^*, $\tilde{\tau}$ of R are defined by:

$$\rho(Y,Z) = \sum_{i=1}^{2n+1} R(e_i, Y, Z, e_i), \quad \tau = \sum_{i,j=1}^{2n+1} R(e_i, e_j, e_j, e_i);$$

$$\rho^*(Y,Z) = \sum_{i=1}^{2n+1} R(e_i, Y, Z, \varphi e_i), \quad \tau^* = \sum_{i,j=1}^{2n+1} R(e_i, e_j, e_j, \varphi e_i);$$

$$\tilde{\rho}(Y,Z) = R(\xi, Y, Z, \xi), \quad \tilde{\tau} = \sum_{i=1}^{2n+1} R(\xi, e_i, e_i, \xi),$$

where $\{e_i, e_{2n+1} = \xi\}$, $(i = 1, 2, \ldots, 2n)$ is an orthonormal basis.

A decomposition of the space of curvature tensors $\mathcal{R}$ over an almost contact vector space with Norden metric into 20 mutually orthogonal factors is obtained in [4]. Let us recall the decomposition $\mathcal{R} = h\mathcal{R} \oplus v\mathcal{R} \oplus w\mathcal{R}$, where the subspaces $h\mathcal{R}$, $v\mathcal{R}$, $w\mathcal{R}$ have the following orthogonal decompositions:

$$\begin{aligned} h\mathcal{R} &= h\mathcal{R}_1 \oplus h\mathcal{R}_1^{\perp} \oplus h\mathcal{R}_2^{\perp} \oplus h\mathcal{R}_3^{\perp} = \omega_1 \oplus \omega_2 \oplus \cdots \oplus \omega_{11}, \\ v\mathcal{R} &= v\mathcal{R}_1 \oplus v\mathcal{R}_1^{\perp} \oplus v\mathcal{R}_2^{\perp} = v_1 \oplus v_2 \oplus v_3 \oplus v_4 \oplus v_5, \\ w\mathcal{R} &= w_1 \oplus w_2 \oplus w_3 \oplus w_4. \end{aligned}$$

2. Lie groups as almost contact manifolds with Norden metric

Let V be a real vector space. We consider the set $V^{\mathbb{C}} = \{X + iY, \ X, Y \in V\}$ with the following two operations on $V^{\mathbb{C}}$:

$$\begin{aligned} (X + iY) + (Z + iT) &= X + Z + i(Y + T); \\ (a + ib)(X + iY) &= aX - bY + i(aY + bX), \quad a, b \in \mathcal{R}. \end{aligned}$$

Then $V^{\mathbb{C}}$ is a complex vector space and $V^{\mathbb{C}}$ is called a complexification of V. It is well known, if $\mathfrak{g}$ is a real Lie algebra then the complexification $\mathfrak{g}^{\mathbb{C}}$ of $\mathfrak{g}$ is a complex Lie algebra with commutator defined by:

$$[X + iY, Z + iT] = [X, Z] - [Y, T] + i\left([Y, Z] + [X, T]\right).$$

Let $\mathfrak{g}$ be a real Lie algebra ($\dim \mathfrak{g} = 2n+1$) supplied with an almost contact structure (φ, ξ, η). Taking into account (2) we have $\mathfrak{g} = D \oplus \{\xi\}$. Then the natural complexification of the real Lie algebra $\mathfrak{g}$ is the direct sum

$$\mathfrak{g}^{\mathbb{C}} = D^{\mathbb{C}} \oplus \{\xi\}.$$

From $\varphi^2 = -id$ on D it follows a decomposition of $D^{\mathbb{C}}$ into a direct sum $D^{\mathbb{C}} = D^{10} \oplus D^{01}$, where D^{10} and D^{01} are he eigenspaces of φ. We have

$$D^{10} = \mathrm{span}\{Z_\alpha = e_\alpha - i\varphi e_\alpha\}_{\alpha\in I}, \quad I = \{1,\dots,n\},$$
$$D^{01} = \mathrm{span}\{Z_{\bar\alpha} = e_\alpha + i\varphi e_\alpha\}_{\bar\alpha\in \bar I}, \quad \bar I = \{\bar 1,\dots,\bar n\},$$
$$\mathfrak{g}^{\mathbb{C}} = \mathrm{span}\{Z_A\}_{A\in I\cup\bar I\cup I_0}, \quad I_0 = \{0\},\ Z_0 = \xi,$$

where $\{e_\alpha, \varphi e_\alpha, \xi\}$ is a φ-basis of $\mathfrak{g}$. Let G be a real connected Lie group with a real Lie algebra $\mathfrak{g}$ ($\dim\mathfrak{g} = 2n+1$) and (φ,ξ,η,g) is an almost contact structure with Norden metric on $\mathfrak{g}$. In this way, the induced manifold (G,φ,ξ,η,g) is an almost contact manifold with Norden metric. The complex-linear continuations of the fundamental tensors for almost contact manifolds with Norden metric in terms of the essential (i.e. non-zero in general) complex components are descriebed in [3]. Characterization of each basic class by the essential complex equations for the fundamental tensors is obtained, too. Further we give some basic facts from [3] which we will need.

The generalized Nijenhuis tensor field of type $(1,2)$ is

$$N = [\varphi,\varphi] + d\eta\otimes\xi,$$

where

$$[\varphi,\varphi](X,Y) = \varphi^2[X,Y] + [\varphi X,\varphi Y] - \varphi[\varphi X,Y] - \varphi[X,\varphi Y]$$

is the usual Nijenhuis tensor field of type $(1,2)$, formed by φ and

$$d\eta = (\nabla_X\eta)\,Y - (\nabla_Y\eta)\,X.$$

The corresponding to N tensor field of type $(0,3)$ is $N(X,Y,Z) = g(N(X,Y),Z)$. The complex components with respect to the complex basis $\{Z_A\}_{A\in I\cup\bar I\cup I_0}$ of the complex-linear continuations of the tensors F, g, $d\eta$, N are denoted by:

- $F_{ABC} = F(Z_A,Z_B,Z_C) = \overline{F_{\bar A\bar B\bar C}}$;
- $g_{AB} = g(Z_A,Z_B) = \overline{g_{\bar A\bar B}}$;
- $\eta_{AB} = d\eta(Z_A,Z_B) = \overline{\eta_{\bar A\bar B}}$;
- $N(Z_A,Z_B) = N_{AB}^C.Z_C$ and $N_{AB}^C = -N_{BA}^C = \overline{N_{\bar A\bar B}^{\bar C}}$;
- $N_{ABC} = N_{AB}^S.g_{SC} = \overline{N_{\bar A\bar B\bar C}}, \quad A,B,C\in\{1,\dots,n,\bar 1,\dots,\bar n,0\}$.

We put $[Z_A,Z_B] = C_{AB}^C.Z_C$ and $C_{AB}^C = -C_{BA}^C = \overline{C_{\bar A\bar B}^{\bar C}}$. We have the following equalities for the commutators of the basis vector fields of $D^{\mathbb{C}}\oplus\{\xi\}$

$$\begin{pmatrix}[Z_\alpha,Z_\beta]\\ [Z_\alpha,Z_{\bar\beta}]\\ [Z_\alpha,Z_0]\end{pmatrix} = \begin{pmatrix} C_{\alpha\beta}^\gamma & -\frac14 N_{\alpha\beta}^{\bar\gamma} & -\frac12 N_{\alpha\beta}^0\\ C_{\alpha\bar\beta}^\gamma & C_{\alpha\bar\beta}^{\bar\gamma} & -\eta_{\alpha\bar\beta}\\ C_{\alpha 0}^\gamma & -\frac12 N_{\alpha 0}^{\bar\gamma} & -N_{\alpha 0}^0\end{pmatrix}\cdot\begin{pmatrix}Z_\gamma\\ Z_{\bar\gamma}\\ Z_0\end{pmatrix}. \tag{6}$$

Let us recall:

Definition 2.1 ([5]). *The almost contact structure (φ, ξ, η) (resp. the almost contact manifolod) is said to be partially integrable if the contact distribution D is involutive and the restriction $\varphi_{|D}$ to each integral manifold $\mathcal{D}$ for D is an itegrable almost complex structure.*

Definition 2.2 ([5]). *The almost contact structure (φ, ξ, η) (resp. the almost contact manifolod) is said to be integrable if (φ, ξ, η) is partially integrable and the components φ_i^j of φ with respect to an adapted local coordinate system $(U_{\mathcal{D}}; \{x^\alpha, y^\alpha = x^{n+\alpha}, t = const, \alpha \in I\})$, for any integral manifold $\mathcal{D}$ of D, immersed in M, does not depend on t.*

It is known:

Proposition 2.1 ([6]). *The almost contact structure (φ, ξ, η) is partially integrable iff $N^0_{\alpha\beta} = N^{\bar{\gamma}}_{\alpha\beta} = \eta_{\alpha\bar{\beta}} = 0$.*

Proposition 2.2 ([6]). *The almost contact structure (φ, ξ, η) is integrable iff $N = 0$, $d\eta = 0$.*

Proposition 2.3 ([3]). *The class of partially integrable almost contact manifolds with Norden metric is exactly $\mathcal{F}_1 \oplus \mathcal{F}_2 \oplus \mathcal{F}_4 \oplus \mathcal{F}_5 \oplus \mathcal{F}_6 \oplus \mathcal{F}_9 \oplus \mathcal{F}_{10} \oplus \mathcal{F}_{11}$ and the class of integrable almost contact manifolds with Norden metric is exactly $\mathcal{F}_1 \oplus \mathcal{F}_2 \oplus \mathcal{F}_4 \oplus \mathcal{F}_5 \oplus \mathcal{F}_6$.*

Having in mind Proposition 2.1, Proposition 2.2 and (6) we have:

Proposition 2.4. *Let $\mathfrak{g}$ be a real Lie algebra, supplied with an almost contact structure (φ, ξ, η) and Norden metric g. Then (φ, ξ, η) is a partially integrable iff for the complexification $\mathfrak{g}^{\mathbb{C}} = D^{\mathbb{C}} \oplus \{\xi\}$ of $\mathfrak{g}$ the following conditions are valid:*

(i) *$D^{\mathbb{C}}$ is a complex subalgebra of $\mathfrak{g}^{\mathbb{C}}$;*
(ii) *the subspaces D^{10} and D^{01} of $D^{\mathbb{C}}$ are subalgebras of $D^{\mathbb{C}}$.*

Proposition 2.5. *If (φ, ξ, η) is an integrable almost contact structure on a real Lie algebra $\mathfrak{g}$ and g is Norden metric on $\mathfrak{g}$ then for the complexification $\mathfrak{g}^{\mathbb{C}} = D^{\mathbb{C}} \oplus \{\xi\}$ of $\mathfrak{g}$ the following conditions are valid:*

(i) *$D^{\mathbb{C}}$ is an ideal of $\mathfrak{g}^{\mathbb{C}}$;*
(ii) *the subspaces D^{10} and D^{01} of $D^{\mathbb{C}}$ are subalgebras of $D^{\mathbb{C}}$.*

3. A Lie group as a 5-dimensional almost contact manifold with Norden metric of the class $\mathcal{F}_9$

Let V be a 5-dimensional real vector space and consider the structure of the Lie algebra defined by the brackets $[E_i, E_j] = C_{ij}^k.E_k$, where $\{E_1, E_2, E_3, E_4, E_5\}$ is a basis of V and $C_{ij}^k \in \mathcal{R}$. Let G be the associated real connected Lie group and $\mathfrak{g}$ be the real Lie algebra of G with a global basis of left invariant vector fields $\{X_1, X_2, X_3, X_4, X_5\}$, induced by the basis of V. Then the Jacobi identity holds

$$\underset{X_i,X_j,X_k}{\mathfrak{S}} \big[[X_i, X_j], X_k\big] = 0. \tag{7}$$

Next we define tensor field φ of type $(1,1)$, vector field ξ, metric g and 1-form η on G by the conditions:

$$\varphi X_i = X_{2+i}, \quad \varphi X_{2+i} = -X_i, \quad \varphi X_5 = 0, \quad (i = 1, 2); \tag{8}$$

$$\begin{aligned} &g(X_i, X_i) = -g(X_{2+i}, X_{2+i}) = g(X_5, X_5) = 1, \quad (i = 1, 2);\\ &g(X_j, X_k) = 0, \quad (j \neq k,\ j, k = 1, 2, 3, 4, 5). \end{aligned} \tag{9}$$

We denote $X_5 = \xi$. Then the structure (φ, ξ, η, g) defined by (8), (9) is an almost contact structure with Norden metric on G and $(G, \varphi, \xi, \eta, g)$ is a 5-dimensional almost contact manifold with Norden metric.

Now, we consider the complexification $\mathfrak{g}^{\mathbb{C}}$ of the 5-dimensional real Lie algebra $\mathfrak{g}$ of G, which is a 5-dimensional complex Lie algebra. Let $\{Z_1, Z_2, Z_{\bar{1}}, Z_{\bar{2}}, Z_0\}$ be a complex basis of $\mathfrak{g}^{\mathbb{C}}$, where $Z_1 = X_1 - iX_3$, $Z_2 = X_2 - iX_4$, $Z_{\bar{1}} = \overline{Z_1}$, $Z_{\bar{2}} = \overline{Z_2}$, $Z_0 = X_5 = \xi$. From (8) and (9) for the complex basis we obtain

$$\varphi Z_j = iZ_j, \quad \varphi Z_{\bar{j}} = -iZ_{\bar{j}}, \quad \varphi Z_0 = 0, \quad (j = 1, 2); \tag{10}$$

$$g_{11} = g_{22} = 2, \quad g_{1\bar{1}} = g_{1\bar{2}} = g_{12} = g_{2\bar{2}} = g_{10} = g_{20} = 0, \quad g_{00} = 1. \tag{11}$$

Using the well known condition

$$\begin{aligned} 2g(\nabla_X Y, Z) = {} & Xg(Y, Z) + Yg(X, Z) - Zg(X, Y) + g([X, Y], Z)\\ & + g([Z, X], Y) + g([Z, Y], X) \end{aligned} \tag{12}$$

for the Levi-Civita connection ∇ of g and (11) we get the following equation for the tensor F defined by (3)

$$\begin{aligned} F(Z_A, Z_B, Z_C) = {} & \frac{1}{2}\{g([Z_A, \varphi Z_B] - \varphi[Z_A, Z_B], Z_C)\\ & + g(\varphi[Z_C, Z_A] - [\varphi Z_C, Z_A], Z_B)\\ & + g([Z_C, \varphi Z_B] - [\varphi Z_C, Z_B], Z_A)\}, \quad A, B, C \in \{1, 2, \bar{1}, \bar{2}, 0\}. \end{aligned} \tag{13}$$

We determine the commutators $[Z_A, Z_B] = C_{AB}^C.Z_C$ of $\mathfrak{g}^{\mathbb{C}}$ such that the manifold $(G, \varphi, \xi, \eta, g)$ is of class $\mathcal{F}_9$ and the Jacobi identity holds. According to Proposition 2.3 the class $\mathcal{F}_9$ belongs to the class of partially integrable almost contact manifolds with Norden metric. The characterization conditions of the class $\mathcal{F}_9$ in local components given in [3] are:

$$\mathcal{F}_9 : F_{\alpha\bar{\beta}0} = -F_{\bar{\beta}\alpha 0} = \frac{i}{2}N_{0\alpha\bar{\beta}}.$$

From the last equalities it follows

$$N_{0\alpha\bar{\beta}} = N_{0\bar{\beta}\alpha} \Longleftrightarrow N_{\alpha 0}^{\bar{\beta}} = N_{\bar{\beta}0}^{\alpha} \tag{14}$$

and the rest components of the Nijenhuis tensor and of the tensor F are zero. Then the commutators (6) become:

$$\begin{pmatrix} [Z_\alpha, Z_\beta] \\ [Z_\alpha, Z_{\bar{\beta}}] \\ [Z_\alpha, Z_0] \end{pmatrix} = \begin{pmatrix} C_{\alpha\beta}^{\gamma} & 0 & 0 \\ C_{\alpha\bar{\beta}}^{\gamma} & C_{\alpha\bar{\beta}}^{\bar{\gamma}} & 0 \\ C_{\alpha 0}^{\gamma} & -\frac{1}{2}N_{\alpha 0}^{\bar{\gamma}} & 0 \end{pmatrix} \cdot \begin{pmatrix} Z_\gamma \\ Z_{\bar{\gamma}} \\ Z_0 \end{pmatrix}, \quad \alpha, \beta, \gamma \in \{1, 2\}. \tag{15}$$

Taking into account that for the class $\mathcal{F}_9$ $F_{11\bar{1}} = F_{11\bar{2}} = F_{12\bar{1}} = F_{12\bar{2}} = F_{21\bar{1}} = F_{21\bar{2}} = F_{22\bar{1}} = F_{22\bar{2}} = 0$ and Jacobi identity we obtain a system for the structural constants $C_{\alpha\beta}^{\gamma}$, $C_{\alpha\bar{\beta}}^{\gamma}$, $C_{\alpha\bar{\beta}}^{\bar{\gamma}}$ from (15). Using (10), (11), (13), (15) we give one solution of this system:

$$C_{1\bar{1}}^{2} = -C_{12}^{1} = -C_{1\bar{1}}^{\bar{2}} = C_{1\bar{2}}^{\bar{1}} = a, \; C_{12}^{2} = C_{1\bar{2}}^{2} = -C_{2\bar{2}}^{1} = -C_{2\bar{2}}^{\bar{1}} = ia, \; a \in \mathcal{R};$$
$$C_{1\bar{1}}^{1} = C_{1\bar{1}}^{\bar{1}} = C_{1\bar{2}}^{1} = C_{1\bar{2}}^{\bar{2}} = C_{2\bar{2}}^{2} = C_{2\bar{2}}^{\bar{2}} = 0.$$

From (14) we have:

$$\begin{aligned} &N_{10}^{\bar{2}} = N_{\bar{2}0}^{1} \in \mathbb{C}; \quad N_{10}^{\bar{1}} = N_{\bar{1}0}^{1} = \overline{N_{10}^{\bar{1}}} \Longrightarrow N_{10}^{\bar{1}}, \; N_{\bar{1}0}^{1} \in \mathcal{R}; \\ &N_{20}^{\bar{2}} = N_{\bar{2}0}^{2} = \overline{N_{20}^{\bar{2}}} \Longrightarrow N_{20}^{\bar{2}}, \; N_{\bar{2}0}^{2} \in \mathcal{R}. \end{aligned} \tag{16}$$

Having in mind that for the class $\mathcal{F}_9$ $F_{110} = F_{120} = F_{210} = F_{220} = F_{001} = F_{002} = F_{01\bar{1}} = F_{01\bar{2}} = F_{02\bar{2}} = 0$ and Jacobi identity we obtain a system for the structural constants $C_{\alpha 0}^{\gamma}$, $N_{\alpha 0}^{\bar{\gamma}}$ from (15). Using (10), (11), (13), (15), (16) we give one solution of this system:

$$C_{10}^{1} = C_{20}^{2} = N_{10}^{\bar{1}} = N_{\bar{1}0}^{1} = N_{20}^{\bar{2}} = N_{\bar{2}0}^{2} = 0; \; C_{20}^{1} = -C_{10}^{2} = m, \; m \in \mathcal{R}.$$

Finally, for the commutators of the basis vector fields of $\mathfrak{g}^{\mathbb{C}}$ we have

$$\begin{aligned} &[Z_1, Z_2] = -aZ_1 + iaZ_2, \quad [Z_1, Z_{\bar{1}}] = aZ_2 - aZ_{\bar{2}}, \\ &[Z_1, Z_{\bar{2}}] = iaZ_2 + aZ_{\bar{1}}, \quad\;\; [Z_2, Z_{\bar{2}}] = -iaZ_1 - iaZ_{\bar{1}}, \\ &[Z_1, \xi] = -mZ_2 + mZ_{\bar{2}}, \quad [Z_2, \xi] = mZ_1 + mZ_{\bar{1}} \quad a, m \in \mathcal{R}. \end{aligned} \tag{17}$$

From the equalities (17) for the commutators of the basis vector fields of $\mathfrak{g}^{\mathbb{C}}$ we obtain the commutators of the basis fields of the real Lie algebra $\mathfrak{g}$

$$\begin{aligned}
&[X_1, X_2] = -[X_1, X_3] = aX_4, \quad [X_2, X_3] = aX_2 + aX_3, \\
&[X_3, X_4] = -[X_2, X_4] = aX_1, \quad [X_2, \xi] = 2mX_1, \\
&[X_3, \xi] = -2mX_4, \quad [X_1, X_4] = [X_1, \xi] = [X_4, \xi] = 0, \quad a, m \in \mathcal{R}.
\end{aligned} \tag{18}$$

The non-zero components of the tensor F are $F_{145} = F_{235} = -F_{325} = -F_{415} = m$, where $F_{ijk} = F(X_i, X_j, X_k)$. Then $F(X, Y, Z) = m(x^1y^4z^5 + x^1y^5z^4 + x^2y^3z^5 + x^2y^5z^3 - x^3y^2z^5 - x^3y^5z^2 - x^4y^1z^5 - x^4y^5z^1)$, where $X = x^iX_i$, $Y = y^iX_i$, $Z = z^iX_i$, $(i = 1, 2, 3, 4, 5)$. We verify that

$$F(X, Y, Z) = F(\varphi X, \varphi Y, Z) + F(\varphi X, Y, \varphi Z) = -F(Y, Z, X) - F(Z, X, Y),$$

which is the characterization condition of the class $\mathcal{F}_9$ in global variables given in [2].

So, for the manifold $(G, \varphi, \xi, \eta, g)$ constructed above, we establish the truthfulness of the following

Theorem 3.1. *Let $(G, \varphi, \xi, \eta, g)$ be a 5-dimensional almost cotact manifold with Norden metric, where G is a connected Lie group with corresponding Lie algebra $\mathfrak{g}$ determined by the global basis of left invariant vector fields $\{X_1, X_2, X_3, X_4, X_5\}$; φ is a tensor of type $(1,1)$ defined by (8), $\xi = X_5$, g is a Norden metric defined by (9) and $\eta(X) = g(X, \xi)$. Then $(G, \varphi, \xi, \eta, g)$ is of class $\mathcal{F}_9$ if and only if the Lie algebra $\mathfrak{g}$ of G belongs to the 2-parametric family determined by (18).*

Let us recall that a Lie algebra $\mathfrak{g}$ is said to be solvable if its derived series given by

$$D^0\mathfrak{g} = \mathfrak{g}, \; D^1\mathfrak{g} = [\mathfrak{g}, \mathfrak{g}], \; \ldots, \; D^{k+1}\mathfrak{g} = [D^k\mathfrak{g}, D^k\mathfrak{g}], \; \ldots$$

vanish for some $k \in \mathbb{N}$. Then, having in mind the equalities (18) it is easy to check that $D^2\mathfrak{g} = \{0\}$ and therefore the Lie algebras $\mathfrak{g}$ constructed by us are 2-step solvable.

4. Curvature properties of the constructed manifold

Let R be the curvature tensor of type $(0,4)$ on the manifold $(G, \varphi, \xi, \eta, g)$ constructed in previous section. We denote its components by $R_{ijks} = R(X_i, X_j, X_k, X_s)$ $(i, j, k, s \in \{1, 2, 3, 4, 5\})$. Using (12), (8), (9), (18) we receive the non-zero components of R as follows

$$\begin{aligned}
&R_{2314} = R_{2413} = R_{3443} = -R_{4554} = R_{1551} = m^2, \\
&R_{2525} = -R_{3535} = 3m^2, \quad m \in \mathcal{R}.
\end{aligned} \tag{19}$$

Taking into account (19) we find

$$R(X,Y,Z,W) = hR(X,Y,Z,W) + wR(X,Y,Z,W).$$

The curvature tensor hR satisfies the following equation

$$R(x,y,z,w) = R(\varphi x,\varphi y,z,w) + R(\varphi x,y,\varphi z,w) + R(\varphi x,y,z,\varphi w),$$

which is the characterization condition of the class

$$h\mathcal{R}_2 \oplus h\mathcal{R}_1 \oplus h\mathcal{R}_1^{\perp} = \omega_5 \oplus \omega_6 \oplus \omega_7 \oplus \omega_8.$$

The curvature tensor $R(\xi,y,z,\xi)$ satisfies the following equation $R(\xi,y,z,\xi) = R(\xi,\varphi y,\varphi z,\xi) - 2m^2 g(y,z)$. Finally

$$R \in \omega_5 \oplus \omega_6 \oplus \omega_7 \oplus \omega_8 \oplus w\mathcal{R}.$$

Further, we give the non-zero components $\rho_{ij} = \rho(X_i,X_j)$, $\rho^*_{ij} = \rho^*(X_i,X_j)$, $\tilde{\rho}_{ij} = \tilde{\rho}(X_i,X_j)$ $(i,j \in \{1,2,3,4,5\})$ of the Ricci-type tensors as follows

$$\rho_{11} = -\rho_{22} = \rho_{33} = -\rho_{44} = 2m^2, \quad \rho_{55} = -4m^2 \Longrightarrow \rho(Y,Z) = \rho(Z,Y);$$

$$\rho^*_{13} = \rho^*_{24} = \rho^*_{31} = \rho^*_{42} = m^2 \Longrightarrow$$

$$\rho^*(\varphi Y,\varphi Z) = -\rho^*(Y,Z), \quad \rho^*(Y,Z) = \rho^*(Z,Y);$$

$$\tilde{\rho}_{11} = -\tilde{\rho}_{44} = m^2, \quad \tilde{\rho}_{33} = -\tilde{\rho}_{22} = 3m^2 \Longrightarrow \tilde{\rho}(Y,Z) = \tilde{\rho}(Z,Y).$$

So, the scalar curvatures are: $\tau = \tilde{\tau} = -4m^2$, $\tau^* = 0$.

References

1. D. Blair, *Contact manifolds in Riemannian geometry*, Lecture Notes in Math., vol. 509, Springer Verlag, Berlin, 1976.
2. G. Ganchev, V. Mihova and K. Gribachev, *Almost contact manifolds with B-metric*, Math. Balkanica **7** (1993), 262–276.
3. G. Nakova, *Some remarks on the integrability conditions for almost contact manifolds with B-metric*, Plovdiv University "Paissii Hilendarski" Scientific Works-Mathematics, vol. 33, no. 3, 2001, 87–94.
4. ———, *Curvature tensors on almost contact manifolds with B-metric*, Trends in Complex Analysis, Differential Geometry and Mathematical Physics, eds. S. Dimiev and K. Sekigawa, World Scientific, Singapore, 2003, 145–158.
5. K. Yano and S. Ishihara, *On integrability conditions of a structure f, satisfying $f^3 + f = 0$*, Quart. J. Math. **15** (1964), 217–222.
6. M. Hristov, *On the integrability conditions for almost contact metric manifolds*, Aspects of Complex Analysis, Differential Geometry and Mathematical Physics, eds. S. Dimiev and K. Sekigawa, World Scientific, Singapore, 1999, 134–147.

COMPLEXIFICATION OF THE PROPAGATOR FOR THE HARMONIC OSCILLATOR

T. NITTA

Department of Education, Mie University
Kurimamachiyamachi, Tsu, 514-8507, Japan
E-mail: nitta@edu.mie-u.ac.jp

In 1999, Ken Loo used Nonstandard Analysis and provided a rigorous computation for the harmonic oscillator Feynman path integral. He computed it without having prior knowledge of the classical path. Usually such kind of calculation is done using the classical behavior. He emphasized that the time is real-valued in his paper . We extend his result to the case the time is valued in the complex number field, that is, we complexify his calculation. The complexified Feynman path integral connects the integral kernel for Schrödinger's equation and the propagator for the heat equation. The complex analyticity in the nonstandard world can be push down to the standard world. We will assume that the reader is familiar with Nonstandard Analysis.

0. Introduction

A heat equation and a Schrödinger equation are formally very similar. If we put $i\,t$ instead of t in the Schrödinger equation, then it becomes a heat equation. Physicians call it a complex time. When the initial function at $t = 0$ is the *delta*-function in the two equations, the solutions are called propagators. In order to calculate the propagators, we have two methods: using a stochastic Wiener measure for the heat equation and a Feynman path integral for the Schroedinger equation. One of the most fundamental examples is the case of a harmonic oscillator, whose potential is a constant multiple of x^2. The propagators are obtained from many kinds of calculations. Using Nonstandard Analysis, Ken Loo provided a rigorous computation for the harmonic oscillator Feynman path integral. Especially we remark the Loo's calculation that does not use the knowledge how the harmonic oscillator moves in the subject of the classical mechanics.

Our purpose is to connect the two propagators using the complexified calculation. We use the same kind of path integral with a complex parameter

t. We do not use Wiener measure for calculating the propagator of heat equation. The convergence is absolutely uniform for the heat equation's case, and it is conditional for the Schrödinger equation's case. We shall extend Loo's calculation to a complex time case. His calculation can be generalized and the propagator for the harmonic oscillator can be treat for a complex number t . Most of Loo's calculations are useful, but some parts are quite different. We mention these different parts from his calculation. We use nonstandard argument, but it is not essential for our calculation.

1. Preliminaries, notations (cf. [2,3])

In Feynman's formulation of quantum mechanics ([1]), the propagator of the one-dimensional harmonic oscillator is the following path integral:

$$K(q, q_0, t) = \lim_{n\to\infty} \int_{\mathbf{R}^n} \left(\frac{m}{2\pi i\hbar\epsilon}\right)^{(n+1)/2} \exp\left(\frac{i\epsilon}{\hbar}\sum_{j=1}^{n+1}\left(\frac{m}{2}\left(\frac{x_j - x_{j-1}}{\epsilon}\right)^2 - \frac{m}{2}\lambda^2 x_j^2\right)\right) dx_1 dx_2 \cdots dx_n,$$

where $x_0 = q_0$, $x_{n+1} = q$, $\epsilon = \frac{t}{n}$.

In nonstandard analysis, it is known that

$$\lim_{n\to\infty} a_n = a \quad \text{iff} \quad {}^*a_w \approx a$$

for any infinite natural number $\omega \in {}^*\mathbf{N} - \mathbf{N}$, where ${}^*a_n$ is the $*$-extension of $\{a_n\}_{n\in\mathbf{N}}$, and $\approx$ means that ${}^*a_w - a$ is infinitesimal, that is, the standard part of a_w is a, usually denoted by $st({}^*a_w) = a$. The path integral is written as

$$st \int_{{}^*\mathbf{R}^w} \left(\frac{m}{2\pi i\hbar\epsilon}\right)^{(\omega+1)/2} \exp\left(\frac{i\epsilon}{\hbar}\sum_{j=1}^{w+1}\left(\frac{m}{2}\left(\frac{x_j - x_{j-1}}{\epsilon}\right)^2 - \frac{m}{2}\lambda^2 x_j^2\right)\right) dx_1 dx_2 \cdots dx_w.$$

By extending t to a complex number, we complexify the path integral to the following:

$$st \int_{{}^*\mathbf{R}^w} \left(\frac{m}{2\pi i\hbar\epsilon}\right)^{(\omega+1)/2} \exp\left(\frac{i\epsilon}{\hbar}\sum_{j=1}^{w+1}\left(\frac{m}{2}\left(\frac{x_j - x_{j-1}}{\epsilon}\right)^2 - \frac{m}{2}\lambda^2 x_j^2\right)\right) dx_1 dx_2 \cdots dx_w,$$

where $\epsilon \in \mathbf{C}$.

Now let T_n be the $(n+2)\times(n+2)$ matrix as $(T_n)_{0,0}=1$, $(T_n)_{i,i}=2-\epsilon^2\lambda^2$, $1\le i\le n$, $(T_n)_{n+1,n+1}=1-\epsilon^2\lambda^2$, $(T_n)_{j,j+1}=(T_n)_{j+1,j}=-1$, $0\le j\le n$, $(T_n)_{i,j}=0$, (i,j): otherwise. Then we have

$$\frac{i\epsilon}{\hbar}\sum_{j=1}^{w+1}\left(\frac{m}{2}\left(\frac{x_j-x_{j-1}}{\epsilon}\right)^2-\frac{m}{2}\lambda^2x_j^2\right)=\frac{im}{2\hbar\epsilon}{}^txT_nx,$$

where $x={}^t(x_0,x_1,\cdots,x_{n+1})$ and we write A_n, B_n for the $n\times n$ matrices as $(A_n)_{i,j}=2\delta_{i,j}-\delta_{i,j+1}-\delta_{i+1,j}$, $(B_n)_{i,j}=\delta_{ij}$ and $A_n-\epsilon^2\lambda^2B_n$ as S_n.

2. Complexification

Since the matrix S_n is parametrized by t, the inversability of S_n depends on the parameter t.

Proposition 2.1. *Let $t\in\mathbf{C}$. For $t\neq\pm\frac{\sqrt{2}n}{\lambda}\sqrt{1-\cos\frac{k\pi}{n+1}}$, $k=1,2,\ldots,n$, S_n is invertible.*

Proof. Since

$$\begin{aligned}&A_n{}^t\left(\sin\frac{k\pi}{n+1},\sin\frac{2k\pi}{n+1},\cdots,\sin\frac{nk\pi}{n+1}\right)\\&=\left(2-2\cos\frac{k\pi}{n+1}\right){}^t\left(\sin\frac{k\pi}{n+1},\sin\frac{2k\pi}{n+1},\cdots,\sin\frac{nk\pi}{n+1}\right),\end{aligned}$$

$k=1,2,\ldots,n$, eigenvalues of the n by n symmetric matrix A_n are $2-2\cos\frac{k\pi}{n+1}$, $k=1,2,\ldots,n$. Hence, the n distinct eigenvalues of S_n are $2-2\cos\frac{k\pi}{n+1}-\lambda^2\left(\frac{t}{n}\right)^2$, $k=1,2,\cdots,n$. Therefore S_n is invertible if and only if t is not equal to $\pm\frac{\sqrt{2}n}{\lambda}\sqrt{1-\cos\frac{k\pi}{n+1}}$, $k=1,2,\ldots,n$. □

We transform Proposition 2.1 into ${}^*\mathbf{C}$. From now on, we put the bar when we represent $*$-transformed objects.

Theorem 2.2. *For any $\omega\in{}^*\mathbf{N}-\mathbf{N}$ and $t\in{}^*\mathbf{C}$, $\bar{S_\omega}$ is invertible if and only if $t\neq\pm\frac{\sqrt{2}\omega}{\lambda}\sqrt{1-\cos\frac{k\pi}{\omega+1}}$, $k=1,2,\ldots,\omega$, where the bar denotes an $*$-transform.*

Proof. $*$-transforming Proposition 2.1 and setting $n=\omega$, we have Theorem 2.2. □

Let $w(s)$ be an arbitrary path for $0\le s\le t$ with $w(0)=q_0$, $w(t)=q$. We separate a path into a classical one and a quantum fluctuation one:

$x_j = w(\frac{jt}{n+1}) + y_j = w_j + y_j$. We fix the boundaries $w(0) = x_0 = q_0$ and $w(t) = x_{n+1} = q$. Because of it, $y_0 = 0, y_{n+1} = 0$. Let T_n be the $(n+2)\times(n+2)$ symmetric matrix defined by $(T_n)_{0,0} = 1$, $(T_n)_{1,0} = (T_n)_{0,1} = (T_n)_{n+1,n} = (T_n)_{n,n+1} = -1$, $(T_n)_{n+1,n+1} = 1-\epsilon^2\lambda^2$, $(T_n)_{i,0} = (T_n)_{0,i} = 0$, $(2 \le n+1)$, $(T_n) = (T_n)_{n+1,j} = (T_n)_{j,n+1} = 0$, $(0 \le j \le n)$, $(T_n)_{k,l} = (S_n)_{k,l}$, $(1 \le k, l \le n)$. For y, w, we define n-dimensional vectors $\hat{y}$, $\hat{w}$, $\hat{\hat{w}}$ as $(\hat{y})_i = y_i$, $(\hat{w})_i = w_i$, $(1 \le i \le n)$, $(\hat{\hat{w}})_1 = w_0$, $(\hat{\hat{w}})_n = w_{n+1}$, $(\hat{\hat{w}})_j = 0$, $(2 \le j \le n-1)$. Let $\rho = S_n\hat{w} - \hat{\hat{w}}$.

Lemma 2.3. *Under the assumption on t, $t \in \mathbf{C}$, whose imaginary part is negative, or $t \in \mathbf{R}$, $t \neq \pm\frac{\sqrt{2}n}{\lambda}\sqrt{1-\cos\frac{k\pi}{n+1}}$, $k = 1, 2, \ldots, n$. Let $\epsilon = \frac{t}{n}$. Then*

$$\int_{\mathbf{R}^n}\Big(\frac{m}{2\pi i\hbar\epsilon}\Big)^{\frac{n+1}{2}}\exp\Big(\frac{i\epsilon}{\hbar}\sum_{j=1}^{n+1}\Big(\frac{m}{2}\Big(\frac{x_j-x_{j-1}}{\epsilon}\Big)^2 - \frac{m}{2}\lambda^2x_j^2\Big)\Big)dx_1dx_2\cdots dx_n$$

$$= \exp\Big(\frac{im}{\hbar\epsilon}({}^t wT_nw - {}^t\rho S_n^{-1}\rho)\Big)\Big(\frac{m}{2\pi i\hbar\epsilon}\Big)^{1/2}\sqrt{\frac{1}{\det S_n}}\,.$$

Proof. By Proposition 2.1, S_n is invertible under the assumption of t. Since S_n is symmetric, the following holds:

$${}^t\hat{y}S_n\hat{y} + 2\,{}^t\rho\hat{y} = {}^t(\hat{y} + S_n^{-1}\rho)S_n(\hat{y} + S_n^{-1}\rho) - {}^t\rho S_n^{-1}\rho.$$

We write $\hat{y}+S_n^{-1}\rho = z$, that is, $y_j + (S_n^{-1}\rho)_j = z_j$, $(1 \le j \le n)$. The integral in the left hand side of the equation is

$$\exp\Big(\frac{im}{\hbar\epsilon}({}^t wT_nw - {}^t\rho S_n^{-1}\rho)\Big)\int_{\mathbf{R}^n}\Big(\frac{m}{2\pi i\hbar\epsilon}\Big)^{\frac{n+1}{2}}\exp\Big(\frac{im}{2\hbar\epsilon}{}^t zS_nz\Big)dz_1dz_2\cdots dz_n.$$

The n distinct eigenvalues of $\frac{im}{2\hbar\epsilon}S_n$ are $\frac{mn}{2\hbar}\frac{i}{t}\big(2\big(1-\cos\frac{k\pi}{n+1}\big) - \frac{\lambda^2}{n^2}t^2\big)$, $k = 1, 2, \ldots, n$. When we write $t = t_1 + it_2$, $t_1, t_2 \in \mathbf{R}$, by a simple calculation, the real part of $\frac{mn}{2\hbar}\frac{i}{t}\big(2\big(1-\cos\frac{k\pi}{n+1}\big) - \frac{\lambda^2}{n^2}t^2\big)$ is $\frac{mn}{2\hbar|t|^2}\big(2\big(1-\cos\frac{k\pi}{n+1}\big) + \frac{\lambda^2}{n^2}|t|^2\big)t_2$ for each k. In this, the terms $\frac{mn}{2\hbar|t|^2}$, $1-\cos\frac{k\pi}{n+1}$ and $\frac{\lambda^2}{n^2}|t|^2$ are positive. Hence the real part of each eigenvalue of $\frac{im}{2\hbar\epsilon}S_n$ is negative if and only if the imaginary part $t_2$of t is negative. When the imaginary part of t is zero, all eigenvalues of $\frac{im}{2\hbar\epsilon}S_n$ are pure imaginary ,and nonzero under the assumption of t. We have the statement after diagonalizing of S_n and doing the integrals. □

We transform Lemma 2.3 into ${}^*\mathbf{C}$.

Corollary 2.4. *Under the previous definition of $w(s)$ and the assumption on t, $t \in \mathbf{C}$, whose imaginary part is negative, or $t \in \mathbf{R}$, $t \neq \pm\frac{\sqrt{2}\omega}{\lambda}\sqrt{1-\cos\frac{k\pi}{\omega+1}}$, $k = 1, 2, \ldots, \omega$, the one-dimensional harmonic oscillator internal functional integral is well-defined. It is equal to* $\exp\left(\frac{im}{\hbar\epsilon}\left({}^t\bar{w}\bar{T}_\omega\bar{w} - {}^t\bar{\rho}\bar{S}_\omega^{-1}\bar{\rho}\right)\right)\left(\frac{m}{2\pi i\hbar\omega}\right)^{1/2}\sqrt{\frac{1}{\det \bar{S}_\omega}}$.

Proof. Lemma 2.3 directly implies the statement through the $*$-transformation. □

We recall that ϵ is $\frac{t}{n}$. We denote $A_{j,n}$ and $C_{j,n}$ with $0 < j \leq n$ to be the j by j matrices defined as $(A_{j,n})_{i,i} = 2$, $(A_{j,n})_{i+1,i} = (A_{j,n})_{i,i+1} = -1$, $1 \leq i \leq j-1$, $(A_{j,n})_{j,j} = 1$, $(A_{j,n})_{i,k} = 0$, (i,k): otherwise and $(C_{j,n})_{i,i} = 1$, $1 \leq i \leq j-1$, $(C_{j,n})_{i,k} = 0$, (i,k): otherwise. We define $D_{j,n} = \det\left(A_{j,n} - \epsilon^2\lambda^2 C_{j,n}\right)$.

We shall calculate the standard part of the obtained internal functional integral in Corollary 2.4.

Proposition 2.5. *After $*$-transforming of $D_{j,n}$, we have that for $\omega \in {}^*\mathbf{N} - \mathbf{N}$, $\bar{D}_{k,\omega} \approx {}^*\cos\frac{kt\lambda}{\omega} = {}^*\cos\frac{k\epsilon}{\lambda}$.*

Proof. By the definition $D_{j,n}$ satisfies

$$D_{j,n} = (2 - \epsilon^2\lambda^2)D_{j-1,n} - D_{j-2,n}, \quad 2 \leq j \leq n.$$

We denote $a_\pm$ by the solutions of $a^2 - (2-\epsilon^2\lambda^2)a + 1$, that is ,

$$a_\pm = \frac{2 - \epsilon^2\lambda^2 \pm \epsilon\lambda\sqrt{4-\epsilon^2\lambda^2}}{2} \approx 1 \pm i\epsilon\lambda \approx e^{\pm i\epsilon\lambda}$$

for $0(\epsilon^2)$. For some constants $A_\pm$, $D_{j,n} = A_+(a_+)^{j-1} + A_-(a_-)^{j-1}$. The initial conditions imply that $D_{1,n} = 1 = A_+ + A_-$, $D_{2,n} = 1 - \epsilon^2\lambda^2 A_+(a_+) + A_-(a_-)$. Hence $A_\pm = \frac{1}{2} \pm i\frac{\epsilon\lambda}{2\sqrt{4-\epsilon^2\lambda^2}} \approx \frac{1}{2}$ for $0(\epsilon^2)$. Let n be ω and let k be in ${}^*\mathbf{N}$. Then $\epsilon = \frac{t}{\omega}$ is infinitesimal. Hence $D_{k,\omega} = A_+(a_+)^{k-1} + A_-(a_-)^{k-1} \approx \frac{1}{2}(e^{i(k-1)\epsilon\lambda} + e^{-i(k-1)\epsilon\lambda}) \approx \frac{1}{2}(e^{ik\epsilon\lambda} + e^{-ik\epsilon\lambda}) = \cos(k\epsilon\lambda)$. □

For $\bar{S}_\omega$, we obtain the following:

Theorem 2.6. *$\epsilon \det \bar{S}_\omega$ is infinitesimally close to $\frac{\sin(\lambda t)}{\lambda}$.*

Proof. Let $S_{j,n} = \det(A_{j,n} - \epsilon^2\lambda^2 B_{j,n})$, where $1 \leq j \leq n$. Then expanding $S_{j,n}$ on the bottom row, we have $S_{j,n} = D_{j,n} + (1-\epsilon^2\lambda^2)S_{j-1,n}$. Hence $S_{n,n} - S_{1,n} = \sum_{j=2}^n D_{j,n} - \epsilon^2\lambda^2\sum_{j=2}^n S_{j-1,n}$, and $\epsilon S_{n,n} = \epsilon S_{1,n} + \epsilon\sum_{j=2}^n D_{j,n} -$

$\epsilon^3\lambda^2\sum_{j-1}^{n} S_{j-1,n}$. We transform n to ω, then $D_{m,\omega} = \cos(m\epsilon\lambda) + h_m$, where h_m is infinitesimal, $1 \leq m \leq \omega$. We remark that ϵ is not only real but complex.

$$\epsilon S_{\omega,\omega} = \epsilon S_{1,\omega} + \sum_{m=1}^{\omega} \epsilon \cos\frac{mt\lambda}{\omega} - \epsilon\cos(\epsilon\lambda) + \epsilon\sum_{m=2}^{\omega} h_m - \epsilon^2\lambda^2\sum_{m=2}^{\omega} \epsilon S_{m-1,\omega}.$$

Now $\epsilon S_{1,n}$ is $\epsilon(1-\epsilon^2\lambda^2)$, which is infinitesimal, and $|\cos(m\epsilon\lambda)| < e^{-t_2\lambda}$, where t_2 is the imaginary part of t, and $|\epsilon\sum_{m=2}^{\omega} h_m| < |\epsilon|\omega\max_{1\leq m\leq\omega}|h_m| = |t|\max_{1\leq m\leq\omega}|h_m| \approx 0$. We have $|D_{k,\omega}| \approx |\cos((k-1)\theta)| = |\cos(k\lambda\epsilon)| < e^{-t_2\lambda}$, and $|\epsilon^2\lambda^2\sum_{m=2}^{\omega}\epsilon S_{m-1,\omega}| \leq \epsilon^3\lambda^2|S_{1,\omega}| + \epsilon^2\lambda^2(\sum_{m=2}^{\omega}|t|e^{-t_2\lambda} + |t||S_{1,\omega}|) \approx 0$. Furthermore, since $\cos(\lambda z)$ is holomorphic, $\sum_{m=1}^{\omega}\epsilon\cos\frac{mt\lambda}{\omega} \approx \int_0^t \cos(\lambda z)dz = \sin(\lambda t)$. □

We calculate the exponential part in Corollary 2.4.

Proposition 2.7. *The exponential in Corollary 2.4 satisfies the following:*

$$\exp\Big(\frac{im}{\hbar\epsilon}({}^t\bar{w}\bar{T}_\omega\bar{w} - t\bar{\rho}\bar{S}_\omega^{-1}\bar{\rho})\Big) \approx \exp\Big(\frac{im}{\hbar}\frac{\lambda}{\sin\lambda t}((q_0^2+q^2)\cos\lambda t - 2qq_0)\Big),$$

where $q_0 = x_0$ *and* $q = x_{\omega+1}$.

Proof. $\frac{1}{\epsilon}({}^t w T_n w - {}^t\rho S_n^{-1}\rho) = \frac{1}{\epsilon\det(S_n)}\big(q_0^2(S_{n,n} - S_{n-1,n}) + \rho^2(S_{n,n} - S_{n-1,n}) - 2qq_0\big) - \epsilon\lambda^2q^2$. When we transform n to ω, and use Theorem 2.6, we obtain $\frac{1}{\hbar\epsilon}({}^t\bar{w}\bar{T}_\omega\bar{w} - {}^t\bar{\rho}\bar{S}_\omega^{-1}\bar{\rho}) = \frac{1}{\epsilon\det(S_\omega)}\big(q_0^2(S_{\omega,\omega} - S_{\omega-1,\omega}) + \rho^2(S_{\omega,\omega} - S_{\omega-1,\omega}) - 2qq_0\big) - \epsilon\lambda^2q^2 \approx \frac{\lambda}{\sin\lambda t}\big((q_0^2+q^2)\cos\lambda t - 2qq_0\big)$. □

We write the standard part of a as $st(a)$. Proposition 2.7 implies the following:

Theorem 2.8. *Let* $t \in \mathbf{R}$, $t \neq st\Big(\pm\frac{\sqrt{2}n}{\lambda}\sqrt{1-\cos\frac{k\pi}{\omega+1}}\Big)$, $k = 1,2,\ldots,\omega$, *or* $t \in \mathbf{C}$, *whose imaginary part is negative. The complexified one-dimensional harmonic oscillator standard functional integral is given by* $\left(\frac{m}{2\pi i\hbar}\right)^{1/2}\sqrt{\frac{\lambda}{\sin(\lambda t)}}\exp\big(\frac{im}{\hbar}\frac{\lambda}{\sin\lambda t}((q_0^2+q^2)\cos\lambda t - 2qq_0)\big)$.

Proof. If $t \neq st\Big(\pm\frac{\sqrt{2}n}{\lambda}\sqrt{1-\cos(\frac{k\pi}{\omega+1})}\Big)$, $k = 1,2,\ldots,\omega$, then $t \neq \pm\frac{\sqrt{2}n}{\lambda}\sqrt{1-\cos(\frac{k\pi}{\omega+1})}$, $k = 1,2,\ldots,\omega$, for arbitrary infinite number ω. The theorem is followed from Corollary 2.4, Theorem 2.6 and Proposition 2.7. □

For the d-dimensional harmonic oscillator, we write $\mathbf{q}_0$, $\mathbf{q}$ are d-dimensional vectors and $|\mathbf{q}_0|^2$, $|\mathbf{q}|^2$ are the square norms and $\mathbf{q}_0\mathbf{q}$ is the inner product of $\mathbf{q}_0$ and $\mathbf{q}$. We have:

Corollary 2.9. *For the complexified d-dimensional harmonic oscillator standard functional integral, we have* $\left(\frac{m}{2\pi i\hbar}\right)^{\frac{d}{2}}\left(\frac{\lambda}{\sin(\lambda t)}\right)^{\frac{d}{2}}\exp\left(\frac{im}{\hbar}\frac{\lambda}{\sin\lambda t}((|\mathbf{q}_0|^2 + |\mathbf{q}|^2)\cos\lambda t - 2\mathbf{q}\mathbf{q}_0)\right)$.

Proof. Factorizing Theorem 2.8 into a product on d dimensional, we have the corollary. □

We shall calculate the trace of the propagator obtained in Theorem 2.8. In order to do it, we prepare the following lemma.

Lemma 2.10. *Let α_n's be a convergent sequence of complex numbers whose limit is α, satisfying the following condition* (i) *or* (ii).

(i) *The real parts of both α_n's and α are negative.*
(ii) *α_n's and α are non-zero pure imaginary.*

Then

$$\lim_{n\to\infty}\int_{-\infty}^{\infty}\exp(\alpha_n q^2)dq = \int_{-\infty}^{\infty}\lim_{n\to\infty}\exp(\alpha_n q^2)dq \; \left(= \int_{-\infty}^{\infty}\exp(\alpha q^2)dq\right).$$

Proof. We denote the real parts of α_n, α by a_n, a. For positive real numbers M, N, the absolute value $\left|\int_{-\infty}^{-N}\exp(\alpha_n q^2) + \int_M^{\infty}\exp(\alpha_n q^2)\right|$ is less than the sum of $\int_{-\infty}^{-N}\exp(a_n q^2) + \int_M^{\infty}\exp(a_n q^2)$. Hence the sequence of the improper integrals $\int_{-\infty}^{\infty}\exp(\alpha_n q^2)dq$ is absolutely and uniformly convergent for α_n to A. Hence $\lim_{n\to\infty}\int_{-\infty}^{\infty}\exp(\alpha_n q^2)dq = \int_{-\infty}^{\infty}\lim_{n\to\infty}\exp(\alpha_n q^2)dq = \int_{-\infty}^{\infty}\exp(\alpha q^2)dq$. □

We obtain the following:

Theorem 2.11. *Let $t \in \mathbf{R}$, $t \neq \pm st\left(\frac{\sqrt{2}\omega}{\lambda}\sqrt{1-\cos\frac{k\pi}{\omega+1}}\right)$, $k = 1, 2, \ldots, \omega$, or $t \in \mathbf{C}$, whose imaginary part is negative. The trace of the complexified one-dimensional harmonic oscillator standard functional integral is given by* $\frac{1}{2i\sin(\lambda t/2)}$.

Proof. We put $q_0 = q$ in $\left(\frac{m}{2\pi i\hbar}\right)^{\frac{1}{2}}\sqrt{\frac{\lambda}{\sin(\lambda t)}}\exp\left(\frac{im}{\hbar}\frac{\lambda}{\sin\lambda t}((q_0^2 + q^2)\cos\lambda t - 2qq_0)\right)$. The trace is the following integral:

$$\int_{-\infty}^{\infty}\left(\frac{m}{2\pi i\hbar}\right)^{\frac{1}{2}}\sqrt{\frac{\lambda}{\sin(\lambda t)}}\exp\left(\frac{im}{\hbar}\frac{\lambda}{\sin\lambda t}2((\cos\lambda t - 1)q^2)\right)dq$$

$$= \frac{1}{2i\sin(\lambda t/2)}. \qquad \square$$

For the d-dimensional harmonic oscillator, we have:

Corollary 2.12. *For the trace of the complexified d-dimensional harmonic oscillator standard functional integral, we have* $\left(\frac{1}{2i\sin(\lambda t/2)}\right)^d$.

Proof. Factorizing Theorem 2.11 into a product on d dimensional, we have the corollary. $\square$

Further problems

In [5] and [6] we construct a theory for path integral for fields. It is a further problem whether the calculation of the d-dimensional harmonic oscillator can be extended to the case of harmonic oscillator for field or not, that is, d can be an infinite number ∞ or not.

Acknowledgement

I would like to thank S. Murai and Y. Imanishi for their beautiful typing and encouragement.

References

1. R.P. Feynman and A.R. Hibbs, *Quantum mechanics and path integrals*, McGrow-Hill Inc. All rights (1965).
2. K. Loo, *A rigorous real time Feynmann path integral*, J. Math.Phys. **40** (1999), 64–70.
3. ______, *Nonstandard Feynmann path integral for the harmonic oscillator*, J. Math.Phys. **40** (1999), 5511–5521.
4. T. Nitta and T. Okada, *Double infinitesimal Fourier transformation for the space of functionals and reformulation of Feynman path integral*, Lecture Note Series in Mathematics, Osaka University Vol.**7** (2002), 255–298 in Japanese.
5. ______, *Infinitesimal Fourier transformation for the space of functionals*, in Topics in almost Hermitian geometry and related fields, World Scientific (2005).
6. ______, *Poisson summation formula for the space of functionals*, Nihonkai Math. J. **16** (2005), 1–21.
7. T. Nitta, T. Okada and A. Tzouvaras, *Classification of non-well-founded sets and an application*, Math. Log. Quart. **49** (2003), 187–200.
8. M. Saito, *Ultraproduct and non-standard analysis*, in Japanese, Tokyo tosho, 1976.

SOME CRITICAL ALMOST KÄHLER STRUCTURES WITH A FIXED KÄHLER CLASS

T. OGURO

Department of Mathematical Sciences,
School of Science and Engineering, Tokyo Denki University,
Saitama, 350-0394, Japan,
E-mail: oguro@r.dendai.ac.jp

K. SEKIGAWA

Department of Mathematics,
Faculty of Science, Niigata University,
Niigata, 950-2181, Japan
E-mail: sekigawa@math.sc.niigata-u.ac.jp

A. YAMADA

Division of General Education,
Nagaoka National College of Technology,
Nagaoka, Niigata 940-8532, Japan
E-mail: ayamada@nagaoka-ct.ac.jp

In this paper, we discuss critical points of the functional $\mathscr{F}_{\lambda,\mu}(J,g) = \int_M(\lambda\tau + \mu\tau^*)dM_g$ (τ is the scalar curvature, τ^* is the $*$-scalar curvature and $(\lambda,\mu) \in \mathbb{R}^2\setminus(0,0)$) on the space $\mathcal{AK}(M,[\Omega])$ of all almost Kähler structures (J,g) whose Kähler classes coincide with $[\Omega]$ on a compact symplectic manifold (M,Ω), and show that (J,g) is a critical point of $\mathscr{F}_{\lambda,\mu}$ if and only if $(\mu-\lambda)\rho$ is J-invariant, where ρ are the Ricci tensor. This result itself can be obtained by the result of Koda and the Moser's stability theorem. In the present article, we shall give another direct proof for the same result without using Moser's theorem taking the variational problems in more general situation into consideration.

Keywords: Almost Kähler manifold.

1. Introduction

Let M be a compact orientable manifold of dimension m. We denote by $\mathcal{M}(M)$ the set of all Riemannian metrics on M and $\mathcal{R}(M)$ the set of all

2000 Mathematics Subject Classification. 53C15, 53C55.

Riemannian metrics of the same volume element. It is well-known that a Riemannian metric $g \in \mathcal{R}(M)$ is a critical point of the functional $\mathscr{A}$ on $\mathcal{R}(M)$ defined by

$$\mathscr{A}(g) = \int_M \tau \, dM_g \tag{1}$$

if and only if g is an Einstein metric, where τ is the scalar curvature of g and dM_g is the volume form of g.

Now, let M be a compact manifold of dimension $m = 2n$ admitting an almost complex structure. We denote by $\mathcal{AH}(M)$ the set of all almost Hermitian structures and $\mathcal{AH}(M, \Omega)$ the set of all almost Hermitian structures with the same Kähler form Ω. An almost Hermitian manifold $M = (M, J, g)$ with the closed Kähler form Ω ($d\Omega = 0$) is called an almost Kähler manifold. Let $M = (M, J, g)$ be a compact almost Kähler manifold and Ω the corresponding Kähler form. Then, we may note that any almost Hermitian structure $(J, g) \in \mathcal{AH}(M, \Omega)$ is an almost Kähler structure on M. In this case, we denote $\mathcal{AH}(M, \Omega)$ by $\mathcal{AK}(M, \Omega)$. In [2], Blair and Ianus studied critical points of the functional $\mathscr{F}$ on $\mathcal{AK}(M, \Omega)$ defined by

$$\mathscr{F}(J, g) = \int_M (\tau^* - \tau) \, dM_g, \tag{2}$$

where τ^* is the $*$-scalar curvature of (M, J, g). They proved that (J, g) is a critical point of $\mathscr{F}$ on $\mathcal{AK}(M, \Omega)$ if and only if the Ricci tensor ρ is J-invariant.

We denote by $\mathcal{AK}(M, [\Omega])$ the set of all almost Kähler structures on M with the same Kähler class $[\Omega]$ in the de Rham cohomology group. It is well-known that $\mathcal{AK}(M, \Omega)$ is a contractible Fréchet space. However, the space $\mathcal{AK}(M, [\Omega])$ might be disconnected in general.

In [5,6], Koda studied critical points of the functional $\mathscr{F}_{\lambda,\mu}$ on $\mathcal{AH}(M, \Omega)$ and $\mathcal{AK}(M, [\Omega])$ defined by

$$\mathscr{F}_{\lambda,\mu}(J, g) = \int_M (\lambda\tau + \mu\tau^*) \, dM_g, \qquad (\lambda, \mu) \in \mathbb{R}^2 \backslash (0, 0). \tag{3}$$

and gave a necessary condition for $(J, g) \in \mathcal{AK}(M, [\Omega])$ to be a critical point of $\mathscr{F}_{\lambda,\mu}$. The purpose of the present paper is to improve his result. Namely, we will give a necessary and sufficient condition for $(J, g) \in \mathcal{AK}(M, [\Omega])$ to be a critical point of $\mathscr{F}_{\lambda,\mu}$.

The diffeomorphism group $\mathrm{Diff}\,(M)$ of a compact symplectic manifold (M, Ω) acts on the space $\mathcal{AK}(M, [\Omega])$ in the natural way. Let $\mathscr{F}$ be a functional on the space $\mathcal{AK}(M, [\Omega])$ which is invariant under the action of the group $\mathrm{Diff}\,(M)$. Then, by applying the Moser's stability theorem, we may easily show that (J,g) is a critical point of the functional $\mathscr{F}$ on the space $\mathcal{AK}(M,[\Omega])$ if and only if (J,g) is a critical one of $\mathscr{F}$ along the space $\mathcal{AK}(M,\Omega)$ corresponding to the Kähler form Ω of (J,g). Now, we may easily check that the functional $\mathscr{F}_{\lambda,\mu}$ is Diff(M)-invariant. Applying the above observation to the functional $\mathscr{F}_{\lambda,\mu}$ on the space $\mathcal{AK}(M,[\Omega])$ and taking account of the result of Koda ([5]), we may easily show that $(J,g) \in \mathcal{AK}(M,[\Omega])$ is a critical point of the functional $\mathscr{F}_{\lambda,\mu}$ if and only if $(\mu-\lambda)\rho$ is J-invariant, where ρ is the Ricci tensor of (M,J,g) (Theorem 3.1).

In the present article, we shall give another direct proof for the same result without using the Moser's stability theorem.

2. Preliminaries

Let $M=(M,J,g)$ be a $2n$-dimensional compact almost Kähler manifold with almost Hermitian structure (J,g) and Ω the Kähler form of M defined by $\Omega(X,Y)=g(X,JY)$ for $X,Y \in \mathfrak{X}(M)$. We assume that M is oriented by the volume form $dM_g=((-1)^n/n!)\Omega^n$. We denote by ∇, R, ρ and τ the Riemannian connection, the curvature tensor, the Ricci tensor and the scalar curvature of M, respectively. The curvature tensor R is defined by $R(X,Y)Z=[\nabla_X,\nabla_Y]Z-\nabla_{[X,Y]}Z$ for $X,Y,Z \in \mathfrak{X}(M)$. A tensor field ρ^* on M of type $(0,2)$ defined by

$$\rho^*(x,y)=\operatorname{trace}(z\mapsto R(x,Jz)Jy)=\frac{1}{2}\operatorname{trace}(z\mapsto R(x,Jy)Jz) \tag{4}$$

is called the Ricci $*$-tensor, where $x,y,z \in T_p(M)$ (the tangent space of M at $p\in M$). We denote by τ^* the $*$-scalar curvature of M which is the trace of the linear endomorphism Q^* defined by $g(Q^*x,y)=\rho^*(x,y)$. We may remark that ρ^* satisfies $\rho^*(JX,JY)=\rho^*(Y,X)$ for any $X,Y\in\mathfrak{X}(M)$. Thus ρ^* is symmetric if and only if ρ^* is J-invariant.

In this paper, for any orthonormal bases (resp. any local orthonormal frame field) $\{e_i\}_{i=1,\dots,2n}$ at any point $p\in M$ (resp. on a neighborhood of p), we shall adopt the following notational convention:

$$\begin{aligned} R_{ijk\ell}=g(R(e_i,e_j)e_k,e_\ell),\dots,R_{\bar{i}\bar{j}\bar{k}\bar{\ell}}&=g(R(Je_i,Je_j)Je_k,Je_\ell),\\ \rho_{ij}=\rho(e_i,e_j),\dots,\rho_{\bar{i}\bar{j}}&=\rho(Je_i,Je_j),\\ \rho^*_{ij}=\rho^*(e_i,e_j),\dots,\rho^*_{\bar{i}\bar{j}}&=\rho^*(Je_i,Je_j),\\ J_{ij}=g(Je_i,e_j),\quad \nabla_iJ_{jk}&=g((\nabla_{e_i}J)e_j,e_k), \end{aligned} \tag{5}$$

and so on, where the Latin indices run over the range $1,\dots,2n$. Then, we have $J_{ij}=-J_{ji}$, $\nabla_iJ_{jk}=-\nabla_iJ_{kj}$ and $\nabla_iJ_{\bar{j}\bar{k}}=-\nabla_iJ_{jk}$. The condition

$d\Omega = 0$ is equivalent to $\mathfrak{S}_{i,j,k} \nabla_i J_{jk} = \nabla_i J_{jk} + \nabla_j J_{ki} + \nabla_k J_{ij} = 0$. Further, since M is a quasi-Kähler manifold and a semi-Kähler manifold, we have $\nabla_i J_{jk} + \nabla_{\bar{i}} J_{\bar{j}k} = 0$ and $\sum_a \nabla_a J_{ai} = 0$. The following curvature identity due to Gray ([4]) is known:

$$2\sum_a (\nabla_a J_{ij}) \nabla_a J_{kl} = R_{ijkl} - R_{ij\bar{k}\bar{\ell}} - R_{\bar{i}\bar{j}kl} + R_{\bar{i}\bar{j}\bar{k}\bar{\ell}} \qquad (6)$$

$$+ R_{\bar{i}j\bar{k}l} + R_{\bar{i}jk\bar{\ell}} + R_{i\bar{j}\bar{k}l} + R_{i\bar{j}k\bar{\ell}}.$$

From this equality, we have

$$\rho^*_{ij} + \rho^*_{ji} - \rho_{ij} - \rho_{\bar{i}\bar{j}} = \sum_{a,b} (\nabla_a J_{ib}) \nabla_a J_{jb}, \qquad (7)$$

and further $\|\nabla J\|^2 = 2(\tau^* - \tau)$. Therefore, M is integrable if and only if $\tau^* = \tau$.

3. Critical point of $\mathscr{F}_{\lambda,\mu}$

Let $M = (M, J, g)$ be a $2n$-dimensional compact almost Kähler manifold. We denote by Ω the Kähler form of M and by $\mathcal{AK}(M, [\Omega])$ the set of all almost Kähler structures on M with the same Kähler class $[\Omega]$. Let $(J(t), g(t))$ be a curve in $\mathcal{AK}(M, [\Omega])$ such that $(J(0), g(0)) = (J, g)$. Since the Kähler form $\Omega(t)$ corresponding to $(J(t), g(t))$ satisfies $[\Omega(t)] = [\Omega]$, there exists a 1-parameter family of 1-forms $\alpha(t)$ satisfying $\Omega(t) = \Omega + d\alpha(t)$. We denote by $\nabla^{(t)}$, $R(t)$, $\rho(t)$, $\rho^*(t)$, $\tau(t)$ and $\tau^*(t)$ the Riemannian connection, the curvature tensor, the Ricci tensor, the Ricci $*$-tensor, the scalar curvature and the $*$-scalar curvature of $(M, J(t), g(t))$, respectively. Let $(U; x_1, \ldots, x_{2n})$ be a local coordinate system of a coordinate neighborhood U of M. With respect to the natural frame $\{\partial_i = \partial/\partial x_i\}_{i=1,\ldots,2n}$, we put

$$g(t)(\partial_i, \partial_j) = g(t)_{ij}, \qquad J(t)(\partial_i) = J(t)_i{}^j \partial_j,$$
$$(\nabla^{(t)}_{\partial_i} J(t))\partial_j = (\nabla^{(t)}_i J(t)_j{}^k)\partial_k, \qquad R(t)(\partial_i, \partial_j)\partial_k = R(t)_{ijk}{}^\ell \partial_\ell,$$
$$\rho(t)(\partial_i, \partial_j) = \rho(t)_{ij}, \qquad \rho^*(t)(\partial_i, \partial_j) = \rho^*(t)_{ij},$$

and let $(g(t)^{ij}) = (g(t)_{ij})^{-1}$. In particular, $g(0)_{ij} = g_{ij}$, $J(0)_i{}^j = J_i{}^j$, $\nabla^{(0)}_i J(0)_j{}^k = \nabla_i J_j{}^k$, $R(0)_{ijk}{}^\ell = R_{ijk}{}^\ell$, $\rho(0)_{ij} = \rho_{ij}$, $\rho^*(0)_{ij} = \rho^*_{ij}$.

Now, put

$$\left.\frac{d}{dt}\right|_{t=0} g(t)_{ij} = h_{ij}, \quad \left.\frac{d}{dt}\right|_{t=0} J(t)_i{}^j = K_i{}^j, \quad \left.\frac{d}{dt}\right|_{t=0} \alpha(t)_i = A_i. \qquad (8)$$

Then, $h = (h_{ij})$ is a symmetric $(0,2)$-tensor on M and

$$\left.\frac{d}{dt}\right|_{t=0} g(t)^{ij} = -h^{ij}, \tag{9}$$

where we may use standard notational convention for tensor analysis: thus h^{ij} means $h^{ij} = g^{ia}g^{jb}h_{ab}$. We denote by $dM_{g(t)}$ the volume form of M with respect to $g(t)$. Then, we have

$$\left.\frac{d}{dt}\right|_{t=0} dM_{g(t)} = \frac{1}{2}(g^{ij}h_{ij})\, dM_g. \tag{10}$$

From (9), we see that the connection coefficients $\Gamma(t)_{ij}^k$ of $\nabla^{(t)}$ satisfy

$$\left.\frac{d}{dt}\right|_{t=0} \Gamma(t)_{ij}^k = \frac{1}{2}g^{ka}(\nabla_i h_{aj} + \nabla_j h_{ia} - \nabla_a h_{ij}).$$

Therefore, the differential of $R(t)_{ijk}{}^{\ell}$ and $\rho(t)_{ij}$ at $t = 0$ are given respectively by the following ([9]):

$$\begin{aligned}\left.\frac{d}{dt}\right|_{t=0} R(t)_{ijk}{}^{\ell} &= \frac{1}{2}(-R_{ijk}{}^{a}h_a{}^{\ell} + R_{ija}{}^{\ell}h_k{}^{a} \\ &\quad + \nabla_i\nabla_k h_j{}^{\ell} - \nabla_j\nabla_k h_i{}^{\ell} - \nabla_i\nabla^{\ell}h_{jk} + \nabla_j\nabla^{\ell}h_{ik}),\end{aligned} \tag{11}$$

$$\begin{aligned}\left.\frac{d}{dt}\right|_{t=0} \rho(t)_{ij} &= \frac{1}{2}(-R_{aij}{}^{b}h_b{}^{a} + \rho_{ia}h_j{}^{a} \\ &\quad + \nabla_a\nabla_j h_i{}^{a} - \nabla_i\nabla_j h_a{}^{a} - \nabla^a\nabla_a h_{ij} + \nabla_i\nabla_a h_j{}^{a}),\end{aligned} \tag{12}$$

$$\left.\frac{d}{dt}\right|_{t=0} \tau(t) = -\rho_{ij}h^{ij} + \nabla^i\nabla^j h_{ij} - \nabla^i\nabla_i h, \tag{13}$$

where $h = h_a{}^a$. Further, since $(J(t), g(t)) \in \mathcal{AK}(M, [\Omega])$, we have

$$K_a{}^iJ_j{}^a + J_a{}^iK_j{}^a = 0, \tag{14}$$

$$h_{ij} = h_{ab}J_i{}^aJ_j{}^b + K_{ia}J_j{}^a + J_{ia}K_j{}^a, \tag{15}$$

$$K_j{}^i = -h_a{}^iJ_j{}^a - A_j{}^i, \tag{16}$$

$$h_{ij} = -h_{ab}J_i{}^aJ_j{}^b + J_i{}^aA_{aj} + J_j{}^aA_{ai}, \tag{17}$$

where $A_{ij} = dA(\partial_i, \partial_j) = \nabla_i A_j - \nabla_j A_i$.

Conversely, let (h, A) be the pair of a symmetric $(0,2)$-tensor h and a 1-form A satisfying (17) and define a $(1,1)$-tensor K by (16), then the equalities (14) and (15) hold ([6]).

From (16) and (17), we have

$$K_j{}^i = h_j{}^aJ_a{}^i - A_b{}^aJ_a{}^iJ_j{}^b. \tag{18}$$

From (16), (17) and (18), we have

$$\left.\frac{d}{dt}\right|_{t=0} J(t)^{ij} = -h^{ia}J_a{}^j + g^{ia}K_a{}^j = A_{ab}J^{ia}J^{jb}, \tag{19}$$

$$\begin{aligned}\left.\frac{d}{dt}\right|_{t=0} \rho^*(t)_{ij} &= \rho^*_{ia}h_j{}^a - \frac{1}{2}R_{iua}{}^b J_j{}^u J^{ac}h_{bc} \\ &\quad - \frac{1}{2}J^{ab}J_j{}^c\nabla_i\nabla_a h_{bc} + \frac{1}{2}J^{ab}J_j{}^c\nabla_c\nabla_a h_{bi} \\ &\quad + \frac{1}{2}\left(2J_j{}^q\rho^{*p}_i - J_j{}^u J^{pa}J^{qb}R_{iuab}\right)A_{pq},\end{aligned} \tag{20}$$

$$\left.\frac{d}{dt}\right|_{t=0} \tau^*(t) = \rho^*_{ab}h^{ab} - J^{ia}J^{jb}\nabla_a\nabla_b h_{ij} - 2J^{ip}\rho^*_{iq}A_p{}^q, \tag{21}$$

$$\begin{aligned}\left.\frac{d}{dt}\right|_{t=0} \nabla_i^{(t)}J(t)_j{}^k &= -h_a{}^k\nabla_i J_j{}^a + \frac{1}{2}J_j{}^a(\nabla_a h_i{}^k - \nabla_i h_a{}^k - \nabla^k h_{ia}) \\ &\quad - \frac{1}{2}J_a{}^k(\nabla_i h_j{}^a + \nabla_j h_i{}^a - \nabla^a h_{ij}) - \nabla_i A_j{}^k.\end{aligned} \tag{22}$$

We are ready to compute the first variation of (2). We shall use the notational convention (5) with respect to a (local) orthonormal frame field $\{e_i\}_{i=1,\ldots,2n}$. From (10), (13) and (21), we have

$$\begin{aligned}&\left.\frac{d}{dt}\right|_{t=0} \mathscr{F}_{\lambda,\mu}(J(t),g(t)) \\ &= \int_M \sum_{i,j}\Big(-\lambda\rho_{ij} + \mu\rho^*_{ij} - \mu\sum_{a,b}\nabla_a\nabla_b(J_{ia}J_{jb}) + \frac{1}{2}(\lambda\tau + \mu\tau^*)\delta_{ij}\Big)h_{ij}\,dM_g \\ &\quad + 2\mu\int_M\sum_{i,j}\rho^*_{\bar{i}j}A_{ij}\,dM_g\end{aligned} \tag{23}$$

Here, we get

$$\sum_{a,b}\nabla_a\nabla_b(J_{ia}J_{jb}) = \sum_{a,b}(\nabla_a\nabla_b J_{ia})J_{jb} + \sum_{a,b}(\nabla_b J_{ia})\nabla_a J_{jb}. \tag{24}$$

Moreover, from (7),

$$\sum_{a,b}(\nabla_a\nabla_b J_{ia})J_{jb} = -\sum_{a,b,s}(R_{abis}J_{sa} + R_{abas}J_{is})J_{jb} = -\rho^*_{ij} + \rho_{\bar{i}\bar{j}},$$

$$\sum_{a,b}(\nabla_b J_{ia})\nabla_a J_{jb} = -\frac{1}{2}\sum_{a,b}(\nabla_i J_{ab})\nabla_j J_{ab} + \rho^*_{ij} + \rho^*_{ji} - \rho_{ij} - \rho_{\bar{i}\bar{j}}.$$

Therefore, from (23), we obtain

$$\left.\frac{d}{dt}\right|_{t=0} \mathscr{F}_{\lambda,\mu}(J(t),g(t)) = \int_M\sum_{i,j}(T_{ij}h_{ij} + s_{ij}A_{ij})dM_g, \tag{25}$$

$$T_{ij} = (\mu - \lambda)\rho_{ij} + \frac{1}{2}(\lambda\tau + \mu\tau^*)\delta_{ij} + \frac{\mu}{2}\sum_{a,b}(\nabla_i J_{ab})\nabla_j J_{ab}, \tag{26}$$

$$s_{ij} = 2\mu\,\rho^{*}_{\bar{i}j}. \tag{27}$$

We note that T_{ij} and s_{ij} define a symmetric and a skew-symmetric (0,2)-tensor fields T and s on M, respectively. Since $A_{ij} = \nabla_i A_j - \nabla_j A_i$, we have

$$\int_M \sum_{i,j} s_{ij}A_{ij}\,dM_g = -2\int_M \sum_i s_i A_i\,dM_g, \quad s_i = \sum_a \nabla_a s_{ai}. \tag{28}$$

Summing up above argument, we obtain

Lemma 3.1. *Let (M,Ω) be a compact symplectic manifold. Then, $(J,g) \in \mathcal{AK}(M,[\Omega])$ is a critical point of the functional $\mathscr{F}_{\lambda,\mu}$ if and only if*

$$\int_M \Big(\sum_{i,j} T_{ij}h_{ij} - 2\sum_i s_i A_i\Big)dM_g = 0, \tag{29}$$

holds for any pair (h, A) of a symmetric $(0,2)$-tensor h and a 1-form $A = (A_i)$ satisfying (17), where $T = (T_{ij})$ and $s = (s_i)$ are the symmetric $(0,2)$-tensor field and the 1-form given by (26) and (28), respectively.

We here recall the following fact due to Blair and Ianus:

Lemma 3.2 ([2]). *Let B be a symmetric $(0,2)$-tensor on M. Then,*

$$\int_M \sum_{i,j} B_{ij}D_{ij}\,dM_g = 0$$

for all symmetric $(0,2)$-tensor D satisfying $DJ + JD = 0$ if and only if B is J-invariant.

Let (M,Ω) be a compact symplectic manifold and suppose $(J,g) \in \mathcal{AK}(M,[\Omega])$ a critical point of the functional $\mathscr{F}_{\lambda,\mu}$. Then, from Lemma 3.1, if $A = 0$, we have $\int_M \sum_{i,j} T_{ij}h_{ij}\,dM_g = 0$ for any symmetric $(0,2)$-tensor h satisfying $hJ + Jh = 0$. Thus, by virtue of Lemma 3.2, we conclude that T is J-invariant. Next, let $A = (A_i)$ be an arbitrary 1-form on M and define a symmetric $(0,2)$-tensor $h = (h_{ij})$ by $h_{ij} = (A_{\bar{i}j} + A_{\bar{j}i})/2$, where $A_{ij} = \nabla_i A_j - \nabla_j A_i$. Then, we have $h_{ij} + h_{\bar{i}\bar{j}} = A_{\bar{i}j} + A_{\bar{j}i}$, and thus the pair (h, A) satisfies (17). Since T_{ij} is symmetric and J-invariant, we have

$$\int_M \sum_{i,j} T_{ij}h_{ij}\,dM_g = \int_M \sum_{i,j} T_{i\bar{j}}A_{ij}\,dM_g = 2\int_M \sum_{i,j}(\nabla_j T_{\bar{j}i})A_i\,dM_g.$$

We put $\hat{t}_i = \sum_a \nabla_a T_{\bar{a}i}$. Then, (29) becomes $2\int_M \sum_i (\hat{t}_i - s_i) A_i \, dM_g = 0$, and hence

$$\hat{t}_i - s_i = 0 \qquad (i = 1, \ldots, 2n). \tag{30}$$

Conversely, for a given $(J, g) \in \mathcal{AK}(M, [\Omega])$, if $T = (T_{ij})$ is a symmetric J-invariant (0,2)-tensor field and (30) is valid for $i = 1, \ldots, 2n$, then we may easily observe that the equality (29) holds for any pair (h, A) satisfying (17).

Therefore, we here conclude that $(J, g) \in \mathcal{AK}(M, [\Omega])$ is critical point of the functional $\mathscr{F}_{\lambda,\mu}$ if and only if the symmetric tensor $T = (T_{ij})$ is J-invariant and the equality (30) holds for each $i = 1, \ldots, 2n$.

Now, we assume that T is J-invariant. Then, we have

$$\hat{t}_i = -t_{\bar{i}}, \qquad t_i = \sum \nabla_a T_{ai}. \tag{31}$$

for $i = 1, \ldots, 2n$. By direct computation, for each i $(1 \leq i \leq n)$, we have

$$\begin{aligned} t_i &= \frac{1}{2}(\mu - \lambda)\nabla_i \tau + \frac{1}{2}(\lambda \nabla_i \tau + \mu \nabla_i \tau^*) \\ &\quad + \frac{\mu}{2} \sum_{a,b,c} (\nabla_a \nabla_a J_{bc}) \nabla_i J_{bc} + \frac{\mu}{2} \sum_{a,b,c} (\nabla_a J_{bc}) \nabla_a \nabla_i J_{bc} \\ &= \mu \nabla_i \tau^* + \frac{\mu}{2} \sum_{b,c} (2\rho^*_{b\bar{c}} - \rho_{b\bar{c}} + \rho_{\bar{b}c}) \nabla_i J_{bc} + \mu \sum_{a,b,c} (\nabla_a J_{bc}) R_{aib\bar{c}}. \end{aligned}$$

Since,

$$\sum_{b,c} \rho^*_{b\bar{c}} \nabla_i J_{bc} = -\sum_{b,c} \rho^*_{bc} \nabla_{\bar{i}} J_{bc}, \quad \sum_{b,c} \rho_{b\bar{c}} \nabla_i J_{bc} = \sum_{b,c} \rho_{\bar{b}c} \nabla_i J_{bc} = 0,$$

we obtain

$$t_{\bar{i}} = \mu \nabla_{\bar{i}} \tau^* + \mu \sum_{a,b} \rho^*_{ab} \nabla_i J_{ab} + \mu \sum_{a,b,c} (\nabla_a J_{bc}) R_{a\bar{i}b\bar{c}}. \tag{32}$$

On one hand, we have

$$\begin{aligned} s_i &= 2\mu \sum_{a,b} J_{ab} \nabla_a \rho^*_{bi} = -\mu \sum_{a,b,c,u,v} J_{ab} \nabla_a (J_{ic} R_{bcuv} J_{uv}) \\ &= -\mu \sum_{a,b} (\nabla_i J_{ab}) \rho^*_{ab} - \mu \nabla_{\bar{i}} \tau^* - \mu \sum_{a,u,v} R_{a\bar{i}u\bar{v}} \nabla_a J_{uv}. \end{aligned}$$

Thus, if T is J-invariant, the condition (30) automatically holds. Further, from (26), we observe that the J-invariance of T and $(\mu - \lambda)\rho$ are equivalent. Therefore, we finally obtain

Theorem 3.1. *Let (M, Ω) be a compact symplectic manifold. Then, $(J, g) \in \mathcal{AK}(M, [\Omega])$ is a critical point of the functional $\mathscr{F}_{\lambda,\mu}$ if and only if the symmetric $(0,2)$-tensor field $(\mu - \lambda)\rho$ is J-invariant.*

This theorem is an extension of the result by Blair and Ianus ([2]).

Corollary 3.1. *Let (M, Ω) be a compact symplectic manifold. Then, the functional $\mathscr{F}_{\lambda,\lambda}$ $(\lambda \neq 0)$ is constant on each connected component of $\mathcal{AK}(M, [\Omega])$.*

Remark 3.1. Let (M, Ω) be a $2n$-dimensional compact symplectic manifold. Then, we have the following formula (cf. [1]):

$$\mathscr{F}_{\frac{1}{2},\frac{1}{2}}(J, g) = \frac{4\pi}{(n-1)!}(c_1 \cdot [\Omega]^{(n-1)})(M) \tag{33}$$

for any $(J, g) \in \mathcal{AK}(M, [\Omega])$, where c_1 is the first Chern class of (M, J). By Corollary 3.1 and (33), we see that $(c_1 \cdot [\Omega]^{(n-1)})(M)$ is constant on each connected component of $\mathcal{AK}(M, [\Omega])$.

Acknowledgment

The authors would like to express their thanks to Professor Apostolov for his valuable advice and suggestion.

References

1. V. Apostolov and T. Draghici, *The curvature and the integrability of almost Kähler manifold: a survey*, Symplectic and contact topology: interactions and perspectives. Toronto, ON/Montreal, QC, 2001, 25–53, Fields Inst. Commun., 35 Amer. Math. Soc., Providence, RI, 2003 (arXiv:math.DG/0302152).
2. D.E. Blair and S. Ianus, *Critical associated metrics on symplectic manifolds*, Contemporary Math. **51** (1986), 23–29.
3. S.I. Goldberg, *Integrability of almost Kähler manifolds*, Proc. Amer. Math. Soc. **21** (1969), 96–100.
4. A. Gray, *Curvature identities for Hermitian and almost Hermitian manifolds*, Tôhoku Math. J. **28** (1969), 601–612.
5. T. Koda, *Critical almost Hermitian structures*, Indian J. Pure Appl. Math. **26** (1995), 679–690.
6. T. Koda, *Almost Kähler structures with a fixed Kähler class*, Math. J. Toyama Univ. **27** (2004), 125–131.
7. J. Moser, *On the volume elements on a manifold*, Trans. Amer. Math. Soc. **120** (1965), 286–294.
8. K. Sekigawa, *On some compact Einstein almost Kähler manifolds*, J. Math. Soc. Japan **39** (1987), 677–684.
9. S. Tanno, Deformations of Riemannian metrics on 3-dimensional manifolds, Tôhoku Math. J **27** (1975), 437–444.

ON THE HYPOELLIPTICITY OF SOME CLASSES OF OVERDETERMINED SYSTEMS OF DIFFERENTIAL AND PSEUDODIFFERENTIAL OPERATORS

P.R. POPIVANOV

Institute of Math. and Informatics
of the Bulgarian Academy of Sciences,
Sofia 1113, Bulgaria
E-mail: popivano@math.bas.bg

This paper deals with the (micro)local C^∞ hypoellipticity of the solutions of some classes of overdetermined systems of pseudodifferential operators having simple and double characteristics. The results in the case of double characteristics do not depend on the lower order terms. As an application we study the hypoellipticity of systems of complex valued vector fields forming locally integrable m-dimensional Frobenius-Lie algebra. We also propose a short survey on some recent results on the hypoellipticity of this system of vector fields.

Keywords: Hypoellipticity; Subellipticity; Overdetermined systems.

1. Introduction

1. In the paper under consideration we study the (micro)local C^∞ hypoellipticity of two classes of pseudodifferential systems with simple and double characteristics. As our systems are consisting of several scalar operators acting on the same complex-valued scalar function u many methods developed in the scalar case can be used. At the beginning we give another proof of the well-known Hörmander's theorems 1.1.5 and 1.2.3 from [1] — sufficiency. The proof is elementary, short and the same idea can be used in other situations, for example, in proving C^∞ hypoellipticity of overdetermined systems with double characteristics. The loss of regularity of the corresponding solutions of the systems under inverstigation is equal to $1/2$, respectively to 1 in both classical and microlocalized Sobolev spaces $H^s_{loc}(\Omega)$, respectively $H^s_{mcl}(\rho)$, $\rho \in \dot{T}^*(\Omega)$. The theorems here proved enable us to study the

MSC Classification: 35H10, 35H20, 35B65.

C^∞ hypoellipticity of complex vector fields integrable system of Frobenius-Lie type having simple and double characteristics. Some of our results in the simple characteristic case — local version — give elementary proofs of theorems due to Bouendi-Trèves [4,5] and Kwong-Song [6]. Our assertions are microlocal in contrast with those in the above mentioned papers and the (micro)local loss of smoothness here obtained /1/2, 1/ is optimal. At the end of the Introduction of this paper and for the sake of completeness a short survey on some recent results on the subject is given. It includes results of J.J. Kohn [15] and Journé-Trépreau [16]. Classical and modern results on the local solvability and hypoellipticity mainly for the scalar case can be found in [2,11,12].

Because of the lack of space we omit the standard definitions of classical homogeneous pseudodifferential operators, C^∞ wave front set $WF(u)$ and microlocalized Sobolev spaces $H^s_{mcl}(\rho^0)$, $u \in D'(\Omega)$. The operator $P(x, D)$ is called C^∞ hypoelliptic iff sing supp Pu = sing supp u and P is microlocally hypoelliptic at the characteristic point $\rho^0 \in \operatorname{Char} P$ iff $\rho^0 \notin WF(Pu) \Rightarrow \rho^0 \notin WF(u)$ /u being assumed a Schwartz distribution, i.e. $u \in D'(\Omega)$/. The corresponding definitions and properties can be found in [7] and [9].

We are going to formulate the main results of this paper.

2. Consider the overdetermined system of pseudodifferential operators /ψdo/:

$$P_j u = f_j, \quad 1 \le j \le d \quad \operatorname{ord} P_j = t, \tag{1}$$

u is a scalar function and $P_j(x, D)$ is a classical ψdo.

In [1] a necessary and sufficient condition for hypoellipticity of (1) with a loss of regularity equal to 1/2 was found. We shall reformulate and prove in an elementary way the sufficiency of the same result. To do this we introduce the characteristic set $\operatorname{Char} P$ of $P = (P_1, \dots, P_d)$:

$$\operatorname{Char} P = \left\{ \rho = (x, \xi) \in \dot{T}^*(\Omega) \,:\, p_j^0(\rho) = 0,\ 1 \le j \le d \right\}, \quad \Omega \subset \mathbf{R}^n.$$

As usual, $\sigma(P) = p^0 = (p_1^0, \dots, p_d^0)$ is the principal symbol of P. Put $p_j^0 = a_j + ib_j$, $a_j = \Re p_j^0$, $b_j = \Im p_j^0$, $1 \le j \le d$. Evidently, $\rho^0 \notin \operatorname{Char} P$ implies the microlocal ellipticity of some symbol: $p_\nu^0(\rho^0) \neq 0$ and therefore we have microlocal hypoellipticity of (1) at the point ρ^0. By $\{\cdot,\cdot\}$ we denote the Poisson bracket of two smooth functions.

Theorem 1.1. *Consider the system* (1) *and assume that* $\rho^0 \in \operatorname{Char} P$. *Then:*

a) *If the following condition holds:*
(NS) *There exist* $1 \le j_0, k_0 \le d$ *and such that* $|\{a_{k_0}, a_{j_0}\}(\rho^0) - \{b_{k_0}, b_{j_0}\}(\rho^0)| + |\{a_{k_0}, b_{j_0}\}(\rho^0) - \{a_{j_0}, b_{k_0}\}(\rho^0)| > 0$,
then the system (1) *consisting of the principal symbols* $p^0(x, D)$ *only is microlocally hypoelliptic at* ρ^0 *with loss of regularity equal to* 1, *i.e.* $Pu = f = (f_1, \dots, f_d) \in H^s_{mcl}(\rho^0) \Rightarrow u \in H^{s+t-1}_{mcl}(\rho^0)$, $\forall u \in D'$, $\forall s \in \mathbf{R}^1$.

b) (S) *The matrix* $\|\{a_j, b_k\}(\rho^0)\|^d_{j,k=1}$ *is symmetric and has at least one positive eigenvalue. Then the system* (1) *is microlocally hypoelliptic at* ρ^0 *with a sharp loss of smoothness equal to* 1/2, *i.e.* $f \in H^s_{mcl}(\rho^0) \Rightarrow u \in H^{s+t-1/2}_{mcl}(\rho^0)$, $\forall s \in \mathbf{R}^1$.

The condition (NS) is equivalent to:

$$\{p^0_{k_0}, p^0_{j_0}\}(\rho^0) \neq 0 \iff \sum_{k \neq j} |\{p^0_k, p^0_j\}(\rho^0)| > 0.$$

Corollary 1.1. *The system* (1) *is hypoelliptic if for each* $\rho^0 \in \operatorname{Char} P$ *one of the following two conditions hold:*

(i) *The matrix* $\|\{a_k, b_j\}(\rho^0)\|$ *is not symmetric.*
(ii) *The matrix* $\|\{a_k, b_j\}(\rho^0)\|$ *is symmetric and possesses at least one positive eigenvalue.*

Certainly, in case (ii) *the result does not depend on the lower order terms /i.e. they can be arbitrary/.*

Following Bouendi-Trèves [5] we consider a locally integrable m-dimensional Frobenius-Lie algebra L of C^∞ smooth complex valued vector fields on an open domain $\Omega \subset R^{m+1}$, $m + 1 = n$. This means that:

c) at each point $z_0 = (t_0, x_0) \in \mathbf{R}^m \times \mathbf{R}^1$ the system L is spanned by m linearly independent vector fields $L_1(z_0), \dots, L_m(z_0)$.
d) L is closed under the operation $[\cdot, \cdot]$ commutator of two vector fields.
e) L is locally integrable, i.e. for each $z_0 \in \Omega$ there exists a neighbourhood U; $z_0 \in U$, and a smooth function Z, such that $LZ = 0$, $dZ(z_0) \neq 0$; Z can be complex-valued.

As Frobenius theorem is not valid in the complex C^∞ case, conditions c), d) do not imply e) as in the real or in the analytical cases.

As it is shown in [5,6] and without loss of generality in studying the hypoellipticity of the system of vector fields L we can assume that we are working in a neighbourhood U of $z_0 = 0$ with local coordinates $z = (t, x)$

for which:

$$L_j = \frac{\partial}{\partial t_j} + \lambda_j(t,x)\frac{\partial}{\partial x}, \quad 1 \le j \le m, \tag{2}$$

$Z(t,x) = x + i\Phi(t,x)$, $\Phi(0,0) = 0$, Φ is real-valued C^∞ function near the origin. Moreover, $[L_j, L_k] = 0$ in U for each $1 \le j,k \le m$. Then one can easily see that

$$\lambda_j = -i\frac{\frac{\partial\Phi}{\partial t_j}}{1 + i\frac{\partial\Phi}{\partial x}}. \tag{3}$$

Put $\operatorname{Char} L = \{\, (t,x;\tau,\xi) \in U \times S^{n-1} : \tau_j = 0,\ \partial\Phi/\partial t_j(t,x) = 0,\ 1 \le j \le m,\ \xi \ne 0 \,\}$, $C = \text{projection}_z \operatorname{Char} L$ and suppose that $C \ne \emptyset, \Omega\ /z = (t,x)/$. Certainly $S^{n-1} = \{\, \xi \in \mathbf{R}^n : |\xi| = 1 \,\}$.

The following result is a generalization of Theorem 2.1 from [6] but with a better result concerning the loss of regularity of the solution /1 in [6], 1/2 here/. The main idea of the proof used here is rather different from this one in [6].

Theorem 1.2. *The system of C^∞ complex-valued vector fields L defined by* c), d), e) *is microlocally hypoelliptic with a loss of smoothness equal to* 1/2 *(i.e. $L_j u \in H^s_{mcl}(\rho^0)$, $1 \le j \le m$, $s \in \mathbf{R}^1$, $\rho^0 \in \operatorname{Char} L \Rightarrow u \in H^{s+1/2}_{mcl}(\rho^0)$) if:*

(iii) $\rho^0 = (t_0, x_0, 0, -1) \in \operatorname{Char} L$ *implies that the partial Hessian of Φ with respect to t: $H_t\Phi(t_0,x_0) = \left\|\partial^2\Phi/\partial t_j \partial t_k(t_0,x_0)\right\|_{j,k=1}^m$ has at least one positive eigenvalue and $\rho^0 = (t_0, x_0, 0, 1) \in \operatorname{Char} L \Rightarrow H_t\Phi(t_0,x_0)$ has at least one negative eigenvalue.*

Corollary 1.2. *The system of C^∞ complex-valued vector fields L is hypoelliptic with sharp loss of regularity* 1/2 *if at each point $z_0 = (t_0, x_0) \in C = \{(t,x) : \partial\Phi/\partial t_j(t,x) = 0\}$ the condition* (iv) *holds:*

(iv) *The Hessian $H_t\Phi(z_0)$ has at least one positive and one negative eigenvalues.*

Remark 1.1. Condition (iv) can be reformulated in the following way.

(iv)′ At each point $z_0 \in C$: $\operatorname{rank} H_t\Phi(z_0) \ge 2$ and $|\operatorname{sgn} H_t\Phi(z_0)| \le \operatorname{rank} H_t\Phi(z_0) - 2$.

At the end of this paper we shall study the C^∞ (micro)hypoellipticity of a second order overdetermined system of complex-valued vector fields belonging to L, namely:

$$L_j L_k u + l_{jk} u = f_{jk}, \quad 1 \le j \le k \le m, \tag{4}$$

$u \in D'(\Omega)$, where L_j are the same as in c), d), e) and l_{jk} are arbitrary C^∞ smooth complex-valued vector fields.

Theorem 1.3. *The system* (4) *is microlocally hypoelliptic at* $\rho^0 \in \mathrm{Char}\, L$ *with a sharp loss of regularity equal to* 1 *if the condition* (iii) *holds.*

Corollary 1.3. *The system* (4) *is locally hypoelliptic with loss of regularity equal to* 1 *if the condition* (iv) *is fulfilled. The hypoellipticity does not depend on the vector fields* l_{jk}.

The proof of the (microlocal) hypoellipticity in Theorem 1.3 relies on the following result proved in [3] /see also [10]/.

Theorem 1.4. *Consider the classical scalar* ψdo P *with full symbol of the form:*

$$p(x,\xi) \sim p_m^2(x,\xi) + p_{2m-1}(x,\xi) + \cdots$$

and suppose that $p_m(\rho^0) = 0$, $\rho^0 \in \dot{T}^*(\Omega)$ *and* $\{\Re p_m, \Im p_m\}(\rho^0) > 0$. *Then* $P(x,D)$ *is microhypoelliptic at* ρ^0 *with sharp loss of smoothness equal to* 1, *i.e.* $Pu \in H^s_{mcl}(\rho^0) \Rightarrow u \in H^{s+2m-1}_{mcl}(\rho^0)$, $\forall s \in \mathbf{R}^1$. *The result does not depend on the lower order terms but only on* p_m.

Remark 1.2. The principal symbol p_m possesses the microlocal form $(\xi_1 + ix_1\xi_n)|\xi|^{m-1}$ near the point $\rho^0 = (x_0 = 0, ; 0, \ldots, 0, 1)$.

3. As we mentioned at the beginning of this paper a short survey on some recent results on the hypoellipticity of several classes of complex valued vector fields forming locally integrable m-dimensional Frobenius-Lie algebra will be proposed. We shall follow closely [16].

Thus, put $L_j = \partial/\partial t_j + i\partial\Phi(t)/\partial t_j \partial/\partial x$, $1 \le j \le m$, i.e. in vector form the system $L_j u = f_j$ takes the form

$$L_\Phi u = f, \tag{5}$$

where $\nabla \equiv \nabla_t = (\partial/\partial t_1, \ldots, \partial/\partial t_m)$, $L_\Phi = \nabla_t + i(\nabla_t \Phi)\partial/\partial x$, $u = u(t,x)$. Evidently, $[L_j, L_k] \equiv 0$.

We shall denote by Σ_Φ the characteristic set of L_Φ, i.e.

$$\Sigma_\Phi = \left\{ (t,x;0,\xi),\ \xi \neq 0,\ \nabla\Phi(t) = 0 \right\}.$$

Certainly, $\nabla\Phi(t_0) \neq 0$ implies that L_Φ is an elliptic system over the point (t_0, x_0), $x_0 \in \mathbf{R}^1$. Consider (5) in a neighbourhood of the point $(\bar{t}, \bar{x})$. The standard changes $x \to -x + \bar{x}$, $t \to t - \bar{t} \Rightarrow \partial/\partial x \to -\partial/\partial x$, $\partial/\partial t \to \partial/\partial t$, $\bar{x} \to 0$, $\bar{t} \to 0$ reduce the study of the microlocal hypoellipticity of (5) to its investigation in a conical neighbourhood of the point $\theta_0^+ = (0,0;0,1) \in \Sigma_\Phi$.

Proposition 1.1 ([16]). *Assume that the real-valued function $\Phi \in C^\infty$ in a neighbourhood of the origin $O \in \mathbf{R}^m$ and the system* (5) *is microlocal hypoelliptic at the characteristic point θ_0^+. Then one can find an open neighbourhood $V \ni O$, $V \subset \mathbf{R}^m$ and such that Φ does not have local maxima in V.*

Contrary to Proposition 1.1 we have the well known result of Maire [14].

Theorem 1.5 ([14]). *Suppose that Φ is real-valued analytic function in an open neighbourhood of the origin in $\mathbf{R}^m$ and Φ does not have local maxima there. Then the system* (5) *is microlocally hypoelliptic at θ_0^+.*

"In general" and for $m \geq 2$ Theorem 1.5 is wrong in the case $\Phi \in C^\infty$ /see [14]/.

The system (5) is called microlocally hypoelliptic at θ_0^+ with loss of regularity /smoothness/ $\delta > 0$ if $L_\Phi u \in H^s_{mcl}(\theta_0^+) \Rightarrow u \in H^{s+1-\delta}_{mcl}(\theta_0^+)$, $\forall s \in \mathbf{R}^1$, $\forall u \in D'(\mathbf{R}^n)$, $n = m + 1$.

Equivalently, we shall say that (5) is microlocally regular of order $\rho < 1$ at θ_0^+ iff $L_\Phi u \in H^s_{mcl}(\theta_0^+) \Rightarrow u \in H^{s+\rho}_{mcl}(\theta_0^+)$, $\forall s$, $\forall u \in D'(\mathbf{R}^n)$.

Certainly, $\rho = 1-\delta$. We shall say that (5) is subelliptic at θ_0^+ if $0 < \delta < 1$ /$\delta = 0 \iff \rho = 1$ corresponds to the elliptic case, i.e. it implies that $\theta_0^+ \notin \Sigma_\Phi$/.

A very interesting theorem — necessary condition for regularity of the system (5) of order $\rho > (-m+1)/4$ is proved in [16].

Theorem 1.6 ([16]). *For each $m \geq 2$ and $\rho > -(m-1)/4$ there exists a real-valued analytic function $\Phi(t)$ without any local maximum in some neighbourhood of the origin in $\mathbf{R}^m$ and such that L_Φ is not regular of order ρ at θ_0^+. The function Φ can be taken in the form: $\Phi(t) = -|t'|^{2r} - |t'|^2 t_m^{2p} + t_m^q$, where $t = (t', t_m)$, $r \geq 1$, $p \geq 0$, $q \geq 2$ are integers.*

This is a more precise version of Theorem 1.5, proved in [16].

Theorem 1.7 ([16]). *The real-valued analytic function $\Phi(t)$ does not possess local maxima in some open neighbourhood V of the origin $O \in \mathbf{R}^m$. Then the system* (5) *is microlocally hypoelliptic at θ_0^+ with (microlocal) order of regularity $\rho = -m/2$, i.e. with loss of regularity $\delta = (m+2)/2$.*

We shall say several words about the proof of Theorem 1.7. It is based on the $L^2 - L^\infty$ subelliptic type estimates. Thus, denote by $\tilde{u}(t,\xi)$ the partial Fourier transform of u with respect to x:

$$\tilde{u}(t,\xi) = \int e^{-ix\xi} u(t,x)\, dx \quad /\hat{u}(\tau,\xi) = \int\int e^{-i(x\xi + t\tau)} u(t,x)\, dt dx/.$$

Then for each $u \in C^1(V, \mathcal{E}'(\mathbf{R}_x))$, $V \ni O$, V — open, $V \subset \mathbf{R}^m$, the system (5) takes the form:

$$\tilde{f}(t,\xi) = (\widetilde{Lu})(t,\xi) = e^{\xi\Phi(t)}\nabla_t(e^{-\xi\Phi(t)}\tilde{u}(t,\xi)). \tag{6}$$

Definition 1.1. Let $s \in \mathbf{R}^1$, $K \subset\subset V$. Then

$$|||u|||^2_{s,K} = \sup_{t\in K}\int_1^\infty \xi^{2s}|\tilde{u}(t,\xi)|^2\,d\xi. \tag{7}$$

Lojasiewicz proved that if the real-valued analytic function Φ does not possess local maxima in some open neighbourhood V_1 of $O \in \mathbf{R}^m$ and $\Phi(0) = d\Phi(0) = 0$, then there exist a sufficiently small neighbourhood V of O, $V \subset\subset V_1$ and a number $\rho \in (0,1]$ such that

$$|\nabla\Phi(t)| \geq |\Phi(t)|^{1-\rho}, \quad \forall t \in V. \tag{8}$$

Therefore, the integral curves of the vector field $\nabla\Phi$ located in V are uniformly bounded.

This is a corollary of Lojasiewicz estimate (8).

Proposition 1.2 ([16]). *Assume that the analytic real valued function $\Phi(t)$ does not possess local maxima in some neighbourhood of the origin $O \in \mathbf{R}^m$. Then one can find two constants $\rho, k > 0$, $\rho \in (0,1]$ and such that for each sufficiently small open neighbourhood V of O and each $t \in V$ there exists a rectifiable curve $c : [0,\sigma] \to \bar{V}$ with the following properties: $c(0) = t$, $c(\sigma) \in \partial V$,*

$$0 \leq s \leq \sigma \Rightarrow \Phi(c(s)) - \Phi(t) \geq ks^{1/\rho}, \tag{9}$$

s being the natural parameter of c /the length of the corresponding arc/. The curve c is piecewise smooth and according to (9) its length is uniformly bounded.

Below we give the proof of the following a-priori estimate.

Proposition 1.3 ([16]). *The analytic function $\Phi(t)$ does not possess local maxima in a neighbourhood of the origin $O \in \mathbf{R}^m$. Then there exists $\rho \in (0,1]$ and such that for each sufficiently small open neighbourhood V of O, each compact $K \subset\subset V$ and every s_0, $s \in \mathbf{R}$ one can find a constant $C(V,K,s,s_0)$ for which*

$$|||u|||^2_{s+\rho/2,K} \leq C(|||u|||^2_{s_0,\bar{V}} + |||L_\Phi u|||^2_{s,\bar{V}}), \quad \forall u \in C^1(\bar{V}, \mathcal{E}'(\mathbf{R})). \tag{10}$$

The proof of (10) is not difficult. In fact, put $t^* = c(\sigma) \in \partial V$ and apply Proposition 1.2. Then

$$v(t) = v(t^*) + \int_\sigma^0 \frac{d}{ds} v(c(s))\, ds, \quad v \in C^1(V) \tag{11}$$

and $d/dsv(c(s)) = \langle \nabla v(c(s)), c'(s) \rangle$, $|c'(s)| = 1$, i.e. $|d/dsv(c(s))| \leq |\nabla v(c(s))|$, $0 \leq s \leq \sigma$.

Multiplying (11) by $e^{\lambda \Phi(t)}$ we obtain $\forall \lambda \geq 1$:

$$\begin{aligned} |(e^{\lambda\Phi} v)(t)| \leq{}& e^{\lambda(\Phi(t) - \Phi(t^*))} |(e^{\lambda\Phi} v)(t^*)| \\ &+ \int_0^\sigma e^{\lambda(\Phi(t) - \Phi(c(s)))} |(e^{\lambda\Phi} \nabla v)(c(s))|\, ds. \end{aligned} \tag{12}$$

Evidently, $\int_0^\infty e^{-\lambda k s^{1/\rho}}\, ds = c/\lambda^\rho$. Suppose now that the compact $K \subset\subset V$, $t \in K$. Then the length σ of the curve c, joining t and t^* is $\geq \sigma_0 > 0$. Combining (12) and (9) we get

$$|(e^{\lambda\Phi} v)(t)| \leq e^{-\lambda k \sigma_0^{1/\rho}} |(e^{\lambda\Phi} v)(t^*)| + \int_0^\sigma e^{-\lambda k s^{1/\rho}} |(e^{\lambda\Phi} \nabla v)(c(s))|\, ds.$$

Put $k\sigma_0^{1/\rho} = \varepsilon/2$ and apply the Cauchy-Schwarz inequality to the previous integral. Thus, with some constant $C > 0$ and for each $\lambda \geq 1$ we have:

$$|(e^{\lambda\Phi} v)(t)|^2 \leq C e^{-\lambda\varepsilon} |(e^{\lambda\Phi} v)(t^*)|^2 + C\lambda^{-\rho} \int_0^\sigma |(e^{\lambda\Phi} \nabla v)(c(s))|^2\, ds. \tag{13}$$

Going back to our system (5) we assume that $u \in C^1(\bar{V}, \mathcal{E}'(\mathbf{R}^1_x))$.

Having in mind (6) we write $v = e^{-\lambda\Phi(t)} \tilde{u}(t, \xi)$, $\lambda = \xi$ in (13) and we get

$$|\tilde{u}(t,\xi)|^2 \leq C e^{-\varepsilon\xi} |\tilde{u}(t^*, \xi)|^2 + C \int_0^\sigma \xi^{-\rho} |\tilde{f}(c(s), \xi)|^2\, ds. \tag{14}$$

We multiply (14) by $\xi^{2s+\rho}$ then we integrate w.r. to $\xi \in [1, \infty)$ and according to (7) and the fact that σ is uniformly bounded we obtain the desired estimate (10).

The proof of Theorem 1.7 relies on Proposition 1.3 /the estimate (10)/ and on the bootstrap arguments. We omit it.

We shall complete our survey by formulating a recent result of J.J. Kohn [15].

Consider the operator $E = \sum_1^d X_j^* X_j$, where $\{X_1, \ldots, X_d\}$ are smooth complex valued vector fields in $\Omega \subset \mathbf{R}^n$.

Theorem 1.8 (**[15]** and see also **[13]**). *If the vector fields* $\{X_i, [X_i, X_j]\}$ *span the complex tangent space at the origin then* E *is hypoelliptic near the origin with a sharp loss of regularity equal to* 1.

2. Proof of the main results

Proof of Theorem 1.1. One can see that (1) implies: The commutator

$$[P_k, P_j]u = P_k f_j - P_j f_k. \tag{15}$$

If we denote by $\sigma(P)$ the principal symbol of the ψdo $P(x, D)$, then $\sigma([A, B]) = 1/i\{\sigma(A), \sigma(B)\}$ and therefore the principal symbol of the classical ψdo participating in the left hand side of (15) is given by:

$$\sigma([p_k^0, p_j^0]) = -i(\{a_k, a_j\} - \{b_k, b_j\}) + (\{a_k, b_j\} - \{a_j, b_k\}). \tag{16}$$

Assuming that condition (NS) holds at the point ρ^0 we prove Theorem 1.1 a).

In order to prove Theorem 1.1 b), i.e. (S) being fulfilled, we are looking for real numbers ε_j, $1 \le j \le d$, $\sum_1^d |\varepsilon_j| > 0$ and such that:

$$Qu \equiv \sum_1^d (\varepsilon_j a_j + i\varepsilon_j b_j)u = \sum_1^d \varepsilon_j f_j. \tag{17}$$

Obviously, $\operatorname{ord} Q = t$ and $\rho^0 \in \operatorname{Char} P \Rightarrow \rho^0 \in \operatorname{Char} Q$. According to Hörmander [1,7], the necessary and sufficient condition for the (microlocal) hypoellipticity with loss of regularity 1/2 of the scalar ψdo Q is the following one: $Q(\rho^0) = 0$, $\{\Re Q, \Im Q\}(\rho^0) > 0$. So we have: $0 < \{\sum_1^d \varepsilon_j a_j, \sum_1^d \varepsilon_j b_j\}(\rho^0) = \sum_{j,k=1}^d \varepsilon_j \varepsilon_k \{a_j, b_k\}(\rho^0)$. Having in mind that $\|\{a_j, b_k\}\|(\rho^0)$ is a symmetric matrix we prove Theorem 1.1 b). In fact, if A is real valued matrix and $A = {}^tA$ then there exists a nondegenerate matrix B s.t. $(A\varepsilon, \varepsilon) = ({}^tBABy, y) = \sum_{j=1}^s \lambda_j y_j^2$, where $s = \operatorname{rank} A \le d$ and say $\lambda_1 > 0$; $\varepsilon = By$. Taking $y_0 = (1, 0, \dots, 0)$, $\varepsilon_0 = By_0$ we get: $(A\varepsilon_0, \varepsilon_0) = \lambda_1 > 0$ etc. □

Proof of Theorem 1.2. If L_j are defined by (2) then $L_j Z = 0 \Rightarrow$ (3). Moreover, $[L_j, L_k] = 0$, $1 \le j, k \le m$ implies that locally near the origin

$$\frac{\partial \lambda_k}{\partial t_j} - \frac{\partial \lambda_j}{\partial t_k} + \lambda_j \frac{\partial \lambda_k}{\partial x} - \lambda_k \frac{\partial \lambda_j}{\partial x} = 0, \quad 1 \le j, k \le m. \tag{18}$$

In order to find out $\operatorname{Char} L = \{ (t, x; \tau, \xi) \in \dot{T}^*(\Omega) : \tau_j + \lambda_j \xi = 0,\ 1 \le j \le m \}$, we compute: $0 = \tau_j + \lambda_j \xi = \tau_j - \frac{\partial \Phi}{\partial t_j} \frac{\partial \Phi}{\partial x} \xi / (1 + \Phi_x^2) - i \frac{\partial \Phi}{\partial t_j} \xi / (1 + \Phi_x^2)$.

Therefore, $\operatorname{Char} L = \{(t, x; \tau, \xi) : \partial\Phi/\partial t_j = 0,\ 1 \le j \le m,\ \tau_j = 0,\ \xi \ne 0\}$. Obviously, $\lambda_j = a_j + ib_j$, $\lambda_{j|\operatorname{Char} L} = 0$, $a_j = \tau_j - \frac{\partial\Phi}{\partial t_j}\frac{\partial\Phi}{\partial x}\xi/(1+\Phi_x^2)$, $b_j = -\frac{\partial\Phi}{\partial t_j}\xi/(1+\Phi_x^2)$. To prove Theorem 1.2 we shall apply Theorem 1.1 b). So we must compute $\|\{a_j, b_k\}(\rho^0)\|_{j,k=1}^d$, $\rho^0 \in \operatorname{Char} L$. Then $\{a_j, b_k\}_{|\operatorname{Char} L} = \{\tau_j + \Re\lambda_j\xi, \Im\lambda_k\xi\}_{|\operatorname{Char} L} = \xi(\partial\Im\lambda_k)/(\partial t_j)|_{\operatorname{Char} L}$ and in a similar way $\{a_k, b_j\}_{|\operatorname{Char} L} = \xi\partial\Im\lambda_j/\partial t_k|_{\operatorname{Char} L}$. According to (18) we get $\{a_k, b_j\}_{|\operatorname{Char} L} = \{a_j, b_k\}_{|\operatorname{Char} L}$ and therefore $\|\{a_k, b_j\}\|(\rho^0)$ is symmetric. We can easily see that $\|\{a_j, b_k\}\|_{|\operatorname{Char} L} = -\xi/(1+\Phi_x^2)\,\left\|\partial^2\Phi/\partial t_k\partial t_j\right\|_{|C}$.

Condition (iii) enables us to apply Theorem 1.1 b) and to conclude that the corresponding system L c), d) e) is (micro)hypoelliptic with sharp loss of regularity $1/2$. □

Corollary 2.1. *Suppose that* $\lambda_j = \lambda_j(t)$, $1 \le j \le m$. *Then according to* (18) $\partial\lambda_k/\partial t_j = \partial\lambda_j/\partial t_k \Rightarrow \partial\Re\lambda_j/\partial t_k = \partial\Re\lambda_k/\partial t_j, \partial\Im\lambda_j/\partial t_k = \partial\Im\lambda_k/\partial t_j$, $1 \le j, k \le m$. *Therefore, there exists in a tiny neighbourhood of the origin a smooth complex-valued function* $\lambda(t)$ *and such that* $\lambda_j = \partial\lambda/\partial t_j$, $1 \le j \le m$. *So we can take in* (3) $\Phi = \Phi(t) \Rightarrow \lambda_j = -i\partial\Phi/\partial t_j = \partial\lambda/\partial t_j \Rightarrow \Phi = i\lambda(t)$. *Then the system* L c), d), e) *is hypoelliptic in some neighbourhood of the origin if at each critical point* t_0 *of the real-valued function* $\Phi(t)$, *i.e.* $\nabla\Phi(t_0) = 0$, *we have that* $\operatorname{rank}\|\Phi''_{tt}(t_0)\| \ge 2$ *and* $|\operatorname{sgn}\|\Phi''_{tt}(t_0)\|| \le \operatorname{rank}\|\Phi''_{tt}(t_0)\| - 2$.

Proof of Theorem 1.3. We are looking for real numbers ε_j, $1 \le j \le d$, $\sum_{j=1}^d |\varepsilon_j| > 0$ for which $\left(\sum_{j=1}^d \varepsilon_j P_j\right)^2 + lu = F(x)$, where $P_j(x, D)$ are classical scalar ψdo of order m and

$$P_j P_k u + l_{jk} u = f_{jk}(x), \tag{19}$$

$\operatorname{ord} l_{jk}(x, D) = 2m - 1$, $\operatorname{ord} l(x, D) = 2m - 1$, $1 \le j \le k \le d$, $x \in \Omega \subset \mathbf{R}^n$. Consider the identity

$$\begin{aligned}
\Big(\sum_1^d \varepsilon_j P_j\Big)^2 u &= \sum_{j,k=1}^d \varepsilon_j\varepsilon_k P_j P_k u = \sum_{j\le k} \varepsilon_j\varepsilon_k P_j P_k u + \sum_{j>k} \varepsilon_j\varepsilon_k P_j P_k u \\
&= \sum_{j\le k} \varepsilon_j\varepsilon_k (f_{jk} - l_{jk}u) + \sum_{j>k} \varepsilon_j\varepsilon_k (P_k P_j u + [P_j, P_k]u) \\
&= \sum_{j,k=1}^d \varepsilon_j\varepsilon_k f_{jk} - lu = F(x) - lu,
\end{aligned} \tag{20}$$

where $l = \sum_{j,k=1}^d \varepsilon_j\varepsilon_k l_{jk} + \sum_{k<j}[P_k, P_j]$, $\operatorname{order} l = 2m - 1$, $F(x) = \sum_{j,k=1}^d \varepsilon_j\varepsilon_k f_{kj}$. Put $\sigma(p_j) = p_j^0 = a_j + ib_j$, $a_j = \Re p_j^0$, $b_j = \Im p_j^0$ and

assume that $\rho^0 \in \text{Char } P = \{\rho \in \dot{T}^*(\Omega) : p_j^0(\rho) = 0, 1 \leq j \leq d\}$. Let us denote $Q = \sum_{j=1}^d \varepsilon_j p_j \Rightarrow \sigma(Q) = \sum_{j=1}^d \varepsilon_j a_j + i\sum_{j=1}^d \varepsilon_j b_j$. Consequently: $\rho^0 \in \text{Char}\, P \Rightarrow \rho^0 \in \text{Char}\, Q$. One can easily see that $\{\Re Q^0, \Im Q^0\}(\rho^0) = \sum_{j,k=1}^d \varepsilon_j \varepsilon_k \{a_j, b_k\}(\rho^0)$. We shall assume further on that $\|\{a_j, b_k\}(\rho^0)\|$ is symmetric at each point $\rho^0 \in \text{Char}\, P$ and possesses at least one positive eigenvalue at ρ^0. Applying Theorem 1.4 we conclude that the scalar operator (20) is microlocally hypoelliptic at ρ^0 with microlocal loss of regularity equal to 1 and without any importance of the lower order terms. On the other hand, $\rho^0 \notin \text{Char}\, P \Rightarrow p_{j_0}(\rho^0) \neq 0$ for some j_0. So $P_{j_0} P_{j_0} u + l_{j_0 j_0} u = f_{j_0 j_0} \in H^s_{mcl}(\rho^0) \Rightarrow u \in H^{s+2m}_{mcl}(\rho^0)$. This way we have proved the (micro)local hypoellipticity of the overdetermined system (19) under the condition (S).

The proof of Theorem 1.3 is now trivial as if $a_j = \Re L_j$, $b_j = \Im L_j$ and (3) holds, then $\|\{a_j, b_k\}\|$ is symmetric at $\text{Char}\, L$ and $\|\{a_j, b_k\}\|_{|\text{Char}\, L} = -\xi/(1 + \Phi_x^2)\|\partial^2 \Phi(t,x)/\partial t_j \partial t_k\|_{|C}$, $\xi \neq 0$. The fulfillment of condition (iii) verifies the stated result on local hypoellipticity. A very delicate problem is the investigation of the local solvability of the system (4). These problems are beyond the scope of this paper. /In the simple characteristic case see [8]/. □

The idea in the proof of Theorems 1.1, 1.2 enables us to propose several examples of hypoelliptic systems of type (2), $\lambda_j(t) = i\partial\Phi(t)/\partial t_j$, which are subelliptic ones. Subelliptic systems of the same type (2) are studied in [17].

Example 2.1. Consider the system of vector fields $L_1 = \partial/\partial t_1 + it_2 \partial/\partial x$, $L_2 = \partial/\partial t_2 + it_1 \partial/\partial x$ in $\mathbf{R}_t^2 \times \mathbf{R}_x^1$. Then the scalar operators L_1, L_2 are not microhypoelliptic at $(0,0;0,\pm 1) = \theta_0^\pm$, while the system (L_1, L_2) is microhypoelliptic at $\theta_0^\pm$ with sharp loss of regularity $\delta = 1/2$. In fact, $\Phi = t_1 t_2$ and we can apply Theorem 1.1, (S).

Example 2.2. Consider the system $L_1 = \partial/\partial t_1 + i(\partial\Phi/\partial t_1)\partial/\partial x$ $L_2 = \partial/\partial t_2 + i(\partial\Phi/\partial t_2)\partial/\partial x$ in $\mathbf{R}^3$, where $\Phi(t_1, t_2) = t_1^3/3 + bt_1 t_2^2 + ct_2^3/3$, $c > 0$, $b < 0$, $1 - \sqrt{1+4c} < 2b < 0$ and a, b, c are real constants. Then L_1, L_2 are not microlocally hypoelliptic at $\theta_0^\pm$, while the system (L_1, L_2) is microhypoelliptic there and consequently, the system (L_1, L_2) is hypoelliptic near $O \in \mathbf{R}^3$. The sharp loss of regularity is $\delta = 2/3$.

It is easy to see that $t = 0$ is the only critical point of Φ and Φ does not have local maximum /minimum/ at O as $\Phi(t_1, 0)$ changes sign at $t_1 = 0$.

Example 2.3. L_1, L_2 are the same as in Example 2 but $b < 0$, $c < 0$, $0 > 2b > 1 - \sqrt{1-4c}$. Then the same results as in Example 2 are valid.

Example 2.4. The system (5) is studied with $\Phi = \pm t_1^{2k+1}/(2k+1) + a(t')$, $t' = (t_2, \ldots, t_m)$, $\partial_{t'} a(0) = 0$. Then (5) is microhypoelliptic at $\theta_0^{\pm}$ with sharp loss of regularity $\delta = 2k/(2k+1)$, $k \in \mathbf{N}_0$.

Example 2.5. Put in (5) $\Phi = \pm t_1^{2k+1}/(2k+1) + t_1^{l+1}/(l+1)a(t')$, $t' = (t_2, \ldots, t_m)$, $k, l \in \mathbf{N}$. Suppose that one of the following conditions hold:

5.1) $l > 2k$

5.2) $l = 2k$, $a(0) = 0$

5.3) l — even, $l < 2k$, $a(0) \neq 0$

Then (5) is microlocally hypoelliptic at $\theta_0^{\pm}$ with sharp loss of regularity $\delta = 2k/(2k+1)$ in the cases 5.1), 5.2) and $\delta = l/(l+1)$ in the case 5.3).

The proofs here are based on Chapter XXVII from [7].

References

1. L. Hörmander, Ann. of Math. **83**, 129 (1966).
2. L. Boutet de Monvel, Comm. Pure Appl. Math. **27**, 585 (1974).
3. P. Popivanov, Serdica **3**, 249 (1977) (in Russian).
4. M.S. Baouendi and F. Treves, Adv. in Math. Suppl. studies **7A**, 245 (1981).
5. ———, Ann. of Math. **113**, 387 (1981).
6. K.H. Kwong and B.C. Song, Journal of the Korean Math. Soc. **26**, 195 (1989).
7. L. Hörmander, *The analysis of linear partial differential operators*, vol. IV, Springer Verlag, Berlin, 1985.
8. P. Cordaro and J. Hounie, Journal of funct. analysis **87**, 231 (1989).
9. M. Mascarello and L. Rodino, *Partial differential equations with multiple characteristics*, Wiley-VCH, Berlin, 1997.
10. T. Gramchev and P. Popivanov, *Partial differential equations. Approximate solutions in scales of functional spaces*, Wiley-VCH, Berlin, 2000.
11. G. De Donno and A. Oliaro, Trans. Amer. Math. Soc. **355**, 3405 (2003).
12. F. Nicola and L. Rodino, J. Math. Pures Appl. **83**, 1067 (2004).
13. L. Hörmander, Math. Ann. **217**, 165 (1975).
14. H.-M. Maire, Comm. in P.D.E. **5**, 331 (1980).
15. J.J. Kohn, Annals of Math. **162**, 943 (2005).
16. J.-L. Journé and J.-M. Trépreau, Séminaire EDP, Ecole Polytechnique, Exposé n XIV, 21.02.2006 (2005-2006).
17. M. Derridj, *Subelliptic estimates for some systems of complex vector fields, Hyperbolic problems and related questions*, Birkhaüser volume (to appear in 2006).

LIE GROUPS AS FOUR-DIMENSIONAL RIEMANNIAN PRODUCT MANIFOLDS

D.K. SHTARBEVA

University of Plovdiv, Faculty of Mathematics and Informatics,
Departament of Geometry, 236 Bulgaria Blvd., Plovdiv 4003, Bulgaria,
E-mail: dobrinka@pu.acad.bg

Two examples of four-dimensional Riemannian product manifolds are constructed by means of Lie groups and Lie algebras. The form of the curvature tensor for each of the examples is obtained.

Introduction

Almost product manifolds are originally introduced in [5], where these manifolds are classified into six basic classes. Equivalent characteristic conditions for each of the classes are obtained in [6] in the case of $2n$-dimensional Riemannian almost product manifolds with $\operatorname{tr} P = 0$. Examples of the class $\mathcal{W}_0$ (Riemannian P-manifolds) are given in [8] and examples of the basic classes $\mathcal{W}_i$ $(i = 1, 2, \dots, 6)$ are presented in [4].

In this paper our purpose is to construct examples of two classes of integrable almost product manifolds. All examples are of four-dimensional manifolds and are obtained by constructing four-parametric (or two-parametric) families of Lie algebras corresponding to real connected Lie groups. The manifolds obtained in this way are characterized geometrically. The form of the curvature tensor for each of the examples is found.

1. Riemannian almost product manifolds

Let (M, P, g) be a $2n$-dimensional Riemannian almost product manifold, i.e. P is an almost product structure and g is a metric on M such that

$$P^2X = X, \qquad g(PX, PY) = g(X, Y) \tag{1}$$

for all differentiable vector fields X, Y on M, i.e. $X, Y \in \mathfrak{X}(M)$.

Further, X, Y, Z, W (x, y, z, w, respectively) will stand for arbitrary differentiable vector fields on M (vectors in T_pM, $p \in M$, respectively).

Let ∇ be the Levi-Civita connection of the metric g. Then, the tensor field F of type $(0,3)$ on M is defined by $F(X,Y,Z) = g\left((\nabla_X P)Y, Z\right)$. It has the following symmetries

$$F(X,Y,Z) = F(X,Z,Y) = -F(X,PY,PZ). \tag{2}$$

Let $\{e_i\}$ $(i = 1,2,\dots,2n)$ be an arbitrary basis of T_pM at a point p of M. The components of the inverse matrix of g are denoted by g^{ij} with respect to the basis $\{e_i\}$. The Lie form α associated with F is defined by

$$\alpha(z) = g^{ij} F(e_i, e_j, z). \tag{3}$$

The Nijenhuis tensor field N of the manifold is given as

$$N(X,Y) = [PX,PY] + [X,Y] - P[PX,Y] - P[X,PY]. \tag{4}$$

It is known [5] that the almost product structure P is product if it is integrable, i.e. if $N = 0$.

A classification of the Riemannian almost product manifolds is introduced in [5], where six classes of these manifolds are characterized according to the properties of F. Three of the basic classes $\mathcal{W}_0$, $\mathcal{W}_2$, $\mathcal{W}_5$ and the class $\mathcal{W}_2 \oplus \mathcal{W}_5$ are given as follows:

$$\begin{aligned}
&\mathcal{W}_0 : F(X,Y,Z) = 0;\\
&\mathcal{W}_2 : F(X,Y,Z) = F(PX,Y,Z), \quad \alpha(Z) = 0;\\
&\mathcal{W}_5 : F(X,Y,Z) = -F(PX,Y,Z), \quad \alpha(Z) = 0;\\
&\mathcal{W}_2 \oplus \mathcal{W}_5 : \alpha(Z) = 0;
\end{aligned} \tag{5}$$

Let R be the curvature tensor of ∇, i.e. $R(X,Y)Z = \nabla_X\nabla_Y Z - \nabla_Y\nabla_X Z - \nabla_{[X,Y]}Z$. The corresponding tensor of type $(0,4)$ is denoted by the same letter and is given by $R(X,Y,Z,W) = g\left(R(X,Y)Z, W\right)$.

The Ricci tensor ρ and the scalar curvatures τ and $\overset{*}{\tau}$ of R are defined by:

$$\rho(y,z) = g^{ij} R(e_i, y, z, e_j), \quad \tau = g^{ij}\rho(e_i,e_j), \quad \overset{*}{\tau} = g^{ij}\rho(e_i, Pe_j). \tag{6}$$

We consider the following curvature-like tensors of type $(0,4)$:

$$\begin{aligned}
\pi_1(X,Y,Z,W) &= g(Y,Z)g(X,W) - g(X,Z)g(Y,W);\\
\pi_2(X,Y,Z,W) &= g(Y,PZ)g(X,PW) - g(X,PZ)g(Y,PW);\\
\pi_3(X,Y,Z,W) &= g(Y,Z)g(X,PW) - g(X,Z)g(Y,PW)\\
&\quad + g(Y,PZ)g(X,W) - g(X,PZ)g(Y,W).
\end{aligned} \tag{7}$$

Let $\alpha = \{x, y\}$ be a two-plane spanned by vectors $x, y \in T_pM$, $p \in M$. Then, the sectional curvatures of α are given by:

$$\nu(\alpha; p) = \frac{R(x,y,y,x)}{\pi_1(x,y,y,x)}, \qquad \overset{*}{\nu}(\alpha; p) = \frac{R(x,y,y,Px)}{\pi_1(x,y,y,x)}. \tag{8}$$

2. A Lie group as a four-dimensional Riemannian P-manifold

Let V be a four-dimensional real vector space and consider the structure of the Lie algebra defined by the brackets $[E_i, E_j] = C_{ij}^k E_k$, where $\{E_1, E_2, E_3, E_4\}$ is a basis of V and $C_{ij}^k \in \mathbb{R}$. Then, the Jacobi identity

$$C_{ij}^k C_{ks}^l + C_{js}^k C_{ki}^l + C_{si}^k C_{kj}^l = 0 \tag{9}$$

holds.

Let G be the associated real connected Lie group and $\{X_1, X_2, X_3, X_4\}$ be a global basis of left invariant vector fields induced by the basis of V. We can define an almost product structure on G by the conditions

$$PX_1 = X_3, \quad PX_2 = X_4, \quad PX_3 = X_1, \quad PX_4 = X_2. \tag{10}$$

Further, let us consider the left invariant metric defined by

$$g(X_i, X_i) = 1 \quad \text{for } i = 1,2,3,4, \quad g(X_i, X_j) = 0 \quad \text{for } i \neq j. \tag{11}$$

The introduced metric is a Riemannian metric, that satisfies (1). In this way, the induced four-dimensional manifold (G, P, g) is a Riemannian almost product manifold.

Definition 2.1 ([1]). *An almost product structure P on a Lie group G is said to be* bi-invariant *if*

$$[X, PY] = P[X, Y] \quad \textit{for all } X, Y \in \mathfrak{g}, \tag{12}$$

where $\mathfrak{g}$ is the Lie algebra of G.

The condition (12) implies $N(X, Y) = 0$ for any $X, Y \in \mathfrak{g}$, i.e. P is a product structure (paracomplex structure because $\operatorname{tr} P = 0$). Therefore (G, P, g) is a Riemannian product manifold.

Let P, defined by (10), be a bi-invariant product structure. Then, by (12) we obtain the following conditions for the commutators of the basic vector fields:

$$\begin{aligned} &[X_1, X_4] = -[X_2, X_3] = P[X_1, X_2] = P[X_3, X_4], \\ &[X_1, X_3] = [X_2, X_4] = 0, \end{aligned} \tag{13}$$

and thus we can put the non-zero Lie brackets:

$$\begin{aligned} [X_1, X_2] &= [X_3, X_4] = aX_1 + bX_2 + cX_3 + dX_4, \\ [X_1, X_4] &= -[X_2, X_3] = cX_1 + dX + aX + bX_4, \end{aligned} \tag{14}$$

where a, b, c, d are real parameters.

By direct computation we prove that the commutators (14) satisfy the Jacobi identity. Therefore, the conditions (14) define a four-parametric family of four-dimensional real Lie algebras $\mathfrak{g}$.

Let us recall that a Lie algebra $\mathfrak{g}$ is said to be *solvable* if its derived series

$$\mathfrak{D}^0\mathfrak{g} = \mathfrak{z},\ \mathfrak{D}^1\mathfrak{g} = [\mathfrak{g}, \mathfrak{g}], \ldots, \mathfrak{D}^p\mathfrak{g} = [\mathfrak{D}^{p-1}\mathfrak{g}, \mathfrak{D}^{p-1}\mathfrak{g}], \ldots$$

vanishes for some $p \in \mathbb{N}$. Then, having in mind (14), it is easy to check that $\mathfrak{D}^2\mathfrak{g} = \{0\}$ and thus the Lie algebras $\mathfrak{g}$ are solvable.

We establish the validity of the following

Theorem 2.1. *Let* (G, P, g) *be the four-dimensional Riemannian product manifold constructed by* (10), (11) *and* (12), *and let* $\mathfrak{g}$ *be the associated Lie algebra of* G *introduced by* (14). *Then,* (G, P, g) *is a Riemannian* P*-manifold, i.e.* $(G, P, g) \in \mathcal{W}_0$.

Proof. Let ∇ be the Levi-Civita connection of g. Then, the following well-known condition is valid

$$\begin{aligned} 2g(\nabla_X Y, Z) &= Xg(Y, Z) + Yg(X, Z) - Zg(X, Y) \\ &\quad + g([X, Y], Z) + g([Z, X], Y) + g([Z, Y], X). \end{aligned} \tag{15}$$

Applying (10), (11) and the fact that P is a bi-invariant product structure to (15), we obtain

$$\begin{aligned} &2g\left((\nabla_{X_i} P)X_j, X_k\right) = g([X_i, PX_j] - P[X_i, X_j], X_k) \\ &+ g\left([X_k, PX_j] - [PX_k, X_j], X_i\right) + g\left([X_k, PX_i] - [PX_k, X_i], X_j\right) = 0 \end{aligned}$$

for all $i, j, k = 1, 2, 3, 4$, i.e. $\nabla P = 0$ on $\mathfrak{g}$ and therefore, by (5), the manifold (G, P, g) belongs to the class $\mathcal{W}_0$. □

Let us remark that the Killing form [3] of the considered Lie algebra $\mathfrak{g}$

$$B(X, Y) = \operatorname{tr}(\operatorname{ad} X \operatorname{ad} Y), \quad X, Y \in \mathfrak{g}, \tag{16}$$

has the following form

$$B = 2\begin{pmatrix} b^2 + d^2 & -ab - cd & 2bd & -ad - bc \\ -ab - cd & a^2 + c^2 & -ad - bc & 2ac \\ 2bd & -ad - bc & b^2 + d^2 & -ab - cd \\ -ad - bc & 2ac & -ab - cd & a^2 + c^2 \end{pmatrix}.$$

It is easy to prove, that $\det B = 0$, i.e. the Killing form is degenerate. Thus, the Killing form B can not be a Riemannian metric.

Next, by (11), (14) and (15) we find the components of the Levi-Civita connection on the considered manifold as follows:

$$\begin{aligned} &\nabla_{X_1}X_1 = \nabla_{X_3}X_3 = -aX_2 - cX_4, \quad \nabla_{X_1}X_2 = \nabla_{X_3}X_4 = aX_1 + cX_3,\\ &\nabla_{X_1}X_3 = \nabla_{X_3}X_1 = -cX_2 - aX_4, \quad \nabla_{X_1}X_4 = \nabla_{X_3}X_2 = cX_1 + aX_3,\\ &\nabla_{X_2}X_1 = \nabla_{X_4}X_3 = -bX_2 - dX_4, \quad \nabla_{X_2}X_2 = \nabla_{X_4}X_4 = bX_1 + dX_3,\\ &\nabla_{X_2}X_3 = \nabla_{X_4}X_1 = -dX_2 - bX_4, \quad \nabla_{X_2}X_4 = \nabla_{X_4}X_2 = dX_1 + bX_3. \end{aligned} \tag{17}$$

Let R be the curvature tensor of (G, P, g). From the condition $\nabla P = 0$ it follows that $R(X,Y)PZ = PR(X,Y)Z$ and thus we have $R(X,Y,PZ,PW) = R(X,Y,Z,W)$ for any $X,Y,Z,W \in \mathfrak{g}$, i.e. R is a *Kähler tensor*. We denote its components by $R_{ijks} = R(X_i, X_j, X_k, X_s)$ $(i,j,k,s = 1,2,3,4)$. Then, by (17) we get the following non-zero components of R:

$$\begin{aligned} &R_{1221} = R_{1441} = R_{2332} = R_{3443} = R_{1243} = R_{1423} = -(a^2+b^2+c^2+d^2),\\ &R_{1223} = R_{1241} = R_{1443} = R_{2334} = -2\,(ac+bd)\,. \end{aligned} \tag{18}$$

By (6), (11) and (18), we obtain the non-zero components $\rho_{ij} = \rho(X_i, X_j)$ of the Ricci tensor and the scalar curvatures τ and $\overset{*}{\tau}$ as follows:

$$\begin{aligned} &\rho_{11} = \rho_{22} = \rho_{33} = \rho_{44} = -2\left(a^2+b^2+c^2+d^2\right),\\ &\rho_{13} = \rho_{24} = -4\,(ac+bd)\,,\\ &\tau = -8\left(a^2+b^2+c^2+d^2\right), \quad \overset{*}{\tau} = -16\,(ac+bd)\,. \end{aligned} \tag{19}$$

Let us consider the characteristic two-planes α_{ij} spanned by the basic vectors $\{X_i, X_j\}$ at an arbitrary point of the manifold:

- totally real two-planes — α_{12}, α_{14}, α_{23}, α_{34};
- P-holomorphic two-planes — α_{13}, α_{24}.

Then, having in mind (7), (8) and (18), we obtain the corresponding sectional curvatures:

$$\begin{aligned} &\nu(\alpha_{12}) = \nu(\alpha_{34}) = \nu(\alpha_{14}) = \nu(\alpha_{23}) = -\left(a^2+b^2+c^2+d^2\right),\\ &\overset{*}{\nu}(\alpha_{12}) = \overset{*}{\nu}(\alpha_{34}) = \overset{*}{\nu}(\alpha_{14}) = \overset{*}{\nu}(\alpha_{23}) = -2\,(ac+bd)\,,\\ &\nu(\alpha_{13}) = \nu(\alpha_{24}) = \overset{*}{\nu}(\alpha_{13}) = \overset{*}{\nu}(\alpha_{24}) = 0, \end{aligned} \tag{20}$$

i.e. (G, P, g) is of constant totally real sectional curvatures.

It has been proved [6] that a Riemannian P-manifold (M, P, g) ($\operatorname{tr} P = 0$, $\dim M = 2n \geq 4$) is of constant totally real sectional curvatures ν and $\overset{*}{\nu}$, i.e.

$\nu(\alpha;p)=\nu(p)$, $\overset{*}{\nu}(\alpha;p)=\overset{*}{\nu}(p)$ for any non-degenerate totally real two-plane α in T_pM, if and only if

$$R=\nu\left\{\pi_1+\pi_2\right\}+\overset{*}{\nu}\pi_3. \tag{21}$$

Both functions ν and $\overset{*}{\nu}$ are constant if M is connected and $\dim M\geq 6$. Then, by the last statement and (20), we obtain

Theorem 2.2. *The curvature tensor R of the Riemannian P-manifold (G,P,g) has the form* (21).

It is clear that the equations (20) and (21) immediately imply $\nabla R=0$, i.e. the manifold (G,P,g) is locally symmetric. From (21) we get

$$\rho=\frac{1}{4}\left\{\tau g+\overset{*}{\tau}\widetilde{g}\right\},\quad \nu=\frac{\tau}{8},\quad \overset{*}{\nu}=\frac{\overset{*}{\tau}}{8}. \tag{22}$$

Theorem 2.2 and the equalities (19) and (22) imply the next

Corollary 2.1. *The following conditions are equivalent for (G,P,g):*

(i) $R=\frac{\tau}{8}(\pi_1+\pi_2)$, $\tau<0$;
(ii) $\overset{*}{\tau}=0$;
(iii) $c=\lambda b,\ d=-\lambda a,\ a,b,\lambda\neq 0$;
(iv) $c=\lambda d,\ \ b=-\lambda a,\ \ a,d,\lambda\neq 0$;
(v) $\rho=\frac{\tau}{4}g$, *i.e. the manifold is Einsteinian.*

3. A Lie group as a four-dimensional Riemannian product $\mathcal{W}_2$-manifold

Let G be a real connected Lie group, and let $\mathfrak{g}$ be its Lie algebra. If $\{X_1,X_2,X_3,X_4\}$ is a global basis of left invariant vector fields of G, we define an almost product structure P and an invariant Riemannian metric g on G by the conditions (10) and (11), respectively. Then, as in the previous section, (G,P,g) is a Riemannian almost product manifold.

Definition 3.1 ([1]). *An almost product structure P on a Lie group G is said to be* abelian *if*

$$[PX,PY]=-[X,Y]\quad \textit{for all } X,Y\in\mathfrak{g}. \tag{23}$$

From (23) we immediately derive that the Nijenhuis tensor vanishes on $\mathfrak{g}$, i.e. P is a product structure. Thus, the manifold (G,P,g) is a Riemannian product manifold

Now, let us consider the Lie algebra $\mathfrak{g}$ of G. If P, given by (10), is an abelian product structure we obtain

Proposition 3.1. *Let (G, P, g) be a four-dimensional Riemannian product manifold admitting an abelian product structure P defined by* (10), (11) *and* (23). *Then, the Lie algebra $\mathfrak{g}$ of G is given as follows:*

$$\begin{aligned} &[X_1, X_2] = -[X_3, X_4], \quad i.e. \quad C_{12}^k = -C_{34}^k, \\ &[X_1, X_4] = [X_2, X_3], \quad i.e. \quad C_{14}^k = C_{23}^k, \\ &[X_1, X_3] = C_{13}^k X_k, \quad [X_2, X_4] = C_{24}^k X_k, \end{aligned} \tag{24}$$

where $C_{ij}^k \in \mathbb{R}$ $(i, j, k = 1, 2, 3, 4)$ must satisfy the Jacobi identity.

It is know that a Lie algebra $\mathfrak{g}$ is said to be *nilpotent* if its descending central series

$$\mathfrak{D}^0\mathfrak{g} = \mathfrak{g}, \mathfrak{D}^1\mathfrak{g} = [\mathfrak{g}, \mathfrak{g}], \ldots, \mathfrak{D}^p\mathfrak{g} = [\mathfrak{g}, \mathfrak{D}^{p-1}\mathfrak{g}], \ldots$$

vanishes for some $p \in \mathbb{N}$ A Lie algebra $\mathfrak{g}$ is said to be two-step nilpotent, if $\mathfrak{D}^2\mathfrak{g} = \{0\}$. If a Lie algebra is nilpotent, then it is solvable.

Let us construct our example by putting $C_{12}^k = C_{34}^k = 0$, $(k = 1, \ldots, 4)$ in (24). In this case, for the non-zero Lie brackets of $\mathfrak{g}$ the Jacobi identity (9) implies

$$\begin{aligned} [X_2, X_4] &= \lambda^{-2}[X_1, X_3] = -\lambda^{-1}[X_1, X_4] \\ &= -\lambda^{-1}[X_2, X_3] = \mu(X_1 + \lambda X_2 + X_3 + \lambda X_4), \end{aligned} \tag{25}$$

where $\lambda, \mu \in \mathbb{R}$. Thus, the conditions (25) define a family of four-dimensional real Lie algebras $\mathfrak{g}$, which is characterized by two parameters. By direct computation we establish that the Lie algebras (25) are two-step nilpotent. In this case the Killing form B, defined by (16), is zero.

Now, let us study the above constructed four-dimensional manifold (G, P, g), where the Lie algebra $\mathfrak{g}$ of G is defined by (25) in the case $\lambda = 1$. Having in mind (11), (17) and (25) we obtain the following non-zero components of the Levi-Civita connection of the manifold (G, P, g):

$$\begin{aligned} &\nabla_{X_1} X_1 = -\nabla_{X_2} X_2 = -\mu X_3 + \mu X_4, \quad \nabla_{X_3} X_3 = -\nabla_{X_4} X_4 = \mu X_1 - \mu X_2, \\ &\nabla_{X_1} X_3 = -\nabla_{X_4} X_2 = \mu X_1 + \mu X_4, \qquad \nabla_{X_3} X_2 = -\nabla_{X_1} X_4 = \mu X_1 + \mu X_3, \\ &\nabla_{X_3} X_1 = -\nabla_{X_2} X_4 = -\mu X_2 - \mu X_3, \quad \nabla_{X_4} X_1 = -\nabla_{X_2} X_3 = \mu X_2 + \mu X_4. \end{aligned} \tag{26}$$

Next, taking into account (2) and (11) we get the non-zero components $F_{ijk} = F(X_i, X_j, X_k)$ of F:

$$
\begin{aligned}
&F_{111} = -F_{133} = F_{311} = -F_{333} = F_{222} = -F_{244} = F_{422} = -F_{444} = 2\mu, \\
&F_{114} = F_{314} = -F_{123} = -F_{323} = F_{223} = F_{423} = -F_{214} = -F_{414} \\
&= -F_{112} = -F_{312} = F_{134} = F_{334} = -F_{212} = -F_{412} = F_{234} = F_{434} = \mu, \\
&\alpha_i = 0, \quad \text{for } i = 1, 2, 3, 4.
\end{aligned} \tag{27}
$$

Let R be the curvature tensor of type $(0, 4)$ of (G, P, g). Having in mind (26), we get the following non-zero components R_{ijks} of R:

$$
\begin{aligned}
&R_{1223} = R_{1224} = R_{1231} = R_{1241} = \mu^2, \\
&R_{1221} = R_{1313} = R_{1414} = R_{2323} = R_{2424} = R_{3443} = 2\mu^2, \\
&R_{1442} = R_{1341} = R_{1332} = R_{2342} = 3\mu^2, \quad R_{1324} = R_{1423} = 4\mu^2, \\
&R_{1334} = R_{1443} = R_{2334} = R_{2443} = \mu^2.
\end{aligned} \tag{28}
$$

Then, according to (10) and (28) we obtain

Theorem 3.1. *The curvature tensor R of the manifold (G, P, g) has the form*

$$
R(X, Y, Z, W) = R(PX, PY, PZ, PW), \quad \textit{if} \quad \lambda = 1. \tag{29}
$$

Proof. Because $\lambda = 1$, the equations (28) imply (29). □

References

1. A. Andrada, M. Barberis, I. Dotti and G. Ovando, *Product structures on four-dimensional solvable Lie algebras*, Homology, Homotopy and Applications **7**(1) (2005), 9–37.
2. A. Andrada and I. Dotti, *Double products and hypersymplectic structures on R^{4n}*, Preprint. Available at arXiv: math. DG/0401294.
3. S. Helgason, *Differential geometry, Lie groups and symmetric spaces*, Pure and Applied Math. **80**, Academic Press (1978).
4. V. Miquel, *Some examples of Riemannian almost product manifolds*, Pac. J. Math. **111**(1) (1984), 163–178.
5. A. Naveira, *A classification of Riemannian almost-product manifolds*, Rend. Di Mat. Di Roma **3** (1983), 577–592.
6. M. Staikova, *Curvature properties of Riemannian P-manifolds*, Plovdiv Univ. Sci. Works – Math. **25**(3) (1986).
7. M. Staikova and K. Gribachev, *Canonical connections and their conformal invariants on Riemannian almost product manifolds*, Serdica, Math. J. **18** (1992), 150–161.

8. M. Staikova, K. Gribachev and D. Mekerov, *Riemannian P-manifolds of constant sectional curvatures*, Serdica, M. J. **17** (1991), 212-219.

DIFFERENTIAL FORMS OF MANY DOUBLE-COMPLEX VARIABLES

P.V. STOEV

University of Architecture, Civil Engeneering and Geodesy
1, bul. Hristo Smirnenski
Sofia, Bulgaria

In this paper we develop a calculus of exterior double complex diferential $(1,1)$-forms. The term double-complex variables is used in [1]. In fact the double-complex algebra $C(1,j)$ is algebraically isomorphic to the algebra of bi-complex numbers B. (see for instance [2]). Our treatment is of elementary character, a pure coordinate method over the algebra $C^n(1,j)$ of double complex n-vectors (for further details see [3]). The operators ∂, ∂^*, d and $\partial\partial^*$ are defined. We study the solutions of the equation $\partial\partial^* f = 0$, i.e. the double-complex pluriharmonic functions, and respectively the solutions of the double-complex Laplace equations $\Delta_n f = 0$, $\Delta_{kk} f = 0$, i.e. the double-complex harmonic functions f and complex harmonic ones. Examples of double-complex harmonic surfaces are given.

Keywords: Double-complex n-vectors; Double-complex $(1,1)$-forms; Separate holomorphicity; Double-complex pluriharmonic and n-harmonic functions; Quadratic geometries; Double-complex harmonic surfaces.

1. Elements of the calculus of double-complex differential 1-forms of many double-complex variables

1.1. We recall differential 1-forms $\omega = \varphi(\alpha)d\alpha + \psi(\alpha)d\alpha^*$ on the double-complex algebra $C(1,j)$, $\alpha \in C(1,j)$. These 1-forms generalizes the formula for the differential of a double-complex function $f(\alpha) = f_0(z,w) + jf_1(z,w)$, $j^2 = i$, $i \in C$, namely

$$df = \partial f/\partial\alpha d\alpha + \partial f/\partial\alpha^* d\alpha^*, \tag{1}$$

where $\alpha^* = z - jw$ (the conjugate of $\alpha = z + jw$) and

$$\partial f/\partial\alpha := 1/2(\partial f/\partial z - ji\partial f/\partial w),\ \partial f/\partial\alpha^* := 1/2(\partial f/\partial z + ji\partial f/\partial w), \tag{2}$$

MS Classification: 30G35

with

$$\partial f/\partial z := \partial f_0/\partial z + j\partial f_1/\partial z, \quad \partial f/\partial w := \partial f_0/\partial w + j\partial f_1/\partial w.$$

Replacing f by $f_0 + jf_1$ we receive too

$$\partial f/\partial \alpha = 1/2(\partial f_0/\partial z + \partial f_1/\partial w) - ji/2(\partial f_0/\partial w + i\partial f_1/\partial z), \tag{2'}$$

$$\partial f/\partial \alpha^* = 1/2(\partial f_0/\partial z - \partial f_1/\partial w) + ji/2(\partial f_0/\partial w - i\partial f_1/\partial z), \tag{2''}$$

$$d\alpha = dz + jdw, \quad d\alpha^* = dz - jdw.$$

The operator of exterior differentiation is defined as ordinary

$$d\omega = d\varphi(\alpha) \wedge d\alpha + d\psi(\alpha) \wedge d\alpha^*. \tag{3}$$

By definition $d\alpha \wedge d\alpha = d\alpha^* \wedge d\alpha^* = 0$, and $d\alpha \wedge d\alpha^* = -d\alpha^* \wedge d\alpha$.

1.2. Now, we take a double-complex valued function of n double-complex variables

$$f(\alpha^1, \dots, \alpha^n) = f_0(\alpha^1, \dots, \alpha^n) + jf_1(\alpha^1, \dots, \alpha^n), \tag{4}$$

where $\alpha := (\alpha^1, \dots, \alpha^n) \in C^n(1, j)$, $f(\alpha) := f(\alpha^1, \dots, \alpha^n) \in C(1, j)$.

In this case we have $df = \sum_k \{\partial f/\partial \alpha^k d\alpha^k + \partial f/\partial (\alpha^*)^k d(\alpha^k)^*\}$, or

$$df = \sum_k \{\partial f/\partial \alpha^k d\alpha^k\} + \sum_k \{\partial f/\partial (\alpha^k)^* d(\alpha^k)^*\}. \tag{5}$$

Setting $\sum_k \{\partial f/\partial \alpha^k d\alpha^k\} =: \partial f$ and $\sum_k \{\partial f/\partial (\alpha^k)^* d(\alpha^k)^*\} =: \partial^* f$ we can write shortly

$$df = \partial f + \partial^* f. \tag{5'}$$

A differential 1-form over $C^n(1, j)$ has the following coordinate representation

$$\omega = \sum_k \varphi_k d\alpha^k + \sum_k \psi_k d(\alpha^k)^*, \tag{6}$$

where $\varphi_k = \varphi_k(\alpha^1, \dots, \alpha^n) = \varphi_k(\alpha)$ and $\psi_k = \psi_k(\alpha^1, \dots, \alpha^n) = \psi_k(\alpha)$ are double-complex valued functions of n double-complex variables. For the differential of the 1-form ω we calculate that

$$d\omega = \sum_k (\partial \psi_k/\partial \alpha^k - \partial \varphi_k/\partial (\alpha^k)^*) d\alpha^k \wedge d(\alpha^k)^*. \tag{7}$$

So, ω is a close double-complex 1-form iff

$$\partial \psi_k/\partial \alpha^k - \partial \varphi_k/\partial (\alpha^k)^* = 0 \quad \text{for all} \quad k = 1, \dots, n. \tag{8}$$

In the case $\omega = dg(\alpha)$, $\alpha \in D$, D being the domain of the function g, ω is an exact double-complex form on D.

Proposition 1.1. *Each exact double-complex 1-form $\omega(\alpha)$, $\alpha = (\alpha^k)$, is a close double complex 1-form.*

Proof. If $\omega(\alpha) = dg(\alpha) = \sum\{\partial g/\partial\alpha^k d\alpha^k + \partial g/\partial(\alpha^k)^* d(\alpha^k)^*\}$ evidently

$$\partial/\partial(\alpha^k)(\partial g/\partial(\alpha^k)^*) - \partial/\partial(\alpha^k)^*(\partial g/\partial(\alpha^k)) = 0.$$

According to (8) the double-complex 1-form $\omega(\alpha)$ is closed. □

Example 1.1. Taking $\varphi_k(\alpha) = \alpha^k$ and $\psi_k(\alpha) = (\alpha^k)^*$ we obtain the form $\omega = 1/2d(\sum_k[(\alpha^k)^2 + ((\alpha^k)^*)^2]) = d(\sum_k[(z^k)^2 + i(w^k)^2])$, which is an exact double-complex form on $C(1,j)$ and, of course, it is a closed one.

In the case

$$\omega = \sum_k \varphi_k(\alpha) d\alpha^k, \tag{9}$$

with coefficients $\varphi_k(\alpha)$ which are double-complex holomorphic functions with respect to each double-complex variable α^k on a common open domain D, $k = 1, \ldots, n$ we say that ω is a holomorphic double-complex 1-form on D. When $df = \sum \partial f/\partial\alpha^k d\alpha^k$ on an open domain D, we say that $f = f(\alpha)$ is a separately holomorphic double-complex function. This means that $\partial^* f = 0$, or

$$\partial f/\partial(\alpha^k)^* = 0, \quad k = 1, \ldots, n,$$

and, respectively, the Cauchy-Riemann equations

$$\partial f_0/\partial z - \partial f_1/\partial w = 0, \quad \partial f_0/\partial w - i\partial f_1/\partial z = 0. \tag{9'}$$

2. The equation $\partial\partial^* f = 0$

According to the above introduced notations we have

$$\partial\partial^* f = \sum_k \partial/\partial\alpha^k \left(\sum_l \partial f/\partial(\alpha^l)^* d(\alpha^l)^*\right) d\alpha^k,$$

where $\alpha^k = z^k + jw^k$, $(\alpha^l)^* = z^l - jw^l$.

It is easy to see that $\partial\partial^* f + \partial^*\partial f = 0$. We see that the $(1,1)$-form $\partial\partial^* f$ is expressed in terms of double-complex partial derivatives as follows

$$\partial\partial^* f = \sum_k \sum_l \partial^2 f/\partial\alpha^k\partial(\alpha^l)^* d\alpha^k \wedge d(\alpha^l)^*, \tag{10}$$

or using the matrices $\|\partial^2 f/\partial\alpha\partial(\alpha^l)^*\|_{k,l}$ we can write

$$\partial\partial^* f = (d\alpha^1, \dots, d\alpha^n)\|\partial^2 f/\partial\alpha\partial(\alpha^l)^*\|_{k,l}\begin{pmatrix} d(\alpha^1)^* \\ \vdots \\ d(\alpha^n)^* \end{pmatrix}. \quad (10')$$

Proposition 2.1. *The* $(1,1)$*-form* $\partial\partial^* f$ *can be represented by partial complex derivatives of the type* $\partial/\partial z^k$, $\partial/\partial w^k$, $\partial^2/\partial z^k\partial w^l$ *etc.*

$$\partial\partial^* f = 1/4\sum\sum\{\Delta_{kl}f_0 + \partial^2 f_1/\partial z^l\partial w^k - \partial^2 f_1/\partial z^k\partial w^l \\ + j(\Delta_{kl}f_1 - i(\partial^2 f_0/\partial z^k\partial w^l - \partial^2 f_0/\partial z^l\partial w^k)\}d\alpha^k \wedge d(\alpha^l)^*, \quad (11)$$

with $d\alpha^k \wedge d(\alpha^l)^* = dz^k \wedge dz^l - idw^k \wedge dw^l - j(dz^l \wedge dw^k + dz^k \wedge dw^l)$, *where* $\Delta_{kl}f_0 = \partial^2 f_0/\partial z^k\partial z^l + i\partial^2 f_0/\partial w^k\partial w^l$, $\Delta_{kl}f_1 = \partial^2 f_1/\partial z^k\partial z^l + i\partial^2 f_1/\partial w^k\partial w^l$, f_0 *is the even part, and* f_1 *is the odd part of the double-complex function* f.

Proof. It is enough to calculate $\partial^2 f/\partial\alpha^k\partial(\alpha^l)^* = 1/2\partial/\partial\alpha^k(\partial f/\partial z^l + ji\partial f/\partial w^l)$, (using (2)). One obtain the formula

$$\partial^2 f/\partial\alpha^k\partial(\alpha^l)^* = 1/4(\partial^2 f/\partial z^k\partial z^l + i\partial^2 f/\partial w^k\partial w^l \\ - ji(\partial^2 f/\partial z^k\partial w^l - \partial^2 f/\partial z^l\partial w^k). \quad (12)$$

It remains to use (2), (2′) and (2″). The expresion for $d\alpha^k \wedge d(\alpha^l)^*$ is obtained directly. □

In the case $l = k$ we receive

$$\partial\partial^* f|_{kk} = 1/4\sum_k \partial^2 f/\partial\alpha^k\partial(\alpha^k)^* d\alpha^k \wedge d(\alpha^k)^*, \quad (13)$$

$$\partial\partial^* f|_{kk} = 1/4\sum_k\{\Delta_{kk}f_0 + j(\Delta_{kk}f_1\}d\alpha^k \wedge d(\alpha^k)^*, \quad (13')$$

where

$$\Delta_{kk}f_0 = \partial^2 f_0/(\partial z^k)^2 + i\partial^2 f_0/\partial(w^k)^2,\ \Delta_{kk}f_1 = \partial^2 f_1/(\partial z^k)^2 + i\partial^2 f_0/\partial(w^k)^2.$$

For a function of two complex variables $h(z,w)$ the operator

$$h(z,w) \to \partial^2 h/(\partial z^k)^2 + i\partial^2 h/\partial(w^k)^2$$

is called a double-complex Laplace operator. The above received operator $f_0 \to \Delta_{kk}f_0$ and $f_1 \to \Delta_{kk}f_1$ is double-complex Laplace operator with respect to the double-complex variable $\alpha^k = z^k + jw^k$.

Setting $\Delta_n := \sum \Delta_{kk}$, we present (13′) as follows

$$\partial\partial^* f = 1/4(\Delta_n f_0 + j\Delta_n f_1) d\alpha^k \wedge d(\alpha^k)^*. \tag{13''}$$

The solutions of the equation $\partial\partial^* f = 0$ on an open domain D are called double-complex pluriharmonic functions on D. The solutions $f = f_0 + jf_1$ of the equation

$$\Delta_n f_0 + j\Delta_n f_1 = 0, \tag{14}$$

or equivalently of the system

$$\Delta_n f_0 = 0, \quad \Delta_n f_1 = 0, \tag{14'}$$

are called double-complex n-harmonic functions. The even part and the odd part are called n-harmonic functions of many complex variables and the corresponding surfaces are complex harmonic surfaces.

Corollary 2.1. *The even part and the odd part of a double-complex holomorphic function are plruiharmonic functions. Clearly, each double-complex pluriharmonic function is a double-complex n-harmonic one. The class of double-complex n-harmonic function is larger.*

An example of a double-complex separately holomorphic function $F(\alpha)$ is given below in double-complex coordinates (α^k) and respectivelly in complex coordinates

$$F(\alpha) = \sum_{k,l} \alpha^k \alpha^l = \sum_{k,l} (z^k z^l + iw^k w^l) + j \sum_{k,l} (w^k z^l + w^l z^k).$$

The even part $\sum_{k,l}(z^k z^l + iw^k w^l)$ and the odd part $\sum_{k,l}(w^k z^l + w^l z^k)$ give examples of double-complex pluriharmonic functions.

Having a given double-complex pluriharmonic function $h(z, w)$ we can construct the corresponding double-complex holomorphic one by integrating a system of two equations. Indeed, for one double-complex variable or, equivalently, for two complex variables, let $f(\alpha) = h(z, w) + jf_1(z, w)$ be the mentioned corresponding double-complex holomorphic function. According to Cauchy-Riemann system we have

$$\partial f_1/\partial w = \partial h/\partial z, \quad \partial f_1/\partial z = -i\partial h/\partial w. \tag{15}$$

Now we take the differential of f_1

$$df_1 = \partial f_1/\partial z dz + \partial f_1/\partial w dw = -i\partial h/\partial w dz + \partial h/\partial z dw, \tag{16}$$

and integrate it. For many double-complex variables we follow the same way.

3. Quadratic geometry on $C^n(1,j)$

A scalar product over the algebra $C^n(1,j)$ is defined first for the units 1 and j, considered as double-complex numbers: $1 = 1 + j0$, $1 \in \mathbf{R}$, and the hyper-complex number j joint to $C \times C$ ($j = 0 + j1$, $j \notin \mathbf{C}$). We have in mind

$$\langle 1,1\rangle = 1, \quad \langle 1,j\rangle = \langle j,1\rangle = 0, \quad \langle j,j\rangle = i, \quad i \in \mathbf{C}. \tag{17}$$

Let α, β be a pair of double-complex numbers. By definition

$$\langle \alpha,\beta\rangle = \sum_{k,l}\langle \alpha^k,\beta^l\rangle, \quad k,l = 1,\ldots,n. \tag{18}$$

According to (17)

$$\langle \alpha^k,\beta^l\rangle = z^k z^l + i w^k w^l. \tag{19}$$

and, we receive that $\langle \alpha,\beta\rangle$ is a complex number, namely

$$\langle \alpha,\beta\rangle = \sum (z^k z^l + i w^k w^l). \tag{20}$$

Especially, the equation of the isotropic cone for this scalar product seems as follows

$$\langle \alpha,\alpha\rangle = \sum_k ((z^k)^2 + i(w^k)^2) = 0. \tag{21}$$

The complex quadratic form in the left side in the above written equation just determines the mentiond quadratic geometry [4] over $C^n(1,j)$.

Proposition 3.1. *We have that*

$$\Delta_n\left(\sum_k ((z^k)^2 + i(w^k)^2\right) = 0,$$

i.e. the surface defined by the quadratic form of isotropic cone $\langle \alpha,\alpha\rangle = 0$ *is a complex harmonic surface.*

Proof. In wiev of $\Delta_n := \sum \Delta_{kk}$ we obtain

$$\sum \Delta_n\Big(\sum ((z^k)^2 + i(w^k)^2)\Big) = \sum \Delta_{kk}((z^k)^2 + i(w^k)^2).$$

Evidently

$$\Delta_{kk}((z^k)^2 + i(w^k)^2) = 2 + i(2i) = 0.$$

□

Remark 3.1. In an other quadratic geometry on $C^n(1,j)$, namely this one defined by the following scalar values for the units 1 and j

$$\langle 1,1\rangle = 1, \quad \langle 1,j\rangle = \langle j,1\rangle = 0, \quad \langle j,j\rangle = 1, \quad 1 \in \mathbf{R},$$

the scalar product is $\langle\alpha,\beta\rangle = \sum(z^k z^l + w^k w^l)$, and the isotropic cone is defined by the equation $\langle\alpha,\alpha\rangle = \sum\big((z^k)^2 + (w^k)^2\big) = 0$. Our remark says that the quadratic form $\omega = \sum\big((z^k)^2 + (w^k)^2\big)$ does not satisfies the equation (14′).

Remark 3.2. If we consider a complex algebra in which the Cauchy-Riemann equation are $\partial u(z,w)/\partial z = \partial v(z,w)/\partial w$, $\partial u(z,w)/\partial v = -\partial v(z,w)/\partial z$ (i.e. they are the same as the known ones $\partial u(x,y)/\partial x = \partial v(x,y)/\partial y$, $\partial u(x,y)/\partial y = -\partial v(x,y)/\partial x$ obtained by replacing the real variables x, y by the complex ones, we see that quadratic form $(z^k)^2-(w^k)^2$ does not satisfies the corresponding Laplace equation, which is $\partial^2 u/\partial(z^2)^2 + \partial^2 u/\partial(w^2)^2 = 0$.

This means that the surface of the isotropic cone is not a harmonic surface in the quadratic geometry defined in Remark 3.1.

4. Complex-hermitian double-complex quadratic forms

We shall consider complex-Hermitian quadratic canonical forms [3]

$$Q = Q(\alpha,\alpha) = \sum \lambda_k \alpha^k (\alpha^k)^* = \sum \lambda_k (z^l + jw^l)(z^k - jw^k).$$

It is easy to calculate the even part $Q_0(\alpha,\alpha)$, and the odd part $Q_1(\alpha,\alpha)$

$$Q_0(\alpha,\alpha^*) = \sum_k \lambda_k((z^k)^2 - i(w^k)^2), \quad Q_1(\alpha,\alpha) = 0.$$

Clearly, the considered form Q is not double-complex holomorphic form. Setting $Q_{0k} = \lambda_k((z^k)^2 - i(w^k)^2)$ and $Q_{1k} = 0$, we can write

$$Q_0(\alpha,\alpha) = \sum_k Q_{0k} \quad \text{and} \quad Q_1(\alpha,\alpha) = \sum_k Q_{1k} = 0.$$

We set also $\Delta_k^- := \partial^2 U/(\partial z^k)^2 - i\partial^2 U/\partial(w^k)^2$, $\Delta_k^+ := \partial^2 U/(\partial z^k)^2 + i\partial^2 U/\partial(w^k)^2$, (In fact, we have $\Delta_k^+ = \Delta_{kk}$), and

$$\Delta^- := \sum_k \Delta_k^-, \quad \Delta^+ := \sum_k \Delta_k^+. \tag{22}$$

Proposition 4.1. *The complex valued functions $Q_0(\alpha,\alpha)$ and $Q_1(\alpha,\alpha) = 0$ satisfy the following equations*

$$\Delta^-(U) = 0, \quad \textit{and also } \Delta^+(U) = 0 \quad \textit{if} \quad \sum \lambda_k = 0.$$

Proof. It is easy to see that $\Delta_k^+(Q_{0k}) = 4\lambda_k$, and $\Delta_k^-(Q_{0k}) = 0$. This implies

$$\Delta^+(Q_0) = 4\sum_k \lambda_k, \quad \Delta_-(Q_0) = 0.$$

It remains to take in view the definitions of Δ^- and Δ^+, (22). □

We shall consider the products $\Delta_k^+\Delta_k^-(U) = (\partial^2 U/(\partial z^k)^2)^2 + (\partial^2 U/\partial(w^k)^2)^2$, $k = 1, \ldots, n$, and

$$\Delta(U) = \prod_k \Delta_k^+\Delta_k^-(U) = \prod_k \{(\partial^2 U/(\partial z^k)^2)^2 + (\partial^2 U/\partial(w^k)^2)^2\}. \quad (23)$$

By $HP(p,q)$ is denoted the space of the homogeneous polynomials of the double-complex variables $\alpha^1, \alpha^2, \ldots, \alpha^n$ and its conjugates $(\alpha^1)^*, (\alpha^2)^*, \ldots, (\alpha^n)^*$, of total degree p for α^k, and respectively, of total degree q for $(\alpha^k)^*$. Clearly, $HP(1,0)$ coincides with the space of double-complex linear forms of n complex variables, and $HP(p,0)$ is the space of double-complex homogeneous polynomials of degree p. Analogically, $H(0,q)$ can be considered as the space of double-complex anti-holomorphic polynomials.

Example: $\lambda_1\alpha^1(\alpha^1)^* + \lambda_2\alpha^2(\alpha^2)^* + \cdots + \lambda_n\alpha^n(\alpha^n)^*$ is an element of $HP(1,1)$, $\lambda_j \in C$. The space $HP(1,1)$ is just the space of complex-Hermitian quadratic forms.

We shall present here some observations about the spaces $HP(2,1)$, $HP(1,2)$ and $HP(2,2)$. Let $P = \sum \lambda_k(\alpha^k)^2(\alpha^k)^*$ be an element of $HP(2,1)$, $P = P_0 + jP_1$. In complex coordinates we have the following two complex ternary forms

$$P_0 = \sum \lambda_k((z^k)^3 - iz^k(w^k)^2) \quad \text{and} \quad P_1 = \sum \lambda_k((z^k)^2 w^k - i(w^k)^3).$$

It is easy to see that P is not a double-complex holomorphic polynomial. We have $\Delta^+(P_0) = 8\sum \lambda_k z^k$, $\Delta^+(P_1) = 8\sum \lambda_k w^k$ and $\Delta^-(P_0) = 4\sum \lambda_k z^k$, $\Delta^-(P_1) = 4\sum \lambda_k w^k$. So, the restriction of the polynomial P_0 on the complex hyperplane $Z = \sum \lambda_k z^k = 0$, defines a double-complex harmonic surface as $\Delta^+(P_0|Z) = 0$ and $\Delta^-(P_0|Z) = 0$. It coincides with the intersection of the complex hyperplane Z with the quadratic complex surface S, defined by the equation $\sum \lambda_k(z^k)^2 = 0$. For the polynomial P_1, restricted on the hyperplane $W = \sum \lambda_k w^k = 0$ the situation is similar.

Analogous statements hold for the polynomials of $HP(1,2)$.

In the space $HP(2,2)$ we have $P = \sum \lambda_k(\alpha^k)^2((\alpha^k)^*)^2$ in double-complex coordinates, and in complex coordinates: $P_0 = \sum \lambda_k((z^k)^2 -$

$i(w^k)^2)^2$, $P_1 = 0$. The polynomial P is not a double-complex holomorphic polynomial. Calculating we receive

$$\Delta^+(P_0) = (3+i)\sum \lambda_k((z^k)^2 - (w^k)^2) \quad \text{and}$$
$$\Delta^-(P_0) = (3+i)\sum \lambda_k((z^k)^2 + (w^k)^2),$$

concluding that $\Delta^+(P_0)$ annihilates on the complex surface $\sum \lambda_k((z^k)^2 - (w^k)^2) = 0$, but $\Delta^-(P_0)$ does not annihilate on the same complex surface, as of course we suppose that $P_0 \neq 0$. This means that the restriction of the polynomial P_0 (of complex degree 4) on the considered complex surface (of degree 2) defines a double-complex harmonic surface.

Analogous statement can be formulated for $\Delta^-(P_0)$ and the complex surface $\sum \lambda_k((z^k)^2 + (w^k)^2) = 0$. As a corollary we obtain that the equation defined by the operator Δ (see (23)), annihilates on the surface $\sum \lambda_k((z^k)^4 - (w^k)^4) = 0$.

Proposition 4.2. *All element of $H(2,2)$ satisfy the equation*

$$\Delta(U) = \prod_k \Delta_k^+ \Delta_k^-(U) = \prod_k \{(\partial^2 U/(\partial z^k)^2)^2 + (\partial^2 U/\partial (w^k)^2)^2\} = 0,$$

on the surface $\lambda_k((z^k)^4 - (w^k)^4) = 0$.

References

1. S. Dimiev, *Double complex analytic geometry*, to be published in the Proceedings of the Conference in Bedlevo, Poland.
2. S. Roon, *Bicomplex algebra and function theory*, Complex variables, 1 (2001) 1–71.
3. B. Kiradjiev, *Double-complex quadratic forms*, to appear in the same book.
4. M.M. Postnikov, *Linear Algebra and Differential Geometry*, Lectures on Geometry II, Moskow, 1979 (in Russian).

ISOMETRIC IMMERSIONS AND EXTRINSIC SHAPES OF SMOOTH CURVES

T. SUGIYAMA

Division of Mathematics and Mathematical Science
Department of Computer Science and Engineering
Graduate School of Engineering, Nagoya Institute of Technology
Gokiso, Nagoya, 466-8555, JAPAN
E-mail: tadashi@zelus.ics.nitech.ac.jp

We study isometric immersions which preserve the "order 2" property or curvature logarithmic derivatives of smooth curves without inflection points. We give a report on characterization of totally umbilic immersions and isotropic immersions from this point of view.

1. Introduction

Let $f : M \to \widetilde{M}$ be an isometric immersion between Riemannian manifolds. For a smooth curve γ on M, we call the curve $f \circ \gamma$ on $\widetilde{M}$ the *extrinsic shape* of γ through f. It is an interesting problem to study how properties of isometric immersions are reflected on properties of extrinsic shapes of curves and how properties of curves characterize isometric immersions. In this paper we study immersions which preserve "order 2" property of curves and immersions which preserve logarithmic derivatives of curvature for some curves. There are many results on characterizing some immersions by properties of extrinsic shapes of "nice" curves (for example see [2, 4] and papers in their references). In our results we do not stick on "nice curves" but we consider what kind of properties of curves are preserved by immersions. In the first half, sections 2, 3, 4, we study immersions by "order 2" properties of curves, and in the second half, sections 5, 6, we study immersions by logarithmic derivatives of curvatures.

2. Totally umbilic immersions and order 2 property

For a smooth curve $\gamma : I \to M$ parameterized by its arclength, we define a positive function κ_γ by $\kappa_\gamma = \|\nabla_{\dot\gamma}\dot\gamma\|$ and call it the (first) *curvature function* of γ. We say this curve γ does not have *inflection points* if κ_γ does not vanish along γ. For such a curve γ, we define a unit vector field Y_γ along γ which is orthogonal to $\dot\gamma$ by $Y_\gamma = (1/\kappa_\gamma)\nabla_{\dot\gamma}\dot\gamma$. We then find it satisfies

$$\nabla_{\dot\gamma}\dot\gamma = \kappa_\gamma Y_\gamma, \quad \nabla_{\dot\gamma}Y_\gamma = -\kappa_\gamma\dot\gamma + Z_\gamma,$$

with some vector field Z_γ along γ which is orthogonal to both $\dot\gamma$ and Y_γ. We say a smooth curve γ without inflection points is of *proper order* 2 at $\gamma(t_0)$ if Z_γ vanishes at this point. When κ_γ is a positive constant function and $Z_\gamma \equiv 0$, this curve is called a *circle* of positive curvature. We say a geodesic to be a circle of null curvature.

We here give a relationship between curvatures of curves and their extrinsic shapes. Let $f : M \to \widetilde{M}$ be an isometric immersion. We denote by ∇ and $\widetilde\nabla$ the Riemannian connections of M and $\widetilde{M}$, respectively. Their relations are given as Gauss and Weingarten formulae

$$\widetilde\nabla_X Y = \nabla_X Y + \sigma(X,Y), \quad \widetilde\nabla_X\xi = -A_\xi X + \nabla_X^\perp\xi,$$

for vector fields X, Y on $\widetilde{M}$ tangent to M and a vector field ξ on $\widetilde{M}$ normal to M. Here, σ is the second fundamental form, A is the shape operator, and $\nabla_X^\perp\xi$ denotes the normal component of $\widetilde\nabla_X\xi$. We define the covariant differentiation $\overline\nabla$ of the second fundamental form σ with respect to the connection of $TM \oplus TM^\perp$ by

$$(\overline\nabla_X\sigma)(Y,Z) = \nabla_X^\perp(\sigma(Y,Z)) - \sigma(\nabla_X Y, Z) - \sigma(X, \nabla_X Z).$$

For a smooth curve γ on M we denote by $\widetilde\kappa_\gamma$ the curvature function of the extrinsic shape $\widetilde\gamma = f\circ\gamma$ of γ. By use of Gauss and Weingarten formulae, we can obtain the following by direct calculation (see [5]).

Lemma 2.1. *Curvature functions of a smooth curve γ on M and its extrinsic shape $f\circ\gamma$ satisfy $\widetilde\kappa_\gamma^2 = \kappa_\gamma^2 + \|\sigma(\dot\gamma,\dot\gamma)\|^2$. Hence $f\circ\gamma$ does not have inflection points if γ dose not have inflection points.*

Lemma 2.2. *If we decompose $Z_{f\circ\gamma}$ along $f\circ\gamma$ into tangential and normal components as $\widetilde Z_\gamma^T + \widetilde Z_\gamma^\perp$, we have the following:*

$$\kappa_\gamma\dot{\widetilde\kappa}_\gamma Y_\gamma - \widetilde\kappa_\gamma^3\dot\gamma + \widetilde\kappa_\gamma^2\widetilde Z_\gamma^T = \widetilde\kappa_\gamma\big(\dot\kappa_\gamma Y_\gamma - \kappa_\gamma^2\dot\gamma + \kappa_\gamma Z_\gamma - A_{\sigma(\dot\gamma,\dot\gamma)}\dot\gamma\big), \tag{2.1}$$

$$\dot{\widetilde\kappa}_\gamma\sigma(\dot\gamma,\dot\gamma) = \widetilde\kappa_\gamma\big\{3\kappa_\gamma\sigma(\dot\gamma,Y_\gamma) + (\overline\nabla_{\dot\gamma}\sigma)(\dot\gamma,\dot\gamma) - \widetilde Z_\gamma^\perp\big\}. \tag{2.2}$$

We denote by $\mathfrak{h}$ the mean curvature vector for an isometric immersion f which is defined as

$$M \ni x \longmapsto \frac{1}{n}\sum_{n=1}^{n}\sigma(e_i,e_i) \in \nu(M),$$

where $\{e_i\}$ is an orthonormal basis of tangent space T_xM and $\nu(M)$ is the normal bundle of M. We say M is umbilic at a point x if $\sigma(u,v) = \langle u,v\rangle\mathfrak{h}$ holds for orthonormal basis $u, v \in T_xM$ at $x \in M$. We call M totally umbilic when all points are umbilic. A Riemannian submanifold is called an *extrinsic sphere* if it is totally umbilic and has parallel mean curvature vector. We here recall a characterization of extrinsic spheres due to Nomizu and Yano [4], which is a foundation of our study.

Nomizu-Yano's Theorem. *A Riemannian submanifold M is an extrinsic sphere in $\widetilde{M}$ if and only if all circles on M are circles on $\widetilde{M}$.*

For a standard sphere S^2 in a Euclidean space $\mathbb{R}^3$, geodesics and circles on S^2 can be seen as circles in $\mathbb{R}^3$. Nomizu-Yano's Theorem shows such a property holds for general extrinsic spheres. We generalize their result in the following manner. For a smooth curve γ without inflection points, we denote by ℓ_γ the logarithmic derivative of κ_γ, that is $\ell_\gamma = \kappa'_\gamma/\kappa_\gamma$.

Theorem 2.1. *An isometric immersion $f : M \to \widetilde{M}$ is umbilic at a point $x \in M$ if for each orthonormal pair (u,v) of tangent vectors of M at x there exist two curves $\gamma_1, \gamma_2 : (-\epsilon,\epsilon) \to M$ parameterized by their arclength which satisfy the following four conditions:*

i) γ_1, γ_2 *do not have inflection points,*
ii) $\gamma_i(0) = x$, $\dot{\gamma}_i(0) = u$, $\nabla_{\dot\gamma_i}\dot\gamma_i = (-1)^{i-1}\kappa_{\gamma_i}(0)v$,
iii) *their extrinsic shapes $f\circ\gamma_1$ and $f\circ\gamma_2$ are of proper order* 2 *at* $f(x)$,
iv) $\ell_{\gamma_1}(0) = \ell_{\gamma_2}(0)$.

Outline of proof of Theorem 2.1. Under the condition of Theorem 2.1, we have

$$\begin{aligned}\Big\{\frac{\kappa'_{\gamma_i}(0)}{\kappa_{\gamma_i}(0)} + \frac{(-1)^i}{\kappa_{\gamma_1}(0)}\langle\sigma(u,u),\sigma(u,v)\rangle\Big\}\sigma(u,u)&\\ = (-1)^{i-1}3\kappa_{\gamma_1}(0)\sigma(u,v) + (\bar\nabla_u\sigma)(u,u),&\qquad (i=1,2)\end{aligned} \tag{2.3}$$

by Lemma 2.1. We hence obtain $\sigma(u,v) = 0$ for every orthonormal pair (u,v) of tangent vectors of M at x, and get the conclusion. □

Our condition that an immersion preserves the order 2 property of curves tells not only umbilic property on this immersion but more on the mean curvature vector. We call a point *geodesic* if the second fundamental form vanishes at this point. We set $H = \|\mathfrak{h}\|$ and call the mean curvature of an immersion f. We say f has *parallel normalized mean curvature vector* if either it is minimal ($H=0$) or it satisfies $H \neq 0$ and $\nabla(\mathfrak{h}/H) = 0$. As a consequence of Theorem 2.1 we can characterize totally umbilic immersions with parallel normalized mean curvature vector by the property that they preserve the order 2 property of curves.

Theorem 2.2. *For an isometric immersion $f : M \to \widetilde{M}$ the following conditions are mutually equivalent:*

1) *f is totally umbilic and $\mathfrak{h}/H$ is parallel on outside of the set M_0 of geodesic points of f;*
2) *For every orthonormal pair (u, v) of tangent vectors of M at an arbitrary point $x \in M$, there exist two curves $\gamma_1, \gamma_2 : (-\epsilon, \epsilon) \to M$ parameterized by their arclength which do not have inflection points and satisfy the four conditions mentioned in Theorem 2.1.*

We here make mention on how the conditions in Theorem 2.1 is related to the geodesic property of an immersion f. For a unit tangent vector $u \in U_xM$, we denote by $\mathcal{F}(u)$ the family of all smooth curves which are parameterized by their arclength, do not have inflection points, and satisfy the following conditions;

i) their initial vectors are u,
ii) their extrinsic shapes are of proper order 2 at $f(x)$.

We put $\mathcal{A}(u) := \{\, \ell_\gamma(0) \mid \gamma \in \mathcal{F}(u) \,\} (\subset \mathbb{R})$. Under the conditions in Theorem 2.1 we see f is umbilic at x. Hence we may suppose f is umbilic at x when we consider those conditions.

Theorem 2.3. *Let $f : M \to \widetilde{M}$ is an immersion which is umbilic at a point $x \in M$. Then f is geodesic at x if and only if the set $\mathcal{A}(u)$ contains at least two distinct numbers for some (hence all) $u \in U_xM$.*

By this theorem we see under the second condition in Theorem 2.2 the set $\mathcal{A}(u)$ consists of a single value for every unit tangent vector u at a point in the outside $M \setminus M_0$ of the set of geodesic points. If we denote as $\mathcal{A}(u) := \{a(u)\}$ and define a 1-form ω on $M \setminus M_0$ by $\omega(v) = a(v/\|v\|)\|v\|$, we find it is closed. We take a function φ with $\omega = d\varphi$. When f is an embedding and $M_0 = \emptyset$, by changing metrices conformally as $e^{2\varphi}\langle\ ,\ \rangle$ we find M is an extrinsic sphere in $\widetilde{M}$ with respect to these new metrices.

3. Kähler isometric immersions and order 2 property

In this section we study totally geodesic Kähler immersions by the order 2 property on curves. We shall say a smooth curve γ on a Kähler manifold

(M, J) is *Kähler* at a point $x = \gamma(0)$ if it is of proper order 2 at x and either $Y_\gamma(0) = J\dot{\gamma}(0)$ or $Y_\gamma(0) = -J\dot{\gamma}(0)$ holds. That is, γ satisfies

$$\nabla_{\dot{\gamma}}\dot{\gamma}(0) = \pm\kappa_\gamma(0)J\dot{\gamma}(0), \quad \nabla_{\dot{\gamma}}Y_\gamma(0) = \mp\kappa_\gamma(0)\dot{\gamma}(0),$$

where double signs take the opposite signatures. For a Kähler immersions between Kähler manifolds we have $\sigma(u, Jv) = J\sigma(u, v)$ for an arbitrary pair (u, v) of tangent vectors. By use of (2.3) we can conclude the following:

Theorem 3.1. *A Kähler isometric immersion $f : (M, J) \to (\widetilde{M}, J)$ between Kähler manifolds is geodesic at a point $x \in M$ if for an arbitrary unit tangent vector $u \in U_xM$ there exist two curves $\gamma_1, \gamma_2 : (-\epsilon, \epsilon) \to M$ parameterized by their arclength which satisfy the following conditions:*

i) *γ_1, γ_2 do not have inflection points,*
ii) *$\gamma_i(0) = x$, $\dot{\gamma}_i(0) = u$, $Y_{\gamma_i}(0) = (-1)^{i-1}Ju$,*
iii) *extrinsic shapes $f \circ \gamma_1$, $f \circ \gamma_2$ are of proper order 2, hence are Kähler, at $f(x)$.*

Corollary 3.1. *A Kähler isometric immersion $f : (M, J) \to (\widetilde{M}, J)$ between Kähler manifolds is totally geodesic if and only if the necessary condition in Theorem 3.1 holds at every point $x \in M$.*

This gives a generalization of Maeda and Tanabe's characterization of totally geodesic immersions by Kähler Frenet curves (see [3] for detail).

4. Immersions of rank one symmetric spaces into real space forms

In this section we study isometric immersions of Kähler and quaternionic Kähler manifolds into real space forms by the order 2 property of curves. We denote by $M^m(c;\mathbb{R})$ a real space form of constant sectional curvature c, which is a standard sphere $S^m(c)$, a Euclidean space $\mathbb{R}^m$ and a real hyperbolic space $\mathbb{R}H^m(c)$ according to c is positive, zero and negative. An isometric immersion $f : M \to \widetilde{M}$ is said to be *isotropic* at $x \in M$ if the norm of the second fundamental form $\|\sigma(u, u)\|$ does not depend on the choice of unit tangent vector $u \in U_xM$. When f is isotropic everywhere on M we just call it *isotropic*. For an isotropic immersion f we denote by λ_f the function of M showing the norm $\|\sigma(u, u)\|$ at each point of x and call it the function of isotropy. When this function is constant, we call f *constant isotropic*. Clearly, if f is umbilic at a point it is isotropic at this point, but not vice versa when $\dim(\widetilde{M}) \geq \dim(M) + 2$.

Theorem 4.1. *Let f be an isometric immersion of a connected Kähler manifold M of complex dimension n into a real space form $M^{2n+p}(\tilde{c};\mathbb{R})$. Suppose for every unit tangent vector $u \in U_xM$ at an arbitrary point $x \in M$ there exist two smooth curves $\gamma_1, \gamma_2 : (-\epsilon, \epsilon) \to M$ which are parameterized by their arclength, do not have inflection points and satisfy the following four conditions:*

i) *γ_1, γ_2 are of proper order 2 at x,*
ii) *$\gamma_i(0) = x$, $\dot{\gamma}_i(0) = u$, $Y_{\gamma_i}(0) = (-1)^{i-1}Ju$,*
iii) *extrinsic shapes $f \circ \gamma_1$, $f \circ \gamma_2$ are of proper order 2 at $f(x)$,*
iv) *$\ell_{\gamma_1}(0) = \ell_{\gamma_2}(0)$.*

Then f is parallel and constant isotropic, and is locally equivalent to one of the following;

1) *a totally geodesic $f : \mathbb{C}^n \to \mathbb{R}^{2n+p}$,*
2) *a totally umbilic $f : \mathbb{C}^n \to \mathbb{R}H^{2n+p}(\tilde{c})$,*
3) *an immersion which is given as a composition $f_2 \circ f_1$ of the first standard minimal immersion*

$$f_1 : \mathbb{C}P^n(c) \to S^{n^2+2n+1}((n+1)c/(2n))$$

and a totally umbilic

$$f_2 : S^{n^2+2n+1}((n+1)c/(2n)) \to M^{2n+p}(\tilde{c};\mathbb{R}),$$

where $c \geq 2n\tilde{c}/(n+1)$.

If we give more information on curvature logarithmic derivatives, we can characterize totally geodesic immersions of complex Euclidean spaces.

Theorem 4.2. *An isometric immersion $f : M \to M^{2n+p}(\tilde{c};\mathbb{R})$ of a Kähler manifold is equivalent to a totally geodesic $\mathbb{C}^n \to \mathbb{R}^{2n+p}$ if and only if for every unit tangent vector $u \in UM$ has two smooth curves γ_1, γ_2 satisfying the conditions in Theorem 4.1 and $\ell_{\gamma_i}(0) \neq 0$.*

These theorems can be extend to quaternionic Kähler manifolds. A quaternionic Kähler structure $\mathcal{J}$ on a Riemannian manifold M of real dimension $4n$ is a rank 3 vector subbundle of the bundle of endomorphisms of the tangent bundle TM with the following properties:

i) For each point $x \in M$ there is an open neighborhood G of x in M and sections J_1, J_2, J_3 of the restriction $\mathcal{J}|_G$ such that
   (a) each J_i is an almost Hermitian structure on G , that is, $J_i^2 = -id$ and $\langle J_iX, Y\rangle + \langle X, J_iY\rangle = 0$ for all vector fields X and Y on G, where $\langle\ ,\ \rangle$ is the Riemannian metric of M;

(b) $J_iJ_{i+1} = J_{i+2} = -J_{i+1}J_i$ (i *mod* 3) for $i = 1, 2, 3$;

ii) The condition that $\nabla_X J$ is a section of $\mathcal{J}$ holds for each vector field X on M and section J of the bundle J, where ∇ denotes the Riemannian connection of M.

We call a Riemannian manifold of real dimension $4n$ equipped with quaternionic Kähler structure a quaternionic Kähler manifold.

Theorem 4.3. *Let f be an isometric immersion of a connected quaternionic Kähler manifold $(M, \mathcal{J})$ of quaternionic dimension n (≥ 2) into a real space form $M^{4n+p}(\tilde{c}; \mathbb{R})$. Suppose for every tangent vector $u \in U_xM$ at an arbitrary point $x \in M$ there exist linearly independent $J^{(1)}, J^{(2)}, J^{(3)} \in \mathcal{J}_x$ with $\|J^{(i)}\| = 1$ $(i = 1, 2, 3)$ and smooth curves $\gamma_j : (-\epsilon, \epsilon) \to M$ $(1 \leq j \leq 6)$ which are parameterized by their arclength, do not have inflection points, and satisfy the following five conditions:*

i) *γ_j is of proper order 2 at x for $1 \leq j \leq 6$,*
ii) *$\gamma_j(0) = x$ and $\dot{\gamma}_j(0) = u$ for $1 \leq j \leq 6$,*
iii) *$Y_{\gamma_i}(0) = J^{(i)}u$ and $Y_{\gamma_{i+3}}(0) = -J^{(i)}u$ for $1 \leq i \leq 3$,*
iv) *the extrinsic shape $f \circ \gamma_j$ is of proper order 2 at $f(x)$ for $1 \leq j \leq 6$,*
v) *$\ell_{\gamma_i}(0) = \ell_{\gamma_{i+3}}(0)$ for $1 \leq i \leq 3$.*

We then find f is a parallel and constant isotropic immersion and is locally equivalent to one of the following;

1) *a totally geodesic $f : \mathbb{H}^n \to \mathbb{R}^{4n+p}$ of a quaternionic Euclidean space,*
2) *a totally umbilic $f : \mathbb{H}^n \to \mathbb{R}H^{4n+p}(\widetilde{c})$,*
3) *an immersion which is given as a composition $f_2 \circ f_1$ of the first standard minimal immersion*

$$f_1 : \mathbb{H}P^n(c) \to S^{2n^2+3n-1}\big((n+1)c/(2n)\big)$$

and a totally umbilic immersion

$$f_2 : S^{2n^2+3n-1}\big((n+1)c/(2n)\big) \to M^{4n+p}(\tilde{c}; \mathbb{R}),$$

where $c \geq 2n\tilde{c}/(n+1)$.

Theorem 4.4. *An isometric immersion $f : (M, \mathcal{J}) \to M^{4n+p}(\tilde{c}; \mathbb{R})$ of a connected quaternionic Kähler manifold is equivalent to a totally geodesic $\mathbb{H}^n \to \mathbb{R}^{4n+p}$ if and only if for every tangent vector $u \in TM$ there exist linearly independent $J^{(1)}, J^{(2)}, J^{(3)} \in \mathcal{J}_x$ with $\|J^{(i)}\| = 1$ $(i = 1, 2, 3)$ and six smooth curves γ_j $(1 \leq j \leq 6)$ satisfying the same conditions in Theorem 4.3 and $\ell_{\gamma_{j_u}}(0) \neq 0$ for some j_u.*

It should be noted that every quaternionic isometric immersion, an immersion of a quaternionic Kähler manifold into a quaternionic Kähler manifold which preserve quaternionic Kähler structure, is totally geodesic because $J_1 J_2 = J_3$.

We can also characterize similar isometric immersions of subsets of a Cayley projective plane into a real space form by use of order 2 property of curves. We denote by $\mathbb{O}P^2(c)$ a Cayley projective plane of maximal sectional curvature c. Being different from the cases of Kähler manifolds and quaternionic Kähler manifolds, we restrict ourselves on subsets of $\mathbb{O}P^2(c)$. Therefore we can reduce the number of test curves in the following manner.

Theorem 4.5. *Let f be an isometric immersion of an open subset M of $\mathbb{O}P^2(c)$ into a real space form $\widetilde{M}^{16+p}(\tilde{c})$. Suppose for every tangent vector $u \in U_xM$ at an arbitrary point $x \in M$ there exist normal tangent vector $v \in U_xM$ and mutually exclusive smooth curves $\gamma_i : (\epsilon, -\epsilon) \to M$ $(i = 1, 2, 3)$ which are parameterized by their arclength do not have inflection points, and satisfy the following four conditions:*

i) $\gamma_i(0) = x$, $\dot\gamma_i(0) = \pm u$ *and it does not hold* $\dot\gamma_1(0) = \dot\gamma_2(0) = \dot\gamma_3(0)$,
ii) $\nabla_{\dot\gamma_i}\dot\gamma_i(0)$ *is parallel to* v, $i = 1, 2, 3$,
iii) *the extrinsic shape* $f \circ \gamma_i$ *is of proper order* 2 *at* $f(x)$,
iv) $\ell_1(0) = \ell_2(0) = \ell_3(0)$.

Then the immersion f is locally congruent to a parallel immersion $f_2 \circ f_1$ which is a composition of the first standard minimal immersion

$$f_1 : \mathbb{O}P^2(c) \to S^{25}(3c/4)$$

and a totally umbilic immersion

$$f_2 : S^{25}(3c/4) \to \widetilde{M}^{16+p}(\tilde{c}),$$

where $3c/4 \geq \tilde{c}$.

5. Isotropic immersions and curvature logarithmic derivatives

In our study for immersions which preserve the order 2 property of curves, curvature logarithmic derivatives of curves play quite important role. In this section we shall concentrate our mind on this quantity and study isotropic immersions. Let $f : M \to \widetilde{M}$ be an isometric immersion. For a smooth curve without inflection points we denote by $\tilde{\ell}_\gamma$ the curvature logarithmic derivative of the extrinsic shape $\tilde{\gamma} = f \circ \gamma$. By Lemma 2.1 we see the following relationship between curvature logarithmic derivatives of curves and their extrinsic shapes.

Lemma 5.1. *For a smooth curve γ which is parameterized by its arclength and does not have inflection points, we have*

$$\begin{aligned}\kappa_\gamma^2(\tilde{\ell}_\gamma - \ell_\gamma) + \tilde{\ell}_\gamma\|\sigma(\dot\gamma,\dot\gamma)\|^2 &= \tilde{\kappa}_\gamma^2(\tilde{\ell}_\gamma - \ell_\gamma) + \ell_\gamma\|\sigma(\dot\gamma,\dot\gamma)\|^2 \\ &= 2\kappa_\gamma\langle\sigma(\dot\gamma,\dot\gamma),\sigma(\dot\gamma,Y_\gamma)\rangle + \langle(\bar\nabla_{\dot\gamma}\sigma)(\dot\gamma,\dot\gamma),\sigma(\dot\gamma,\dot\gamma)\rangle.\end{aligned} \tag{5.1}$$

For an orthonormal pair (u,v) of tangent vectors of M, we denote by $\mathcal{G}(u,v)$ a family of smooth curves defined by the following condition: A curve $\gamma:(-\epsilon,\epsilon)\to M$ parameterized by its arclength is an element of $\mathcal{G}(u,v)$ if it does not have inflection points and satisfies the following conditions;

i) $\dot\gamma(0)=u$ and $Y_\gamma(0)=v$,
ii) the curvature logarithmic derivative at x is preserved by f, (i.e. $\tilde{\ell}_\gamma(0) = \ell_\gamma(0)$).

We consider curvature logarithmic derivatives of such curves. We put

$$\begin{aligned}\mathcal{B}(u,v) &:= \{\,\ell_\gamma(0)\mid\gamma\in\mathcal{G}(u,v)\,\},\\ \mathcal{B}(u) &:= \bigcup\{\,\mathcal{B}(u,v)\mid v \text{ is a unit vector orthogonal to } u\,\}.\end{aligned}$$

We can characterize isotropic immersions by the property that some curvature logarithmic derivatives are preserved.

Theorem 5.1. *An isometric immersion $f: M\to\widetilde{M}$ is isotropic at a point $x\in M$ if and only if $\mathcal{B}(u,v)\cap\mathcal{B}(u,-v)\neq\phi$ for every orthonormal pair $(u,v)\in T_xM\times T_xM$ of tangent vectors.*

Theorem 5.2. *An isometric immersion $f: M\to\widetilde{M}$ is geodesic at a point $x\in M$ if and only if the following two conditions hold:*

i) *$\mathcal{B}(u,v)\cap\mathcal{B}(u,-v)\neq\phi$ for every orthonormal pair $(u,v)\in U_xM\times U_xM$,*
ii) *$\mathcal{B}(u)$ contains at least two distinct numbers for every $u\in U_xM$.*

If we restrict ourselves on Kähler immersions between Kähler manifolds, we can weaken the conditions. We put

$$\mathcal{B}_c(u) := \{\,\ell_\gamma(0)\mid\gamma\in\mathcal{G}(u,Ju)\cup\mathcal{G}(u,-Ju)\,\}.$$

Theorem 5.3. *A Kähler isometric immersion $f: M\to\widetilde{M}$ is geodesic at x if and only if the set $\mathcal{B}_c(u)$ contains at least two distinct numbers for every unit tangent vector $u\in U_xM$.*

6. Veronese embeddings

As an application of Theorem 5.1, we can characterize Veronese embeddings. We define a Kähler full isometric immersion $f_k : \mathbb{C}P^n(\tilde{c}/k) \to \mathbb{C}P^N(\tilde{c})$ of a complex projective space of constant holomorphic sectional curvature $\tilde{c}/k$ into a complex projective space of holomorphic sectional curvature $\tilde{c}$ by

$$[z_i]_{0\leq i\leq n} \mapsto \left[\sqrt{k!/(k_0!\cdots k_n!)}\, z_0^{k_0}\cdots z_n^{k_n}\right]_{k_0+\cdots+k_n=k},$$

where $N = N(n,k) := (n+k)!/(n!k!) - 1$ and $[*]$ denotes homogeneous coordinates. We call this the k-th Veronese embedding. It is well-known that it is constant isotropic with isotropy constant $\lambda_f \equiv \tilde{c}(k-1)/(2k)$. We denote by $M^n(\tilde{c};\mathbb{C})$ a complex space form of constant holomorphic sectional curvature $\tilde{c}$, which is a complex projective space $\mathbb{C}P^n(\tilde{c})$, a complex Euclidean space $\mathbb{C}^n$ and a complex hyperbolic space $\mathbb{C}H^n(\tilde{c})$ according to $\tilde{c}$ is positive, zero and negative.

Theorem 6.1. *Let $f : M \to M^N(\tilde{c};\mathbb{C})$ be a non-totally geodesic Kähler isometric full immersion of a Kähler manifold M of complex dimension $n \geq 2$. Then the following conditions are equivalent:*

1) *There is a positive integer k satisfying that $N = N(n,k)$, the ambient space $M^N(\tilde{c};\mathbb{C})$ is $\mathbb{C}P^N(\tilde{c})$, the submanifold M is locally congruent to $\mathbb{C}P^n(\tilde{c}/k)$ and f is locally equivalent to the k-th Veronese embedding f_k;*
2) *For every orthonormal pair (u,v) of tangent vectors of M, we have $\mathcal{B}(u,v)\cap\mathcal{B}(u,-v)\neq\emptyset$.*

Since every holomorphic curve on $M^N(\tilde{c};\mathbb{C})$ is isotropic, the same result does not hold when $n=1$. But we can say the following (c.f. [2]):

Proposition 6.1. *Let $f : M \to \mathbb{C}P^N(\tilde{c})$ be a full isometric immersed holomorphic curve. The set $\mathcal{B}(u,Ju)\cap\mathcal{B}(u,-Ju)$ contains zero for every $u\in UM$ if and only if there is a positive integer k satisfying that $N = N(1,k)$, the submanifold M is locally congruent to $\mathbb{C}P^1(\tilde{c}/k)$ and f is locally equivalent to f_k.*

We should note that in these results in this section we only need the ambient space only to be of constant holomorphic sectional curvature.

References

1. T. Adachi and T. Sugiyama, *A characterization of isotropic immersions by extrinsic shapes of smooth curves.*

2. S. Maeda, *A characterization of constant isotropic immersions by circles*, Arch. Math. (Basel) **81** (2003), 90–95.
3. S. Maeda and H. Tanabe, *Totally geodesic immersions of Kähler manifolds and Kähler Frenet curves*, Math. Z. **252** (2006), 787–795.
4. K. Nomizu and K. Yano, *On circles and spheres in Riemannian geometry*, Math. Ann. **210** (1974), 163–170.
5. T. Sugiyama and T. Adachi, *Totally umbilic isometric immersions and curves of order two*, to appear in Monat. math..
6. ———, *Logarithmic derivatives of geodesic curvature for smooth curves and isometric immersions*, Far East J. Math. Sci. **23** (2006), 181–192.
7. T. Sugiyama, *Totally geodesic Kähler immersions in view of curves of order two*.
8. H. Tanabe, *Quaternionic Frenet curves and totally geodesic immersions*, Yokohama Math. J. **52** (2005), 1–10.
9. ———, *A note on isometric immersions of the Cayley projective plane and Frenet curves*, Proc. Japan Acad. **81** (2005),12–16.

LIE GROUPS AS FOUR-DIMENSIONAL CONFORMAL KÄHLER MANIFOLDS WITH NORDEN METRIC

M. TEOFILOVA

Faculty of Mathematics and Informatics, University of Plovdiv, 236 Bulgaria Blvd., Plovdiv 4003, Bulgaria
E-mail: mar@gbg.bg

An example of a four-dimensional conformal Kähler manifold with Norden metric is constructed on a Lie group. The form of the curvature tensor is obtained and the isotropic-Kähler properties of the manifold are studied.

Introduction

Almost complex manifolds with Norden metric are originally introduced in [7] as generalized B-manifolds. These manifolds are classified into eight classes in [3], and equivalent characteristic conditions for each of the classes are obtained in [4]. Examples of the basic classes of the integrable almost complex manifolds with Norden metric are given in [1]. An example of the only basic class of the considered manifolds with a non-integrable almost complex structure is introduced in [6].

In this paper we present an example of a four-dimensional conformal Kähler manifold with Norden metric which is obtained by constructing a four-parametric family of Lie algebras. We obtain the form of the curvature tensor and we study the conditions the given manifold to be isotropic Kählerian.

1. Almost complex manifolds with Norden metric

Let (M, J, g) be a $2n$-dimensional almost complex manifold with Norden metric, i.e. J is an almost complex structure and g is a metric on M such that

$$J^2X = -X, \qquad g(JX, JY) = -g(X, Y) \tag{1}$$

for all differentiable vector fields X, Y on M, i.e. $X, Y \in \mathfrak{X}(M)$.

The associated metric $\widetilde{g}$ of g, given by $\widetilde{g}(X,Y) = g(X, JY)$, is a Norden metric, too. Both metrics are necessarily neutral, i.e. of signature (n, n).

Further, X, Y, Z, W (x, y, z, w, respectively) will stand for arbitrary differentiable vector fields on M (vectors in T_pM, $p \in M$, respectively).

If ∇ is the Levi-Civita connection of the metric g, the tensor field F of type $(0,3)$ on M is defined by $F(X,Y,Z) = g\left((\nabla_X J)Y, Z\right)$ and has the following symmetries

$$F(X,Y,Z) = F(X,Z,Y) = F(X,JY,JZ). \tag{2}$$

Let $\{e_i\}$ ($i = 1, 2, \ldots, 2n$) be an arbitrary basis of T_pM at a point p of M. The components of the inverse matrix of g are denoted by g^{ij} with respect to the basis $\{e_i\}$. The Lie forms θ and θ^* associated with F, and the Lie vector Ω, corresponding to θ, are defined by, respectively

$$\theta(z) = g^{ij} F(e_i, e_j, z), \qquad \theta^* = \theta \circ J, \qquad \theta(z) = g(z, \Omega). \tag{3}$$

The Nijenhuis tensor field N is given as $N(X,Y) = [JX, JY] - [X,Y] - J[JX,Y] - J[X,JY]$. It is known that the almost complex structure J is complex, if and only if $N = 0$ ([8]).

A classification of the almost complex manifolds with Norden metric is introduced in [3], where eight classes of these manifolds are characterized according to the properties of F. The three basic classes and the class $\mathcal{W}_1 \oplus \mathcal{W}_2$ of the complex manifolds with Norden metric are given by:

$$\begin{aligned}
&\mathcal{W}_1 : F(X,Y,Z) = \frac{1}{2n}\big[g(X,Y)\theta(Z) + g(X,Z)\theta(Y) \\
&\qquad\qquad\qquad + g(X,JY)\theta(JZ) + g(X,JZ)\theta(JY)\big]; \\
&\mathcal{W}_2 : F(X,Y,JZ) + F(Y,Z,JX) + F(Z,X,JY) = 0, \quad \theta = 0; \\
&\mathcal{W}_3 : F(X,Y,Z) + F(Y,Z,X) + F(Z,X,Y) = 0; \\
&\mathcal{W}_1 \oplus \mathcal{W}_2 : F(X,Y,JZ) + F(Y,Z,JX) + F(Z,X,JY) = 0.
\end{aligned} \tag{4}$$

The class $\mathcal{W}_0$ of the Kähler manifolds with Norden metric is given by $F = 0$.

Let R be the curvature tensor of ∇, i.e. $R(X,Y)Z = \nabla_X \nabla_Y Z - \nabla_Y \nabla_X Z - \nabla_{[X,Y]} Z$ and $R(X,Y,Z,W) = g\left(R(X,Y)Z, W\right)$.

The Ricci tensor ρ and the scalar curvatures τ and $\overset{*}{\tau}$ of R are given by:

$$\rho(y,z) = g^{ij} R(e_i, y, z, e_j), \quad \tau = g^{ij} \rho(e_i, e_j), \quad \overset{*}{\tau} = g^{ij} \rho(e_i, Je_j). \tag{5}$$

It is well known that the Weyl tensor W on a $2n$-dimensional pseudo-Riemannian manifold ($2n \geq 4$) is determined by

$$W = R - \frac{1}{2n-2}\Big\{\psi_1(\rho) - \frac{\tau}{2n-1}\pi_1\Big\}, \tag{6}$$

where

$$\begin{aligned}\psi_1(\rho)(X,Y,Z,W) &= g(Y,Z)\rho(X,W) - g(X,Z)\rho(Y,W)\\ &\quad + g(X,W)\rho(Y,Z) - g(Y,W)\rho(X,Z),\\ \pi_1(X,Y,Z,W) &= g(Y,Z)g(X,W) - g(X,Z)g(Y,W).\end{aligned} \tag{7}$$

The Weyl tensor vanishes if and only if the manifold is conformally flat.

Let $\alpha = \{x, y\}$ be a non-degenerate two-plane spanned by the vectors $x, y \in T_pM$, $p \in M$. Then, the sectional curvature of α is given by:

$$\nu(\alpha; p) = \frac{R(x,y,y,x)}{\pi_1(x,y,y,x)}. \tag{8}$$

We consider the following basic sectional curvatures in T_pM with respect to the structures J and g: *holomorphic sectional curvatures* if $J\alpha = \alpha$ and *totally real sectional curvatures* if $J\alpha \perp \alpha$ with respect to g.

The square norm $\|\nabla J\|^2$ of ∇J is introduced in [5] by

$$\|\nabla J\|^2 = g^{ij}g^{kl}g\left((\nabla_{e_i}J)e_k, (\nabla_{e_j}J)e_l\right). \tag{9}$$

Then, the definition of F, (2) and (9) imply

$$\|\nabla J\|^2 = g^{ij}g^{kl}g^{pq}F_{ikp}F_{jlq}, \quad F_{ikp} = F(e_i, e_k, e_p). \tag{10}$$

Definition 1.1 ([6]). *An almost complex manifold with Norden metric, satisfying the condition* $\|\nabla J\|^2 = 0$, *is said to be* isotropic Kählerian.

It is known ([9]) that the curvature tensor R on any almost complex manifold with Norden metric satisfies the identity

$$(\nabla_X F)(Y,Z,JW) - (\nabla_Y F)(X,Z,JW) = R(X,Y,Z,W) + R(X,Y,JZ,JW). \tag{11}$$

Further, by (2) and (3) we obtain the following properties:

$$\begin{aligned}&(\nabla_X F)(Y,Z,W) = (\nabla_X F)(Y,W,Z);\\ &(\nabla_X F)(Y,JZ,W) = -(\nabla_X F)(Y,Z,JW) - g((\nabla_X J)Z, (\nabla_Y J)W)\\ &\qquad\qquad - g((\nabla_X J)W, (\nabla_Y J)Z);\\ &(\nabla_X \theta^*)Y = (\nabla_X \theta)JY + F(X,Y,\Omega);\\ &\theta(\Omega) = g^{ik}g^{jl}g((\nabla_{e_i}J)e_k, (\nabla_{e_j}J)e_l).\end{aligned} \tag{12}$$

Let us denote $\overset{**}{\tau} = g^{il}g^{jk}R(e_i, e_j, Je_k, Je_l)$. If R is a Kähler tensor, i.e. if $R(X,Y,JZ,JW) = -R(X,Y,Z,W)$, we have $\overset{**}{\tau} = -\tau$.

Let (M, J, g) be in $\mathcal{W}_1 \oplus \mathcal{W}_2$. Then, by (4) and (9) we get

$$2g^{il}g^{jk}g\big((\nabla_{e_i}J)e_k, (\nabla_{e_j}J)e_l\big) = \|\nabla J\|^2. \tag{13}$$

Theorem 1.1. *On a complex manifold with Norden metric it is valid*

$$\tau + \overset{**}{\tau} + \theta(\Omega) - 2\mathrm{div}(J\Omega) = \frac{1}{2}\|\nabla J\|^2, \tag{14}$$

where $\mathrm{div}(J\Omega) = \nabla_i J_k^i \Omega^k$.

Proof. By the properties (12), from (11) we obtain

$$\begin{aligned}&(\nabla_X F)(Y, Z, JW) + (\nabla_Y F)(X, W, JZ) + g((\nabla_X J)Z, (\nabla_Y J)W)\\&+ g((\nabla_X J)W, (\nabla_Y J)Z) = R(X, Y, Z, W) + R(X, Y, JZ, JW).\end{aligned} \tag{15}$$

Then, taking into account (12), (13) and $\nabla g = 0$, the total trace of (15) implies (14). □

It has been proved that on a $\mathcal{W}_1$-manifold with Norden metric it is valid $\|\nabla J\|^2 = \frac{2}{n}\theta(\Omega)$ ([10]). Then, Theorem 1.1 induces

Corollary 1.1. *On a $\mathcal{W}_1$-manifold with Norden metric we have*

$$\tau + \overset{**}{\tau} - 2\mathrm{div}(J\Omega) = -\frac{n-1}{2}\|\nabla J\|^2. \tag{16}$$

The equality (16) and Definition 1.1 immediately imply

Corollary 1.2. *A $\mathcal{W}_1$-manifold with Norden metric is isotropic Kählerian if and only if* $\tau + \overset{**}{\tau} = 2\mathrm{div}(J\Omega)$.

Further, let us consider the class $\mathcal{W}_2$. By (4) and (14) it follows

Corollary 1.3. *On a $\mathcal{W}_2$-manifold with Norden metric it is valid*

$$2(\tau + \overset{**}{\tau}) = \|\nabla J\|^2. \tag{17}$$

Then, Corollary 1.3 and Definition 1.1 give rise to

Corollary 1.4. *A $\mathcal{W}_2$-manifold with Norden metric is isotropic Kählerian if its curvature tensor R is Kählerian.*

2. A Lie group as a four-dimensional conformal Kähler manifold with Norden metric

Let $\mathfrak{g}$ be a real four-dimensional Lie algebra corresponding to a real connected Lie group G. If $\{X_1, X_2, X_3, X_4\}$ is a global basis of left invariant vector fields on G and $[X_i, X_j] = C_{ij}^k X_k$, then the Jacobi identity is valid:

$$C_{ij}^k C_{ks}^l + C_{js}^k C_{ki}^l + C_{si}^k C_{kj}^l = 0. \tag{18}$$

We define an almost complex structure on G by the conditions:

$$JX_1 = X_3, \quad JX_2 = X_4, \quad JX_3 = -X_1, \quad JX_4 = -X_2. \tag{19}$$

Let us consider the left-invariant metric given by

$$\begin{aligned} &g(X_1, X_1) = g(X_2, X_2) = -g(X_3, X_3) = -g(X_4, X_4) = 1, \\ &g(X_i, X_j) = 0 \text{ for } i \neq j. \end{aligned} \tag{20}$$

The introduced metric is Norden because of (19). Hence the induced 4-dimensional manifold (G, J, g) is an almost complex manifold with Norden metric.

It is known ([2]) that an almost complex structure J on a Lie group G is said to be *abelian* if

$$[JX, JY] = [X, Y] \quad \text{for al} \quad X, Y \in \mathfrak{g}. \tag{21}$$

From (21) we derive that the Nijenhuis tensor vanishes on $\mathfrak{g}$, i.e. J is a complex structure. Thus, (G, J, g) is a complex manifold with Norden metric.

The well-known equality

$$\begin{aligned} 2g(\nabla_X Y, Z) = &\, Xg(Y, Z) + Yg(X, Z) - Zg(X, Y) \\ &+ g([X, Y], Z) + g([Z, X], Y) + g([Z, Y], X) \end{aligned} \tag{22}$$

implies

$$\begin{aligned} &2F(X_i, X_j, X_k) = g([X_i, JX_j] - J[X_i, X_j], X_k) \\ &+ g([X_k, JX_i] - [JX_k, X_i], X_j) + g(J[X_k, X_j] - [JX_k, X_j], X_i). \end{aligned} \tag{23}$$

Let (G, J, g) be a $\mathcal{W}_1$-manifold. Then, by (3), (4), (21) and (23) we get

Lemma 2.1. *If (G, J, g) is a four-dimensional $\mathcal{W}_1$-manifold, admitting an Abelian complex structure, the Lie algebra $\mathfrak{g}$ of G is given by:*

$$\begin{aligned} &C_{13}^1 = C_{14}^2 - C_{12}^4,\ C_{13}^2 = C_{12}^3 - C_{14}^1,\ C_{13}^3 = C_{12}^2 + C_{14}^4,\ C_{13}^4 = -C_{12}^1 - C_{14}^3, \\ &C_{24}^1 = -C_{12}^4 - C_{14}^2,\ C_{24}^2 = C_{12}^3 + C_{14}^1,\ C_{24}^3 = C_{12}^2 - C_{14}^4,\ C_{24}^4 = C_{14}^3 - C_{12}^1, \end{aligned} \tag{24}$$

where $C_{ij}^k \in \mathbb{R}$ $(i, j, k = 1, 2, 3, 4)$ must satisfy the Jacobi identity (18).

One solution to the equations (18) and (24) is the four-parametric family of Lie algebras $\mathfrak{g}$ defined by

$$\begin{aligned} &[X_1, X_4] = [X_2, X_3] = \lambda_1 X_1 + \lambda_2 X_2 + \lambda_3 X_3 + \lambda_4 X_4, \\ &[X_1, X_3] = -[X_2, X_4] = \lambda_2 X_1 - \lambda_1 X_2 + \lambda_4 X_3 - \lambda_3 X_4, \end{aligned} \tag{25}$$

where $\lambda_i \in \mathbb{R}$ $(i = 1, 2, 3, 4)$. Thus, by (25) we obtain a four-parametric family of four-dimensional $\mathcal{W}_1$-manifolds with Norden metric.

It has been proved that if a Lie algebra admits an abelian complex structure, then it is solvable ([2]). Therefore, the Lie algebras (25) are solvable.

By (20), (22) and (25) we obtain the non-zero components of the Levi-Civita connection of (G, J, g):

$$\begin{array}{ll} \nabla_{X_1}X_1 = \nabla_{X_2}X_2 = \lambda_2X_3 + \lambda_1X_4, & \nabla_{X_3}X_3 = \nabla_{X_4}X_4 = -\lambda_4X_1 - \lambda_3X_2, \\ \nabla_{X_1}X_3 = \nabla_{X_4}X_2 = \lambda_2X_1 - \lambda_3X_4, & \nabla_{X_1}X_4 = -\nabla_{X_3}X_2 = \lambda_1X_1 + \lambda_3X_3, \\ \nabla_{X_2}X_4 = \nabla_{X_3}X_1 = \lambda_1X_2 - \lambda_4X_3, & \nabla_{X_2}X_3 = -\nabla_{X_4}X_1 = \lambda_2X_2 + \lambda_4X_4. \end{array} \tag{26}$$

Then, by (19), (20) and (23) we get the following essential non-zero components $F_{ijk} = F(X_i, X_j, X_k)$ of the tensor F:

$$\begin{array}{ll} \frac{1}{2}F_{222} = F_{112} = F_{314} = \lambda_1, & \frac{1}{2}F_{111} = F_{212} = -F_{414} = \lambda_2, \\ \frac{1}{2}F_{422} = -F_{114} = F_{312} = -\lambda_3, & \frac{1}{2}F_{311} = F_{214} = F_{412} = -\lambda_4. \end{array} \tag{27}$$

Having in mind (1), (3) and (27), we compute the components $\theta_i = \theta(X_i)$ and $\theta_i^* = \theta^*(X_i)$ of the Lie forms θ and θ^*, respectively:

$$\theta_1 = -\theta_3^* = 4\lambda_2, \quad \theta_2 = -\theta_4^* = 4\lambda_1, \quad \theta_3 = \theta_1^* = 4\lambda_4, \quad \theta_4 = \theta_2^* = 4\lambda_3. \tag{28}$$

A $\mathcal{W}_1$-manifold with closed forms θ and θ^* is called *a conformal Kähler manifold with Norden metric*. The subclass of these manifolds is denoted by $\mathcal{W}_1^0$. Such manifolds are conformally equivalent to Kähler manifolds ([1]).

We establish that the Lie form θ^* is closed on (G, J, g). Thus, we have

Proposition 2.1. *The manifold (G, J, g) is conformal Kählerian if and only if the Lie form θ is closed, i.e. if and only if one of the conditions holds: $\lambda_1 = \lambda_4$, $\lambda_2 = -\lambda_3$ or $\lambda_1 = -\lambda_4$, $\lambda_2 = \lambda_3$.*

Next, by (2), (10) and (27) we get the square norm of ∇J

$$\|\nabla J\|^2 = 16(\lambda_1^2 + \lambda_2^2 - \lambda_3^2 - \lambda_4^2). \tag{29}$$

Proposition 2.2. *The manifold (G, J, g) is isotropic Kählerian if and only if the condition $\lambda_1^2 + \lambda_2^2 - \lambda_3^2 - \lambda_4^2 = 0$ holds.*

Taking into account (20) and (26), we compute the non-zero components $R_{ijkl} = R(X_i, X_j, X_k, X_l)$ of the curvature tensor R as follows:

$$
\begin{aligned}
&R_{1221} = \lambda_1^2 + \lambda_2^2, \quad R_{1331} = \lambda_4^2 - \lambda_2^2, \quad R_{1441} = \lambda_4^2 - \lambda_1^2,\\
&R_{2332} = \lambda_3^2 - \lambda_2^2, \quad R_{2442} = \lambda_3^2 - \lambda_1^2, \quad R_{3443} = -\lambda_3^2 - \lambda_4^2,\\
&R_{1341} = R_{2342} = -\lambda_1\lambda_2, \qquad R_{2132} = -R_{4134} = -\lambda_1\lambda_3,\\
&R_{1231} = -R_{4234} = \lambda_1\lambda_4, \qquad R_{2142} = -R_{3143} = \lambda_2\lambda_3,\\
&R_{1241} = -R_{3243} = -\lambda_2\lambda_4, \quad R_{3123} = R_{4124} = \lambda_3\lambda_4.
\end{aligned} \tag{30}
$$

Let us denote $\rho_{ij} = \rho(X_i, X_j)$. Then, by (1), (5) and (30) we obtain the components of the Ricci tensor ρ:

$$
\begin{aligned}
&\rho_{11} = 2(\lambda_1^2 + \lambda_2^2 - \lambda_4^2), \quad \rho_{12} = -2\lambda_3\lambda_4, \quad \rho_{23} = 2\lambda_1\lambda_4,\\
&\rho_{22} = 2(\lambda_1^2 + \lambda_2^2 - \lambda_3^2), \quad \rho_{13} = -2\lambda_1\lambda_3, \quad \rho_{24} = -2\lambda_2\lambda_4,\\
&\rho_{33} = 2(\lambda_4^2 + \lambda_3^2 - \lambda_2^2), \quad \rho_{14} = 2\lambda_2\lambda_3, \quad \rho_{34} = -2\lambda_1\lambda_2,\\
&\rho_{44} = 2(\lambda_4^2 + \lambda_3^2 - \lambda_1^2).
\end{aligned} \tag{31}
$$

By (26) and (31) we get $(\nabla_{X_i}\rho)(X_j, X_k) = 0$ for all $i, j, k = 1, 2, 3, 4$.

Proposition 2.3. *The manifold (G, J, g) is Ricci-symmetric.*

Next, by (1), (5) and (31) we obtain the the scalar curvatures as:

$$
\tau = 6(\lambda_1^2 + \lambda_2^2 - \lambda_3^2 - \lambda_4^2), \qquad \overset{*}{\tau} = -4(\lambda_1\lambda_3 + \lambda_2\lambda_4). \tag{32}
$$

Then, (32), Propositions 2.1 and 2.2 imply

Proposition 2.4. *The considered manifold (G, J, g) has the properties:*

(i) *(G, J, g) is isotropic Kählerian if and only if $\tau = 0$;*

(ii) *(G, J, g) is conformal Kählerian if and only if $\tau = \overset{*}{\tau} = 0$.*

Let us consider the Weyl tensor of (G, J, g). Taking into account (6), (7), (30), (31) and (32), we get $W_{ijkl} = 0$ for all $i, j, k, l = 1, 2, 3, 4$.

Proposition 2.5. *The Weyl tensor of (G, J, g) vanishes. Thus, the curvature tensor has the form $R = \frac{1}{2}\left\{\psi_1(\rho) - \frac{\tau}{3}\pi_1\right\}$.*

By Propositions 2.4 and 2.5 we obtain

Proposition 2.6. *If (G, J, g) is a conformal Kähler manifold, then its curvature tensor has the form $R = \frac{1}{2}\psi_1(\rho)$.*

Further, (7), $\nabla g = 0$, Propositions 2.3 and 2.5 imply

Proposition 2.7. *The manifold (G, J, g) is locally symmetric, i.e. $\nabla R = 0$.*

Let us consider the characteristic two-planes α_{ij} spanned by the basic vectors $\{X_i, X_j\}$ at an arbitrary point of the manifold: totally real two-planes: α_{12}, α_{14}, α_{23}, α_{34} and holomorphic two-planes: α_{13}, α_{24}.

Then, by (7), (8), (20) and (30) we obtain

$$\begin{array}{lll} \nu(\alpha_{12}) = \lambda_1^2 + \lambda_2^2, & \nu(\alpha_{13}) = \lambda_2^2 - \lambda_4^2, & \nu(\alpha_{14}) = \lambda_1^2 - \lambda_4^2, \\ \nu(\alpha_{23}) = \lambda_2^2 - \lambda_3^2, & \nu(\alpha_{24}) = \lambda_1^2 - \lambda_3^2, & \nu(\alpha_{34}) = -\lambda_3^2 - \lambda_4^2. \end{array} \tag{33}$$

Proposition 2.8. *If (G, J, g) is of vanishing holomorphic sectional curvatures, then it is isotropic Kählerian.*

Finally, Propositions 2.2, 2.4 and 2.5 induce

Theorem 2.1. *The following conditions are equivalent:*

(i) *(G, J, g) is isotropic Kählerian;*
(ii) *the condition $\lambda_1^2 + \lambda_2^2 - \lambda_3^2 - \lambda_4^2 = 0$ holds;*
(iii) *the scalar curvature τ vanishes;*
(iv) *the curvature tensor has the form $R = \frac{1}{2}\psi_1(\rho)$.*

References

1. R. Castro, L.M. Hervella and E. García-Río, *Some examples of almost complex manifolds with Norden metric*, Riv. Math. Univ. Parma **15** (1989), 133–141.
2. I. Dotti, *Hypercomplex nilpotent Lie groups*, Contemporary Math. **288** (2001), 310–314.
3. G. Ganchev and A. Borisov, *Note on the almost complex manifolds with a Norden metric*, Compt. Rend. Acad. Bulg. Sci. **39** (1986), 31–34.
4. G. Ganchev, K. Gribachev and V. Mihova, *B-connections and their conformal invariants on conformally Kähler manifolds with B-metric*, Publ. Inst. Math. (Beograd) (N.S.) **42** (1987), 107–121.
5. E. García-Río and Y. Matsushita, *Isotropic Kähler structures on Engel 4-manifolds*, J. Geom. Phys. **33** (2000), 288–294.
6. K. Gribachev, M. Manev and D. Mekerov, *A Lie group as a four-dimensional quasi-Kähler manifold with Norden metric*, JP J Geom. Topol. **6** (2006), 55–68.
7. K. Gribachev, D. Mekerov and G. Djelepov, *Generalized B-manifolds*, Compt. Rend. Acad. Bulg. Sci. **38** (1985), 299–302.
8. A. Newlander and L. Nirenberg, *Complex analytic coordinates in almost complex manifolds*, Ann. Math. **65** (1957), 391–404.
9. M. Teofilova, *Complex connections on complex manifolds with Norden metric*, Contemporary Aspects of Complex Analysis, Differential Geometry and Mathematical Physics, eds. S. Dimiev and K. Sekigawa, World Sci. Publ., Singapore (2005), 326–335.
10. ———, *Curvature properties of conformal Kähler manifolds with Norden metric*, Math. Educ. Math., Proc. of 35th Spring Conference of UBM, Borovec (2006), 214–219.

BÄCKLUND TRANSFORMATIONS AND RIEMANN-HILBERT PROBLEM FOR N WAVE EQUATIONS WITH ADDITIONAL SYMMETRIES

T. VALCHEV

Institute for Nuclear Research and Nuclear Energy,
Bulgarian Academy of Sciences,
72 Tzarigradsko chaussèe, 1784 Sofia, Bulgaria,
E-mail: valtchev@inrne.bas.bg

We construct new singular solutions to a matrix Riemann-Hilbert problem by making use of the equivalence between it and the inverse scattering problem. The Riemann-Hilbert problem under consideration allows to solve N-wave systems related to orthogonal algebras and possesing $\mathbb{Z}_2 \times \mathbb{Z}_2$ reductions. The method we apply is a special type of an auto-Bäcklund transformation, namely the dressing Zakharov-Shabat method.

Keywords: Dressing procedure; Riemann-Hilbert problem; N-wave equations with reductions.

1. Preliminaries

1.1. *Bäcklund transformations and inverse scattering method*

One of the basic and fruitful concepts in the modern theory of integrable systems is the notion of a transformation which allows one to generate new solutions of a nonlinear evolution equation (NLEE) starting from a known one. Originally introduced by A. V. Bäcklund at the end of 19-th century when he studied surfaces with a constant negative Gaussian curvature (for more detailed information on this interesting topic we recommend the books [1]). This transformation bears now his name — Bäcklund transformation (BT). Roughly speaking the Bäcklund transformation is a map which transforms a solution q of a differential equation order into a solution of another differential equation. An example of a Bäcklund transformation is the well known Miura transformation which maps a solution of the modified Korteveg-de Vries equation into a solution of Korteveg-de

Vries equation.

An important subclass of BT relates to the situation when a given solution of a NLEE is transformed into another solution of the same NLEE (in this case they usually talk of an auto-Bäcklund transformation). A common particular case of an auto-BT occurs when NLEE admits a Lax representation

$$L\psi(x,t,\lambda) := (i\partial_x + [J, q(x,t)] - \lambda J)\,\psi(x,t,\lambda) = 0, \tag{1}$$

$$M\psi(x,t,\lambda) := (i\partial_t + V(q(x,t),\lambda))\,\psi(x,t,\lambda) = 0, \tag{2}$$

i.e. NLEE for $q(x,t)$ is equivalent to the formal integrability condition for the linear systems (1) and (2) which holds identically with respect to the spectral parameter λ. The class of auto-BT related to such type of NLEE is derived in [2,3]. We shall restrict ourselves with vanishing boundary conditions, i.e. $\lim_{x\to\pm\infty} q(x,t) = 0$ is fulfilled. In addition, we assume that $U(x,t,\lambda)$ and $V(x,t,\lambda)$ belong to a certain semisimple Lie algebra $\mathfrak{g}$ and J is a constant, real and regular element of its Cartan sublagebra $\mathfrak{h} \subset \mathfrak{g}$ (hence the fundamental solution $\psi(x,t,\lambda)$ is an element of its Lie group G).

Example 1.1. Let $V(x,t,\lambda) := [I, q(x,t)] - \lambda I$ and I is a constant element of $\mathfrak{h}$. Then NLEE looks as follows

$$i[J, q_t(x,t)] - i[I, q_x(x,t)] + [[I,q],[J,q]](x,t) = 0.$$

For every choice of $\mathfrak{g}$, J and I one obtains an N-wave equation. Such type of equations occurs in nonlinear optics [4], differential geometry [5] etc.□

Let $\psi_0(x,t,\lambda)$ be a fundamental set of solutions of the linear problem

$$i\partial_x\psi_0(x,t,\lambda) + ([J, q_0(x,t)] - \lambda J)\psi_0(x,t,\lambda) = 0,$$

$$i\partial_t\psi_0(x,t,\lambda) + (V_0(q_0(x,t),\lambda))\psi_0(x,t,\lambda) = 0.$$

generated by some known solution $q_0(x,t)$. Then construct another fundamental solution $\psi(x,t,\lambda) = g(x,t,\lambda)\psi_0(x,t,\lambda)$ of (1)–(2) by using a special transformation $g(x,t,\lambda)$. This transformation is called a dressing factor and as an immediate consequence of its definition it follows that it satisfies the linear system

$$i\partial_x g(x,t,\lambda) + [J, q(x,t)]g(x,t,\lambda) - g(x,t,\lambda)[J, q_0(x,t)] - \lambda[J, g(x,t,\lambda)] = 0,$$

$$i\partial_t g(x,t,\lambda) + V(x,t,\lambda)g(x,t,\lambda) - g(x,t,\lambda)V_0(x,t,\lambda) = 0.$$

One possible ansatz for $g(x,t,\lambda)$ proposed by Zakharov and Shabat reads

$$g(x,t,\lambda) = 1 + \frac{A(x,t)}{\lambda - \lambda^+} + \frac{B(x,t)}{\lambda - \lambda^-}.$$

The auto-Bäcklund transformation can be represented by using the diagram

$$q_0(x,t) \to \psi_0(x,t,\lambda) \xrightarrow{g} \psi(x,t,\lambda) \to q(x,t).$$

In particular if we start from the trivial solution $q_0(x,t) \equiv 0$ we will obtain the 1-soliton solution.

Zakharov and Shabat [6] proposed a purely algebraic method for construction of the dressing factor by using so-called fundamental analytic solutions (FAS) of the linear problem (1)–(2). The concept of FAS was originally introduced by Shabat [7]. One can prove that the constraints on J and $q(x,t)$ (zero boundary conditions) result in existence of FAS $\chi^+(x,t,\lambda)$ (resp. $\chi^-(x,t,\lambda)$) with analytical properties in the upper half plane $\mathbb{C}_+$ (resp. the lower half plane $\mathbb{C}_-$). There is a standard procedure to derive FAS starting from Jost solutions. Jost solutions $\psi_\pm(x,t,\lambda)$ are defined for $\lambda \in \mathbb{R}$ as fundamental solutions whose assymptotics for $x \to \pm\infty$ are just plane waves

$$\lim_{x\to\pm\infty} \psi_\pm(x,t,\lambda)e^{i(\lambda Jx - f(\lambda)t)} = 1, \quad f(\lambda) := \lim_{x\to\pm\infty} V(x,t,\lambda). \tag{3}$$

In the case of N wave equations we have $f(\lambda) = -\lambda I$. The transition matrix $T(\lambda) = \psi_+^{-1}(x,t,\lambda)\psi_-(x,t,\lambda)$ bears the name a scattering matrix. Thus using Gauss decomposition of the scattering matrix, namely

$$T(\lambda) = T^\mp(\lambda)D^\pm(\lambda)(S^\pm(\lambda))^{-1}$$

where $S^+(\lambda)$ and $T^+(\lambda)$ (resp. $S^-(\lambda)$ and $T^-(\lambda)$) are upper (resp. lower) triangular with unit diagonal elements, while the matrices $D^+(\lambda)$ and $D^-(\lambda)$ are diagonal, one can derive that

$$\chi^\pm(x,t,\lambda) = \psi_-(x,t,\lambda)S^\pm(\lambda) = \psi_+(x,t,\lambda)T^\mp(\lambda)D^\pm(\lambda).$$

Let $\psi_{\pm,0}(x,t,\lambda)$ be Jost solutions of the "bare" linear problem, i.e. that one corresponding to $q_0(x,t)$. Then the dressing procedure affects $\psi_{\pm,0}(x,t,\lambda)$ in the follows way

$$\psi_\pm(x,t,\lambda) = g(x,t,\lambda)\psi_{0,\pm}(x,t,\lambda)g_\pm^{-1}(\lambda), \quad g_\pm(\lambda) := \lim_{x\to\pm\infty} g(x,t,\lambda).$$

The factor $g_\pm(\lambda)$ guarranties that $\psi_\pm(x,t,\lambda)$ have the proper assymptotics (3). It can be checked that the bare scattering matrix $T_0(\lambda)$ and the bare FAS $\chi_0^\pm(x,t,\lambda)$ transform into

$$T(\lambda) = g_+(\lambda)T_0(\lambda)g_-^{-1}(\lambda), \quad \chi^\pm(x,t,\lambda) = g(x,t,\lambda)\chi_0^\pm(x,t,\lambda)g_-^{-1}(\lambda).$$

A very important role in the theory of integrable systems is played by the so-called reduction group originally introduced by Mikhailov [8]. Let G_R be a discrete group acting on G by group automorphisms and on the complex plane of the spectral parameter by conformal transformations. The induced action $\mathcal{K}_h$ of the group G_R on the set of fundamental solutions $\{\psi(x,t,\lambda)\}$ of the linear problem reads

$$\mathcal{K}_h : \chi(x,t,\lambda) \mapsto \tilde{\chi}(x,t,k_h(\lambda)) = \mathbf{K}_h(\chi(x,t,\lambda)),$$

where $K_h \in \mathrm{Aut}(G)$ for any $h \in G_R$. We require that $\{\psi(x,t,\lambda)\}$ stays invariant under this action. The dressing factor $g(x,t,\lambda)$ must be invariant under the action of G_R, i.e.

$$\mathbf{K}_h\left(g(x,t,k^{-1}(\lambda))\right) = g(x,t,\lambda).$$

The imposed requirements lead to certain symmetry conditions for $U(x,t,\lambda)$ and $V(x,t,\lambda)$. Hence G_R reduces the number of independent fields (components of $q(x,t)$) in NLEE and that is why it bears the name a reduction group. Many integrable equations are derived by imposing certain reduction on them.

1.2. *Riemann-Hilbert problem*

One of the classical problems in the theory of analytic functions is the problem for analytic factorization of a function known also as a local Riemann-Hilbert problem (RHP). It reads: let $\gamma \subset \mathbb{C}$ is a smooth, closed curve, $G : \mathbb{C} \to \mathrm{GL(n)}$ is an analytic function on γ, then we are searching for matrix-valued functions ψ_{in} and ψ_{out} analytic on $\mathbb{D}_{\mathrm{in}}$ and $\mathbb{D}_{\mathrm{out}}$ respectively and satisfying

$$\psi_{\mathrm{out}}(\lambda) = \psi_{\mathrm{in}}(\lambda)G(\lambda), \qquad \forall\,\lambda \in \gamma.$$

If $\psi_{\mathrm{in}}(\lambda)$ and $\psi_{\mathrm{out}}(\lambda)$ are solutions then $C\psi_{\mathrm{in}}(\lambda)$ and $C\psi_{\mathrm{out}}(\lambda)$ (C is a constant element of $GL(n)$) are solutions too. Thus to ensure the uniqueness of the solution we have to impose a normalization. One common possibility is the normalization $\lim_{\lambda\to\infty}\psi_{\mathrm{out}}(\lambda) = 1$ which is called canonical.

If $\det\psi(\lambda)$ has poles in the λ-plane then the solution ψ is called a singular solution, otherwise (i.e. $\det\psi(\lambda) \neq 0$, $\forall\lambda \in \mathbb{D} \subset \mathbb{C}$) it is a regular one.

It has been shown (see [6,7]) that there exists a deep connection between RHP and the inverse scattering method. In order to illustate this let us introduce another set of FAS $\eta^{\pm}(x,t,\lambda) := \chi^{\pm}(x,t,\lambda)e^{i(\lambda Jx - f(\lambda)t)}$. Then it can be easily checked that

$$\eta^{+}(x,t,\lambda) = \eta^{-}(x,t,\lambda)G(x,t,\lambda),$$

where

$$G(x,t,\lambda) = e^{-i(\lambda Jx - f(\lambda)t)}(S^-(\lambda))^{-1}S^+(\lambda)e^{i(\lambda Jx - f(\lambda)t)},$$

i.e. $\eta^\pm(x,t,\lambda)$ are solutions of RHP ($\gamma \equiv \mathbb{R}$). It turns out [6] that the opposite holds true as well, namely

Theorem 1.1 (Zakharov-Shabat). *If $\eta^\pm(x,t,\lambda)$ are solutions of RHP with a sewing function $G(x,t,\lambda)$ satisfying*

$$i\partial_x G(x,t,\lambda) - \lambda[J, G(x,t,\lambda)] = 0,$$
$$i\partial_t G(x,t,\lambda) + [f(\lambda), G(x,t,\lambda)] = 0,$$

then $\eta^\pm(x,\lambda)$ are FAS for the linear problem.

Let $\eta_0^\pm(x,t,\lambda)$ be known solutions of RHP (for example, they can be regular ones). Then we are able to obtain new singular solutions just by dressing $\eta_0^\pm(x,t,\lambda)$ as follows

$$\eta_0^\pm(x,t,\lambda) \mapsto \eta^\pm(x,t,\lambda) = g(x,t,\lambda)\eta_0^\pm(x,t,\lambda)g_-^{-1}(\lambda). \tag{4}$$

2. Singular solutions of RHP with additional symmetries

In this section we will outline a method to obtain singular solutions to RHP in the case when they take values in the orthogonal group $SO(n,\mathbb{C})$ and represent FAS to a $\mathbb{Z}_2 \times \mathbb{Z}_2$ reduced linear problem. This problem corresponds to the problem for calculating the quadruplet soliton solutions to a $\mathbb{Z}_2 \times \mathbb{Z}_2$-reduced N-wave equation related to orthogonal algebras which is discussed by the authors in [9]. Our work represents an conceptual continuation of the works [10,11] where the $\mathbb{Z}_2$ reductions and the soliton solutions of N wave equations are studied in detail. We are modifying the methods proposed in [12,13].

We recall that $SO(m,\mathbb{C})$ is the group of all isometries in the complex vector space $\mathbb{C}^m$ with a metric S and the Lie algebra of its generators is defined by

$$so(m,\mathbb{C}) := \left\{\mathfrak{c} \in \mathrm{End}(\mathbb{C}^m) \mid \quad \mathfrak{c}^T S + S\mathfrak{c} = 0\right\}.$$

For the sake of convenience we shall use a basis in $\mathbb{C}^m$ such that S reads

$$S = \sum_{k=1}^{n} (-1)^{k-1}(E_{k,2(n+1)-k} + E_{2(n+1)-k,k}) + (-1)^n E_{n+1,n+1}, \qquad \text{for } \mathfrak{so}(2n+1)$$

$$S = \sum_{k=1}^{n} (-1)^{k-1}(E_{k,2n+1-k} + E_{2n+1-k,k}), \quad \text{for } \mathfrak{so}(2n),$$

where $(E_{kl})_{pq} = \delta_{kp}\delta_{lq}$ is the Weyl basis. Hence the Cartan subalgebra $\mathfrak{h} \subset \mathfrak{so}(n)$ consists of diagonal matrices in the form

$$\mathfrak{h} = \{J = \mathrm{diag}(J_1, \dots, J_n, 0, -J_n, \dots, -J_1)\}, \quad \text{for}\, \mathfrak{so}(2n+1),$$
$$\mathfrak{h} = \{J = \mathrm{diag}(J_1, \dots, J_n, -J_n, \dots, -J_1)\}, \quad \text{for}\, \mathfrak{so}(2n).$$

Let it be given the following action of the group $\mathbb{Z}_2 \times \mathbb{Z}_2$ on FAS

$$\chi^-(x,t,\lambda) = K_1 \left((\chi^+)^\dagger(x,t,\lambda^*)\right)^{-1} K_1^{-1}$$
$$\chi^-(x,t,\lambda) = K_2 \left((\chi^+)^T(x,t,-\lambda)\right)^{-1} K_2^{-1}.$$

K_1 and K_2 belong to the Cartan subgroup of $SO(n)$ and obey the restriction $K_1^2 = K_2^2 = 1$. As a consequence of that reduction the potential $U(x,t,\lambda)$ satisfies the following symmetry conditions

$$K_1 U^\dagger(x,t,\lambda^*)K_1^{-1} = U(x,t,\lambda), \quad K_2 U^T(x,t,-\lambda)K_2^{-1} = -U(x,t,\lambda).$$

The dressing factor $g(x,t,\lambda)$ must be $\mathbb{Z}_2 \times \mathbb{Z}_2$-invariant, i.e.

$$K_1 \left(g^\dagger(x,t,\lambda^*)\right)^{-1} K_1^{-1} = g(x,t,\lambda), \quad K_2 \left(g^T(x,t,-\lambda)\right)^{-1} K_2^{-1} = g(x,t,\lambda),$$

that is why we choose it in the form

$$\begin{aligned} g(x,t,\lambda) = 1 + \frac{A(x,t)}{\lambda - \lambda^+} + \frac{K_1 S A^*(x,t)(K_1 S)^{-1}}{\lambda - (\lambda^+)^*} \\ - \frac{K_2 S A(x,t)(K_2 S)^{-1}}{\lambda + \lambda^+} - \frac{K_1 K_2 A^*(x,t)(K_1 K_2)^{-1}}{\lambda + (\lambda^+)^*}, \end{aligned} \tag{5}$$

where $\lambda^+ = \mu + i\nu$ lies into the upper half plane $\mathbb{C}_+$ and the matrix-valued function $A(x,t)$ is to be found. The requirement for orthogonality of $g(x,t,\lambda)$ (i.e. $g^{-1} = S^{-1}g^T S$) leads to certain algebraic conditions for $A(x,t)$, namely

$$A(x,t)SA^T(x,t) = 0, \quad A(x,t)S\omega^T(x,t) + \omega(x,t)SA^T(x,t) = 0. \tag{6}$$

The factor $\omega(x,t)$ is defined by following expression

$$\omega(x,t) := 1 - \frac{K_2 S A(x,t) S K_2}{2\lambda^+} + \frac{K_1 S A^*(x,t) S K_1}{2i\nu} - \frac{K_1 K_2 A^*(x,t) K_2 K_1}{2\mu}.$$

From the first algebraic condition in (6) it follows that $A(x,t)$ is degenerate. Therefore it can be presented as a product of two rectangular $n \times k$ matrices $X(x,t)$ and $F^T(x,t)$. In terms of these matrix factors the algebraic restrictions (6) read

$$F^T(x,t)SF(x,t) = 0, \qquad \omega(x,t)SF(x,t) = X(x,t)\alpha(x,t),$$

where α is a $k \times k$ skew-symmetric matrix.

Another type of constraints occurs as an effect of the natural prerequisite for λ-independence of the "dressed" solution $q(x,t)$. These constraints are equivalent to some linear differential equations for $F(x,t)$ and $\alpha(x,t)$ which after solving them lead to

$$\alpha(x,t) = F_0^T(\chi_0^+(x,t,\lambda^+))^{-1}\partial_\lambda \chi_0^+(x,t,\lambda^+)SF_0 + \alpha_0,$$

$$F(x,t) = \left(\left(\chi_0^+\right)^T(x,t,\lambda^+)\right)^{-1} F_0, \quad F_0 = \text{const}, \quad \alpha_0 = \text{const}.$$

In the soliton case when $\chi_0^+(x,t,\lambda) = e^{-i\lambda(Jx+It)}$ this result turns into

$$F(x,t) = e^{i\lambda^+(Jx+It)}F_0, \quad \alpha(x,t) = iF_0^T(Jx+It)SF_0 + \alpha_0.$$

To find the factor $X(x,t)$ we derive the auxiliary linear system shown below

$$(SF,\, SG,\, SH,\, SN)\,(x,t) = (X,\, Y,\, Z,\, W)\begin{pmatrix} \alpha & a & b & c \\ a & \alpha & c & b \\ b^* & c^* & \alpha^* & a^* \\ c^* & b^* & a^* & \alpha^* \end{pmatrix}(x,t).$$

The extra entities above are defined as follows

$$\begin{aligned} Y(x,t) &:= K_2SX(x,t), & Z(x,t) &:= K_1SX^*(x,t), \\ W(x,t) &:= K_1K_2X^*(x,t), & G(x,t) &:= K_2SF(x,t), \\ H(x,t) &:= K_1SF^*(x,t), & N(x,t) &:= K_1K_2F^*(x,t), \\ a(x,t) &:= \frac{F^TSG}{2\lambda^+}(x,t), & b(x,t) &:= \frac{F^TSH}{2i\nu}(x,t), \\ c(x,t) &:= \frac{F^TSN}{2\mu}(x,t). \end{aligned}$$

In the simplest case possible when $\text{rank}(X) = \text{rank}(F) = 1$ (i.e. $\alpha(x,t) \equiv 0$) we have

$$\begin{pmatrix} SF \\ SG \\ SH \\ SN \end{pmatrix}(x,t) = \begin{pmatrix} 0 & a & -b & c \\ a & 0 & c & -b \\ b & c & 0 & a^* \\ c & b & a^* & 0 \end{pmatrix}\begin{pmatrix} X \\ Y \\ Z \\ W \end{pmatrix}(x,t).$$

Consequently we obtain that

$$X = \frac{1}{\Delta}\left(a^*SG + bSH - cSN\right), \qquad \Delta := |a|^2 + b^2 - c^2.$$

Thus inserting the result for $X(x,t)$ into (5) we can find $g(x,t,\lambda)$ and after that construct the singular solution applying formula (4). Repeating the same procedure, namely dressing the singular solutions of RHP themselves, we are able to construct another singular solutions and so on.

3. Conclusion

We have demonstrated an approach to derive new singular solutions to RHP by using the equivalence between it and the invrse scattering method. In our case RHP corresponds to N wave equations related to $\mathfrak{so}(m)$ and possessing $\mathbb{Z}_2 \times \mathbb{Z}_2$ symmetries. To achieve our purpose we have modified the classical Zakharov-Shabat dressing method. The singular solution we have obtained that way is tightly connected with the quadruplet to a N wave system with a $\mathbb{Z}_2 \times \mathbb{Z}_2$ reduction. The method can be applied also for the case of symplectic algebras.

Acknowledgement

The author would like to thank prof. Vladimir Gerdjikov for setting the problem and for the useful discussions.

This work is partially supported by a contract 1410 with the National Science Foundation of Bulgaria.

References

1. C. Rogers and W. Schief, *Bäcklund and Darboux Transformations*, Cambridge Univ. Press, Cambridge, 2002.
2. F. Calogero and A. Degasperis, Lett. Nuovo Cimento B **32**, 201–242 (1976); Lett. Nuovo Cimento B **39**, 1–54 (1976).
3. V. Gerdjikov and Khristov, Bul. J. Phys. **7**, n. 2, 119–133, (1980) (In Russian).
4. D. Kaup, A. Reiman and A. Bers, Rev. Mod. Phys. **81**.
5. E. Ferapontov, Diff. Geom and Appl. **5**, 335–369 (1995).
6. V. Zakharov and A. Shabat, Funkts. Anal. Prilozhen **13**, 13–22 (1979).
7. A. Shabat, Diff. Equations, **15**, 1824–1834 (1979) (In Russian).
8. A. Mikhailov, Physica D **3**, 73–117 (1981).
9. V. Gerdjikov and T. Valchev, to be published in: Proc. of 8-th International Conference on Geometry, Integrability and Quantization, June 9–14, 2006, Varna, Bulgaria, Eds. Ivailo Mladenov and Manuel de Leon.
10. V. Gerdjikov, G. Grahovski and N. Kostov, J. Phys. A **34**, 9425–9461 (2001).
11. V. Gerdjikov, G. Grahovski, R. Ivanov and N. Kostov, Inv. Problems **17**, 999–1015 (2001).
12. R. Ivanov, Nuclear Physics B **694**, 509-524 (2004).
13. V. Zakharov and A. Mikhailov, Commun. Math. Phys. **74**, 21–40 (1980).

AUTHOR INDEX